UNSOLVED PROBLEMS
IN ECOLOGY

UNSOLVED PROBLEMS IN ECOLOGY

EDITED BY

ANDREW DOBSON, ROBERT D. HOLT,
AND DAVID TILMAN

PRINCETON UNIVERSITY PRESS
PRINCETON AND OXFORD

Requests for permission to reproduce material from this work
should be sent to permissions@press.princeton.edu

Published by Princeton University Press
41 William Street, Princeton, New Jersey 08540
6 Oxford Street, Woodstock, Oxfordshire OX20 1TR

press.princeton.edu

Library of Congress Cataloging-in-Publication Data

Names: Dobson, Andrew, editor. | Holt, Robert D., editor. |
Tilman, David, 1949– editor.
Title: Unsolved problems in ecology / edited by Andrew Dobson,
Robert D. Holt, and David Tilman.
Description: Princeton : Princeton University Press, [2020] |
Includes bibliographical references and index.
Identifiers: LCCN 2019056794 (print) | LCCN 2019056795 (ebook) |
ISBN 9780691199832 (hardback) | ISBN 9780691199825 (paperback) |
ISBN 9780691195322 (ebook)
Subjects: LCSH: Ecology. | Population biology. | Coexistence of species. |
Ecosystem management. | Biodiversity conservation.
Classification: LCC QH541.145 U57 2020 (print) | LCC QH541.145 (ebook) |
DDC 577–dc23
LC record available at https://lccn.loc.gov/2019056794
LC ebook record available at https://lccn.loc.gov/2019056795

British Library Cataloging-in-Publication Data is available

Editorial: Alison Kalett and Abigail Johnson
Production Editorial: Kathleen Cioffi
Text and Cover Design: Carmina Alvarez
Production: Brigid Ackerman
Publicity: Matthew Taylor and Amy Stewart
Copyeditor: Gregory W. Zelchenko

Cover image courtesy of Andrew Dobson

This book has been composed in Charis SIL

Printed on acid-free paper. ∞

Printed in the United States of America

1 3 5 7 9 10 8 6 4 2

We would like to dedicate this volume to
John Tyler Bonner,
Henry Horn,
and
Robert "Bob" May.

They each had a huge influence on each of us
and we can see signs of their huge intellectual influence
in every chapter in this volume.

Contents

PART III
COEXISTENCE

PART IV
ECOLOGICAL COMMUNITIES AND ECOSYSTEMS

PART V
ECOLOGY AND HEALTH

PART VI
CONSERVATION BIOLOGY AND NATURAL
RESOURCE MANAGEMENT

PART VII
FINAL THOUGHTS

Preface

Andrew Dobson, Robert D. Holt, and David Tilman

The centenary of the Ecological Society of America inspired us to ask ecologists their thoughts about the next century, specifically on the broad question of "What are the Unsolved Problems in Ecology?" We imagined that they might identify two classes of problems: (1) Those people have wrestled with, but where solutions have remained elusive and (2) problems that someone may have just recognized as being potentially huge yet unexamined. The motivation for the book stems from a deep conviction that ecology will be a central defining science of the twenty-first century, just as physics defined the twentieth, and chemistry the nineteeth. Consequently, we put our authors in the position of defining what they think the key agenda for ecology will be within their area of research for the next decades to a full century. Sutherland et al. (2013) honored the centenary of the British Ecological Society by compiling a list of key unanswered questions—in effect, a series of bullet points aiming at future progress in the discipline. We, instead, asked authors to provide a more discursive reflection on open, important questions in the form of essays, providing a more expansive vista across possible future intellectual landscapes.

A strong motivation for the book was a previous volume of essays published in the 1970s that simply asked "What are the unsolved problems for the 20th Century" (Duncan and Weston-Smith 1977); there were only two biological chapters, including one by John Maynard Smith, who astutely pointed out that we did not know why sex had evolved. Curious as it seems, no one had explicitly realized that this was a problem prior to Maynard Smith's explication of the inherent "cost of sex" (1971, 1977; see Bell 1982); although Darwin as early as 1862 presciently remarked "we do not even in the least know why new beings should be produced by the union of two sexual elements, instead of by . . . parthenogenesis" (cited in Kirk 2006), and Bonner (1958) and others do adumbrate some aspects of the issue. This book chapter helped spark the genesis of a whole subdiscipline of studies within evolution, behavioral ecology, and epidemiology. We are ambitious enough to hope that at least one of our chapters in this volume can likewise unearth an intellectual goldmine that transforms thinking within ecology and the broader disciplines of evolution and environmental science. The other biological essay in the 1977 compilation was by Peter Grubb, who pointed out that our knowledge of leaf structure and function at the time was woefully inadequate. This chapter also led to multiple developments in plant physiology and ecology. We were delighted when Peter accepted our invitation to write a chapter for the current book, and doubly so, when he decided to write a chapter that describes how much we still need to know about leaf structure and function, some four decades after his initial distillation of this question.

Some unsolved questions that the authors in this volume bring up are radically new, but others are longstanding. Robert MacArthur towards the end of his life sketched an array of outstanding problems in ecology (MacArthur 1972), focused around the theme of species coexistence, many of which are still with us and touched on in the current volume, including for instance the need for network perspectives, and the importance of understanding "why are some species able to adjust niche widths rapidly when put in a new situation while others are rigid?"; the latter question foreshadowed current concerns with themes such as niche conservatism and evolutionary rescue. MacArthur argued for intellectual pluralism and suggested that ecologists needed to get beyond the biological sciences (including in particular, he notes, the earth sciences) to really come to grips with the issue of species coexistence. These insights resonate today.

We initially planned to obtain three temporal perspectives on the unsolved problems identified by the authors, corresponding roughly to different stages in the trajectories of careers. To this end, we split the set of authors we invited into three broad and overlapping categories: (1) We asked younger researchers whose careers are expanding rapidly as to what they see as the major conceptual challenges facing their research, (2) we asked midcareer scientists to describe what they plan to focus on as the major targets of opportunity in their own careers, and (3) we asked individuals who have helped to define the study of ecology over the last 30 to 50 years to describe the problems they have found intractable or continually challenging, given available techniques and methodology. The skeleton of this structure is faintly discernible within the chapters we received for the final volume, although we perceive two distortions, one of which can fairly readily be dealt with, the other of which presents a significant "unsolved problem in ecology." The first distortion is that we tended to ask people whom we knew personally to write chapters. Although we have all been active in the Ecological Society of America, the British Ecological Society, the American Society of Naturalists, and the Society for Conservation Biology (among others) for more than 30 years, we surely (if unconsciously) are biased in asking friends and colleagues, rather than a broader array of people we may have admired from a distance in these and other ecological societies. This is partly because it is so much easier to pressure and cajole friends and colleagues to deliver manuscripts and forebear with us when the editorial process slows down.

We could have assembled and organized the book in different ways, for instance by soliciting chapters that focused on specific areas of current or historical controversy, or by dividing the contributions papers into those that focus on specific issues, versus more general scientific and societal problems. We again resisted this, partly because such approaches would reflect our own personal knowledge and biases. One notable feature of our collection of essays is that they are all are each by one to two authors. Yet many of the problems identified by these authors will require collaborative efforts among many scientists, and indeed the ecological literature is becoming dominated by multiauthor publications.

In past years, the National Center for Ecological Analysis and Synthesis (NCEAS) in Santa Barbara) was an excellent forum for pushing synthesis in ecological theory and practice. NCEAS supported working groups, postdoctoral fellows, and visiting scholars working on themes such as coexistence theory, ecological networks, and phylogenetic perspectives on community structure. After the ending of core National Science Foundation (NSF) funding, NCEAS metamorphosed into an entity more tightly focused on critical applied issues, such as global food systems sustainability, ocean health, conservation practice, and also providing a venue synthesis across the Long Term Ecological Research Network (LTER) network of sites. These are of course all very important issues to address, but this change in focus means that ecology does not have a think tank where groups of ecologists—often with dissenting opinions—can convene to identify common ground and hatch new perspectives on key conceptual problems in the ecological sciences. The traditional forums of symposia and talks at annual conferences do not at all fill this niche. It is too easy for opposing parties to posture and defend their position rather than work with each other cooperatively and constructively. The long time delays inherent in production allow differences of opinion to fester, slowing the development of new and vital knowledge. Current debates in the ecological literature, ranging from subtleties in coexistence theory, to articulating the biodiversity consequences of habitat fragmentation, to the dilution versus amplification effects in host–parasite ecology, to priority effects and alternative stable states, could all benefit from the working group environment provided by the original avatar of NCEAS.

As pointed out by one of the referees of this volume, "Few if any significant debates in ecology have ever been resolved. People either die or get tired of arguing them. This is not a good thing!" This was much less of a problem when we had NCEAS as a facility to host discussions of areas of controversy which often led to an emerging consensus. The absence of such a concrete center—a think tank for the basic ecological sciences—is in our view a major unsolved problem in ecology.

Just as each essay reflects the personal stance of the author, the selection we have ended up with reflects our own collective vision as to important directions for future research in ecology. A different set of editors might well have ended up with a different suite of unanswered questions in our discipline.

The second distortion is harder to deal with and reflects a broader problem in science. Our initial list of potential authors was well-balanced by subdiscipline and as well-balanced as we could between male and female authors. Most of the more senior women we approached felt themselves too overcommitted with other work to be able to contribute a chapter. Most expressed frustration at the limited amount of time they had available to write primary research papers or even grant proposals, given the heavy loads they experience in terms of being asked to participate in a broad range of administrative—but not directly scientific—tasks. This is a parlous state of affairs that excludes important and insightful voices from not just our compilation of thought pieces, but broader discussions of ecology and other academic disciplines. We hope this situation can be resolved over the next decade, as the different ecological societies, academic institutions, and funding agencies nurture and mentor the next generation of younger scholars. Nonetheless, it is a major unsolved problem that we still need to address with increasing vigor in both the scientific and policy arenas of ecology and the environmental sciences.

We hope the book will appeal to at least three different groups of ecologists: (1) Graduate students at early stages of their careers, who are looking for new and exciting areas in which to develop their research careers. (2) Established ecologists, who are thinking about different directions to take their research, or simply inquisitive about new ideas to include in their courses and symposia. (3) Historians of science who are interested in the forces that shape the development of new ideas within different scientific disciplines.

We thank the authors of each chapter for their contributions, particularly those who also acted as referees for chapters other than their own and provided insightful comments that further enhanced the quality of these chapters. We humbly also thank the two anonymous referees who read the complete volume for Princeton University Press. Their vital insights are reflected in the expanded title, in some of the threads of this introduction and in a modified organization of the order of the chapters. As is inevitable, the task took longer than we first assumed; grappling with the task of compiling and editing this lively set of essays greatly increased our respect for the editors, reviewers, and authors of the ecological journals that keep our discipline vibrant and rigorous. We finally thank everyone involved for their patience, and hope that the final product matches our and your expectations. We have learned a lot and thoroughly enjoyed reading these contributions, and hope that you, and the broader readership of our community, may likewise profit from careful perusal of these essays.

References

Bell, G. 1982. The masterpiece of nature: The evolution and genetics of sexuality. University of California Press.

Bonner, J. M. 1958. The relation of spore formation to recombination. American Naturalist 92:193–200.

Duncan, R., and M. Weston-Smith, eds. 1977. The Encyclopaedia of Ignorance: Everything you ever wanted to know about the unknown. Oxford: Pergamon Press.

Kirk, D. L. 2006. Oogamy: Inventing the sexes. Current Biology 16:R1028–R1030.

MacArthur, R. 1972. Coexistence of species. In: J. A. Behnke, ed. Challenging biological problems: Directions toward their solutions. New York: Oxford University Press, 253–259.

Maynard Smith, J. 1971. What use is sex? Journal of Theoretical Biology 30:319–335.

Sutherland, W. J., et al. 2013. Identification of 100 fundamental ecological questions. Journal of Ecology 101:58–67.

Contributors

Stefano Allesina, Department of Ecology and Evolution, University
of Chicago, and Northwestern Institute on Complex Systems,
Northwestern University

Julien F. Ayroles, Department of Ecology and Evolutionary Biology
and Lewis Sigler Institute, Princeton University

Elizabeth T. Borer, Department of Ecology, Evolution, and Behavior,
University of Minnesota

Tim Caro, Department of Wildlife, Fish and Conservation Biology and
Center for Population Biology, University of California, Davis

Tim Coulson, Department of Zoology, University of Oxford

Andrew Dobson, Department of Ecology and Evolutionary Biology,
Princeton University

Peter J. Grubb, Department of Plant Sciences, University of
Cambridge

Simon P. Hart, Institute of Integrative Biology, ETH Zurich

Ian Hatton, Department of Ecology and Evolutionary Biology, Princeton University

Michael E. Hochberg, CNRS, Institut des Sciences de l'Evolution de
Montpellier, Université de Montpellier, and the Santa Fe Institute

Robert D. Holt, Department of Biology, University of Florida

Marcel Holyoak, Department of Environmental Science and Policy,
University of California, Davis

Kevin D. Lafferty, Western Ecological Research Center, US Geological
Survey

Egbert Giles Leigh Jr., Smithsonian Tropical Research Institute

Simon A. Levin, Department of Ecology and Evolutionary Biology,
Princeton University

Jonathan M. Levine, Department of Ecology and Evolutionary
Biology, Princeton University; and Institute of Integrative Biology,
ETH Zurich

Michel Loreau, Centre for Biodiversity Theory and Modelling, Theoretical and Experimental Ecology Station, CNRS and Paul Sabatier
University

Pablo A. Marquet, Departamento de Ecología and Laboratorio Inter-
nacional en Cambio Global (LINCGlobal) y Centro de Cambio

Global UC, Facultad de Ciencias Biológicas, Pontificia Universidad Católica de Chile; Instituto de Ecología y Biodiversidad (IEB); the Santa Fe Institute; and Instituto de Sistemas Complejos de Valparaíso (ISCV)

Robert M. May, Department of Zoology, University of Oxford

C. Jessica E. Metcalf, Department of Ecology and Evolutionary Biology and Office of Population Research, Princeton University

Helene C. Muller-Landau, Smithsonian Tropical Research Institute

Stephen W. Pacala, Department of Ecology and Evolutionary Biology, Princeton University

Mercedes Pascual, Department of Ecology and Evolution, University of Chicago; and the Santa Fe Institute

Robert M. Pringle, Department of Ecology and Evolutionary Biology, Princeton University

Andrew F. Read, Center for Infectious Disease Dynamics, Departments of Biology and Entomology, Pennsylvania State University

Rolando Rebolledo, Instituto de Ingeniería Matemática, Facultad de Ingeniería, Universidad de Valparaíso

Christina Riehl, Department of Ecology and Evolutionary Biology, Princeton University

Mauricio Tejo, Departamento de Matemática, Universidad Tecnológica Metropolitana

Andrew R. Tilman, Department of Biology, University of Pennsylvania

David Tilman, Ecology, Evolution, and Behavior, University of Minnesota; and Bren School of Environmental Science and Management, University of California, Santa Barbara

Ross A. Virginia, Environmental Studies Program, Dartmouth College

Diana H. Wall, Department of Biology and School of Global Environmental Sustainability, Colorado State University

William C. Wetzel, Department of Entomology, Michigan State University

Rae Winfree, Ecology, Evolution, and Natural Resources, Rutgers University

PART I

Populations, Variability, and Scaling

Ecological Scaling in Space and Time

A New Tool in Plain Sight?

Elizabeth T. Borer

Rapid changes in climate and nutrient deposition in regions around Earth are inducing equally rapid changes in the biosphere (Schimel et al. 1997, Ellis et al. 2010, Running 2012). These abiotic factors are not changing at the same rate or in the same direction in all locations, so organisms are increasingly experiencing novel combinations of precipitation, temperature, and nutrient deposition (Williams et al. 2007, Rockström et al. 2009). The reliance of humans on processes provided by organisms and their interactions within the local biotic and abiotic environment, such as carbon fixation, nutrient cycling, disease transmission, and the quantity, quality, and persistence of freshwater, provides a pressing reason for ecologists to develop a mechanistic understanding of the links between organisms, ecosystem processes, and regional and global cycles (Worm et al. 2006, Kareiva 2011, Cardinale et al. 2012). Yet, the combination of climate and nutrient supply experienced by organisms also results from feedbacks between the physiology of individual organisms and global biogeochemical cycles (Ehleringer and Field 1993, Vitousek et al. 1997, Arrigo 2005, Hooper et al. 2005), making clear our need to better understand processes that span temporal scales from minutes to millennia and spatial scales from suborganismal to Earth's atmosphere to predict future effects of the changing environments on Earth.

The problem of scaling in ecology is not new. Three decades ago, scaling occupied the minds of many ecologists. For example, three decades ago John Wiens wrote an essay calling for a shift to multiscale thinking, development of new theory, and greater focus on collecting data to resolve discontinuities in processes across spatial scales (Wiens 1989). A few years later, Simon Levin published another excellent synthesis of the state of ecological knowledge of scaling (Levin 1992), arguing that knowledge of both large-scale constraints and the aggregate behavior of organisms will be necessary for achieving a predictive, mechanistic understanding of the feedbacks between organisms and ecosystem fluxes. A key topic raised by both Wiens and Levin also was addressed in a book published at nearly the same time (Ehleringer and Field 1993) in which many authors tackled the issue of scaling and cross-scale feedbacks from organismal physiology to global climate and back again.

In the decades since these papers were written, ecologists have continued to develop an understanding of long-term feedbacks, heterogeneity, and links across spatial scales. For example, the effects of forest warming over the short term have been demonstrated to stimulate soil respiration, whereas turnover in microbial composition can increase the carbon use efficiency of the community, leading to attenuation of soil respiration under continuous long-term warming (Melillo et al. 2002, Frey et al. 2013). The effects of diversity on productivity also function via long-term feedbacks. For example, long-term, chronic nutrient addition causes productivity to increase initially, but these effects attenuate over multiple decades because of ongoing loss of species diversity (Isbell et al. 2013). In a different subfield of ecology, research using metagenomic tools is highlighting the links and feedbacks among spatial scales that determine the resident microbial composition, or microbiome, of a host. For example, the identity and relative abundance of microbial species inhabiting an individual is determined at the regional scale by the composition and relative transmission ability of microbial species and at the local scale by the relative abundance of hosts and microbial competitive ability and fitness within individual host species (reviewed in Borer et al. 2013). Many factors, including the abiotic environment (Fenchel and Finlay 2004), host quality (Smith et al. 2005), and host behavior (Lombardo 2008), can play a role in these interactions and feedbacks across spatial scales.

In spite of the forward progress of this field, the fundamental issue of effectively using information about processes at one scale in predictions about outcomes at another scale remains unsolved. In 2011, the Macrosystems Biology program at the National Science Foundation (NSF) was launched to stimulate research and advance greater mechanistic understanding of processes spanning spatial scales (Dybas 2011). Although the availability of funding is certainly a key constraint on intellectual progress, identifying and collecting the types of data that will be useful for making predictions that span scales also represents a major challenge (Levin 1992, Ehleringer and Field 1993, Leibold et al. 2004, Elser et al. 2010, Nash et al. 2014). Perhaps most importantly, ecologists studying feedbacks and linkages across spatial scales are faced with tradeoffs in our capacity to gather data about the biosphere at any scale: the spatial extent versus the temporal extent of a study, the local replication versus the spatial extent of a study, or site-based experimental work versus large-scale observation (Soranno and Schimel 2014).

A New Tool Hiding in Plain Sight

Over the past several decades, a fairly continuous stream of publications has identified conceptual areas of spatial scaling where our ignorance remains vast (e.g., Wiens 1989, Levin 1992, Ehleringer and Field 1993, Peters et al. 2007, Borer et al. 2013). However, ecological science has changed a great deal during this time, giving us a range of new tools and more highly resolved data to study ecological scaling relationships. Meta-analysis has become an accepted tool for quantitative synthesis of the ecological literature and has been used, for example, to examine support for a range of hypotheses about the key determinants of species diversity across spatial scales (Field et al. 2009). Sequencing technology and metagenomics is rapidly extending the conceptual realm and spatial scales being actively considered by ecologists (Borer et al. 2013). Electronic technology also has changed our ability to tackle questions about scaling in myriad ways, including computerization of data acquisition and access, satellite imagery, remote sensing, drone technology, and interpolation of a wide array of environmental data (Campbell et al. 2013). One example of the exciting cutting-edge of technology to examine scaling in ecology is research that is advancing our ability to use remotely sensed spectral variation as a tool for estimating local and regional biodiversity, and concurrently documenting leaf-level traits and functional differences among taxa (Cavender-Bares et al. 2016).

However, the change in the past 30 years that is perhaps most underappreciated for its potential to advance this field is neither statistical nor technological; it is the shift in the culture of ecological science from a field dominated by single investigator projects to one of collaboration (Hampton and Parker 2011, Goring et al. 2014).

Distributed Experimental Networks

Most ecological research is conducted by one or a few scientists over relatively short time scales and small spatial scales (Heidorn 2008), and whereas large-scale, multi-investigator collaborations have become increasingly common in ecology over the past several decades (Nabout et al. 2015), the vast majority of these collaborations generate, share, and analyze observational data (e.g. Baldocchi et al. 2001, Weathers et al. 2013). Although observations of ecological systems represent an exceptionally important tool for characterizing and comparing among systems, manipulative experiments are a far more powerful tool for forecasting a system's behavior under novel environmental conditions. Given the pressing need to effectively forecast ecological responses in a changing global environment, multifactorial experiments measuring responses and feedbacks spanning spatial and temporal scales will be a key tool to complement meta-analyses, large-scale observations, and models.

Although most experiments in ecology are conceived of and performed by single investigators, large-scale, grassroots distributed experimental collaborations are rethinking ecological experimentation and are overcoming the historical tradeoffs in our capacity to gather long-term experimental data across multiple spatial scales (Borer et al. 2014a). By replicating the same experimental treatments and sampling protocols and openly sharing data with each other, ecologists collaborating in distributed experimental networks are able to replicate experiments and directly compare biological and abiotic responses across spatial scales ranging from centimeters to continents. Depending on the question, sampling can occur at multiple scales within sites (e.g., within individual, within plot, plot, block, site) to quantify a plethora of responses to experimental treatments that map onto future scenarios (e.g., multiple nutrients, herbivory, high-latitude warming, drought, and loss of biodiversity; see Arft et al. 1999, Borer et al. 2014b, Duffy et al. 2015, Fay et al. 2015).

This emerging approach to network science is requiring a rethinking of collaboration and a change in scientific culture (Guimerà et al. 2005, Hampton and Parker 2011, Borer et al. 2014a). By using common experimental treatment and sampling protocols and sharing data openly among collaborators, every site improves the dataset through contribution and each investigator benefits from the opportunity to contribute data and ideas as a result of their efforts (Borer et al. 2014a). As with any effective collaboration, careful fostering of a culture of trust and sharing means that contributors have confidence that their efforts will be included and rewarded (Hampton and Parker 2011). In this model, participation is voluntary, and for most distributed experiments organized as grassroots efforts, investigators at each site shoulder the cost of implementing the treatments and collecting the data rather than funding such efforts through a single centralized grant. This pay-to-play funding model means that participation, particularly by international collaborators in understudied regions of the world, is increased when costs are low. And, for a field that seems to perpetually struggle with physics envy, this model of egalitarian collaboration was once called "the dream" by the Director General of the European Center for Particle Research (CERN), Dr. Robert Aymar (Ford 2008).

To forecast future scenarios for ecological responses and feedbacks in nonanalog environmental conditions (Williams and Jackson 2007, Rockström et al. 2009), we need experiments that manipulate multiple global change factors over long periods of time, and we need to understand how novel conditions influence the resulting spatial patterns and processes across multiple scales. Without multifactorial experiments replicated across many sites, it remains difficult to effectively estimate interactions among factors and contingencies in responses associated with, for instance, climate, evolutionary, or geological history. Distributed experimental networks provide such an opportunity.

The benefits of a distributed experiment for tackling questions about processes spanning and feeding back across spatial scales are enormous. This widespread collaboration among scientists dramatically expands the spatial extent of observation while retaining resolution (grain) at the scale of individuals, but also generating data that can be aggregated to capture patterns at larger grain such as block or site. The spatial replication generated by a network with many collaborators allows clear quantification of responses that are shared among sites as well as responses that are contingent on site characteristics (e.g., climate, soils, or evolutionary history). The replication of experimental treatments across many sites and conditions also allows investigation into the patterns and feedbacks resulting from multiple interacting factors by breaking up the colinear and confounded variables that plague single-site studies. By working as a widespread collaborative team to establish multiple treatments and sample at locations spanning regions and continents, distributed experiments overcome the tradeoff between the spatial and temporal scales of sampling that has caused ecologists to rely so heavily on models and meta-analysis for which interactions among treatments and site variables are difficult (and usually impossible) to disentangle.

We provide a few case studies to develop how we envision that this type of approach, harnessing the intellectual and data collection power of scientists spanning regions and continents, could interlink with existing approaches (e.g., modeling, streaming data) to generate a predictive understanding of how biological processes will change and feed back across scales in response to changing environments on Earth.

Case Study 1: Plant Productivity

As we move across spatial and temporal scales of observation, the key controls on the processes and resulting patterns in primary productivity shift (Wiens 1989, Ehleringer and Field 1993, Polis 1999, Peters et al. 2007). For example, roots foraging for soil resources may occur at the scale of millimeters, inducing organismal constraints on productivity (Tian and Doerner 2013). At the scale of meters, intraspecific and interspecific interactions among organisms seeking the same resources may generate webs of direct and indirect interactions that may determine the net carbon fixation and annual productivity of a plant community. For example, concurrent changes and feedbacks in plant quality and composition in response to grazing (Zheng et al. 2012) or chronic nutrient addition (Isbell et al. 2013) can lead to long-term declines in productivity within fields. At regional scales, solar radiation, precipitation, nutrients, or other physical factors may impose the most important constraints on productivity (Polis 1999, Del Grosso et al. 2008). Although local, long-term patterns of evapotranspiration can predict the dominant flora, and thus biome, of a region, direct measurements of leaf-scale transpiration or small-scale measurements of local plant communities may fail to predict the larger-scale pattern (Wang and Dickinson 2012). Thus, we remain limited in our ability to use observed responses at the scale of roots and stomata to interpret satellite information or predict regional climate, although we believe that these changes are important pieces of the puzzle.

The use of meta-analysis has advanced our understanding of the role and interactions among climate, plant chemistry, and vegetation type on regional-scale patterns of plant productivity (Del Grosso et al. 2008). However, in spite of the important insights arising from synthesis across studies, such studies have relied on interpolation and derived metrics of production that may underestimate the role of local-scale processes and overestimate the role of regional climatic drivers (Shoo and Ramirez 2010). They also fail to provide a strong estimation of trajectories of productivity under future scenarios of climate and nutrient deposition. Thus, our ability to predict productivity responses to multiple interacting factors (e.g., concurrent changes in the supply rates of multiple nutrients or climate factors) and feedbacks from plant productivity to climate and nutrient cycles remains limited by the lack of simultaneous, direct manipulations of the environment and measurements of the rates of primary productivity within and among sites.

A coordinated, long-term experiment spanning a wide range of climate and nutrient supply could produce data to test the multiscale hypotheses generated with meta-analysis. By concurrently manipulating factors most likely to determine productivity within sites, regions, and across continents (e.g., climate, local nutrient supply, herbivory; Milchunas and Lauenroth 1993, Del Grosso et al. 2008, Fay et al. 2015), such a study could generate data to clarify the likely trajectories of change in productivity in future, nonanalog environments. These direct estimates of primary productivity, under a wide variety of natural and manipulated environments, produced through large-scale collaboration among scientists, would generate data to clarify the interactions among factors, spatial and temporal feedbacks, and spatial scales at which each factor most strongly constrains primary productivity. Far from supplanting other approaches to studying ecological systems (e.g. observations, meta-analysis, models), this is a complementary approach that takes advantage of the collective power of the research community to generate directly comparable data spanning unprecedented spatial scales.

Case Study 2: The Microbiome

Developments in metagenomics over the past decade have shown that most of the genes and approximately half of the carbon in a human is of microbial origin (Shively et al. 2001, Nelson et al. 2010, Brüls and Weissenbach 2011), leading to a fundamental reassessment, among other things, of what it means to be an individual. Metagenomic studies have demonstrated that an individual's microbiome, the identity and relative abundance of microbial species inhabiting an individual, plays many important functional roles for animal and plant hosts, including digestion and nutrient acquisition, production of anti-inflammatory compounds, and resistance to pathogens (van der Heijden et al. 1998, Gill et al. 2006, DiBaise et al. 2008, Rodriguez et al. 2009, Fraune and Bosch 2010). Thus, the accumulation of microbes and turnover of species within the microbial community of a host are fundamentally important processes that define the composition and function of each host's microbiome. Although what we do know suggests that these processes span and feed back across spatial scales from biotic interactions at microscopic scales within hosts to regional drivers of the abiotic environment, our understanding of the spatial scaling and feedbacks across scales that control host–microbe interactions remains poorly developed (Medina and Sachs 2010).

Recent syntheses of this body of empirical work demonstrate that there are many links and feedbacks from local microbial interactions within a host to larger-scale distributions of microbes (Borer et al. 2013, Borer et al. 2016). Studies of microbes have demonstrated that some taxa are capable of extremely long-distance dispersal, leading to the increasingly debated hypothesis that microbes lack dispersal limitation, and local microbial communities are determined solely through environmental tolerance and selection (Baas-Becking 1934, Cho and Tiedje 2000, Fenchel and Finlay 2004, Antony-Babu et al. 2008, Peay et al. 2010). In addition to regional-scale selection, the abiotic environment also can determine the outcome of competition among microbes within a host (Yatsunenko et al. 2012, Lacroix et al. 2014) and alter the composition of a host's microbiome through feedbacks that alter the nutritional quality of host tissues, from a microbe's perspective, as well as the relative abundance of conspecific hosts (Smith et al. 2005, Keesing et al. 2006, Clasen and Elser 2007, Borer et al. 2010). Another key finding is that hosts are not vessels, but rather play a role in sanctioning and turnover of microbes to favor more beneficial species or strains (Kiers and Denison 2008), thereby feeding back to alter the local and regional composition of microbial taxa. Related to this, recent work has revealed that the composition and relative abundance of the microbes that make up a host's microbiome is constantly changing, likely determined by processes such as host sanctioning, competition, and succession of microbial taxa that feed back across spatial scales (Yatsunenko et al. 2012, Copeland et al. 2015).

However, most studies of the microbiome within hosts are observational, not experimental, and are performed at single sites, focused on single microbial species, examine only a single host species, and do not characterize the regional microbial pool (but see U'Ren et al. 2010, U'Ren et al. 2012). Thus, our knowledge of the relative importance of processes operating at different scales is lacking. Because of this, our ability to predict the response of within-host microbial community diversity and function in a changing biotic and abiotic world is limited by the lack of simultaneous, direct manipulations of the environment and measurements of within-host microbial communities across sites.

Sampling the microbiome of hosts within a distributed experimental network could lay the foundation for predicting how global changes will alter the function of microbial communities inhabiting hosts and feed back to determine the relative abundance of hosts, themselves. For example, by quantifying the effects of experimentally manipulated global change factors on the identity, diversity, and relative abundance of microbes among host plant tissues (scale of millimeters), individual host plants (centimeters to meters), among plots (meters), among species and treatments within a site, among sites (kilometers or greater), and as a function of regional and experimentally imposed environmental gradients, we could better characterize dispersal distances and the role of environmental filtering and, importantly, understand the conditions and scales at which this community filtering is a dominant process controlling the microbiome of individual hosts.

The microbiome is a community of species and interacting individuals; ecological metacommunity paradigms (Leibold et al. 2004, Borer et al. 2016) can help us sort through patterns and responses by the microbiome to experimental treatments spanning spatial scales. For example, a distributed experiment would allow us to determine whether within-host microbial richness increases as a saturating function with increasing microbial taxon pool size (Fukami 2004) and whether this consistently differs by experimental treatment among sites. If niche-based processes (e.g., host chemistry, environmental nutrients) primarily determine microbial composition at the local scale, we expect a strong correlation between host microbial composition and the local environment (Cottenie et al. 2003, Leibold et al. 2004, Chase 2007). Thus, by directly measuring the response of host-associated microbes to multiple concurrent global change factors across a globally relevant range of conditions, a distributed experimental network could generate critical empirical data about the interactions and feedbacks among factors controlling the microbiome. By harnessing the capacity of the research community deeply invested in these questions, these data could effectively complement insights from metagenomic observations, single-site (or lab) studies, and models, providing insights about generality and contingencies determining a host's microbiome at an unprecedented range of spatial scales.

Conclusions

Perhaps this will simply be another essay pointing out our need for progress in understanding the mechanisms underlying ecological relationships spanning spatial and temporal scales. If so, it will be an essay in venerable company. However, as a discipline, we have an ever richer and more diverse set of young scientists spanning the globe. This growth and diversity of ecologists can become a direct asset that can position our field to rethink how we work as a society of scientists. We can harness the collective skills and knowledge of our amazing colleagues to create the newest tool in our own toolbox for generating previously unattainable experimental data documenting processes and feedbacks across scales. More generally, innovation and progress can come in many forms, including rethinking our approach to science. By rethinking how we study the world, redefining how we collect data, and pursuing avenues outside the range of conventional approaches, ecologists may be able to push this field further in the coming decades than we have in the preceding ones.

Acknowledgments

These ideas and examples were developed through many conversations—with a particular thanks to Eric Seabloom for many long and helpful discussions—and as part of a variety of projects funded by the National Science Foundation, including NSF-EF 12-41895, NSF-DEB 1556649.

References

Antony-Babu, S., J.E.M. Stach, and M. Goodfellow. 2008. Genetic and phenotypic evidence for *Streptomyces griseus* ecovars isolated from a beach and dune sand system. Antonie Van Leeuwenhoek International Journal of General and Molecular Microbiology 94:63–74.

Arft, A. M., M. D. Walker, J. Gurevitch, J. M. Alatalo, M. S. Bret-Harte, M. Dale, et al. 1999. Responses of tundra plants to experimental warming: Meta-analysis of the International Tundra Experiment. Ecological Monographs 69:491–511.

Arrigo, K. R. 2005. Marine microorganisms and global nutrient cycles. Nature 437: 349–355.

Baas-Becking, L.G.M. 1934. Geobiologie of inleiding tot de milieukunde. The Hague: Van Stockum and Zoon.

Baldocchi, D., E. Falge, L. Gu, R. Olson, D. Hollinger, S. Running, et al. 2001. FLUXNET: A new tool to study the temporal and spatial variability of ecosystem–scale carbon dioxide, water vapor, and energy flux densities. Bulletin of the American Meteorological Society 82:2415–2434.

Borer, E. T., A.-L. Laine, and E. Seabloom. 2016. A multiscale approach to plant disease using the metacommunity concept. Annual Review of Phytopathology 54:397–418.

Borer, E. T., W. S. Harpole, P. B. Adler, E. M. Lind, J. L. Orrock, E. W. Seabloom, and M. D. Smith. 2014a. Finding generality in ecology: A model for globally distributed experiments. Methods in Ecology and Evolution 5:65–73.

Borer, E. T., L. L. Kinkel, G. May, and E. W. Seabloom. 2013. The world within: Quantifying the determinants and outcomes of a host's microbiome. Basic and Applied Ecology 14:533–539.

Borer, E. T., E. W. Seabloom, D. S. Gruner, W. S. Harpole, H. Hillebrand, E. M. Lind, et al. 2014b. Herbivores and nutrients control grassland plant diversity via light limitation. Nature 508:517–520.

Borer, E. T., E. W. Seabloom, C. E. Mitchell, and A. G. Power. 2010. Local context drives infection of grasses by vector-borne generalist viruses. Ecology Letters 13:810–818.

Brüls, T., and J. Weissenbach. 2011. The human metagenome: our other genome? Human Molecular Genetics 20:R142–R148.

Campbell, J. L., L. E. Rustad, J. H. Porter, J. R. Taylor, E. W. Dereszynski, J. B. Shanley, et al. 2013. Quantity is nothing without quality: automated qa/qc for streaming environmental sensor data. Bioscience 63:574–585.

Cardinale, B. J., J. E. Duffy, A. Gonzalez, D. U. Hooper, C. Perrings, P. Venail, et al. 2012. Biodiversity loss and its impact on humanity. Nature 486:59–67.

Cavender-Bares, J., J. E. Meireles, J. J. Couture, M. A. Kaproth, C. C. Kingdon, A. Singh, et al. 2016. Associations of leaf spectra with genetic and phylogenetic variation in oaks: Prospects for remote detection of biodiversity. Remote Sensing 8:221.

Chase, J. M. 2007. Drought mediates the importance of stochastic community assembly. Proceedings of the National Academy of Sciences of the United States of America 104:17430–17434.

Cho, J. C., and J. M. Tiedje. 2000. Biogeography and degree of endemicity of fluorescent *Pseudomonas* strains in soil. Applied and Environmental Microbiology 66:5448–5456.

Clasen, J. L., and J. J. Elser. 2007. The effect of host *Chlorella NC64A* carbon: Phosphorus ratio on the production of *Paramecium bursaria Chlorella Virus-1*. Freshwater Biology 52:112–122.

Copeland, J. K., L. Yuan, M. Layeghifard, P. W. Wang, and D. S. Guttman. 2015. Seasonal community succession of the phyllosphere microbiome. Molecular Plant-Microbe Interactions 28:274–285.

Cottenie, K., E. Michels, N. Nuytten, and L. De Meester. 2003. Zooplankton metacommunity structure: Regional vs. local processes in highly interconnected ponds. Ecology 84:991–1000.

Del Grosso, S., W. Parton, T. Stohlgren, D. L. Zheng, D. Bachelet, S. Prince, K. Hibbard, and R. Olson. 2008. Global potential net primary production predicted from vegetation class, precipitation, and temperature. Ecology 89:2117–2126.

DiBaise, J. K., H. Zhang, M. D. Crowell, R. Krajmalnik-Brown, G. A. Decker, and B. E. Rittmann. 2008. Gut microbiota and its possible relationship with obesity. Mayo Clinic Proceedings 83:460–469.

Duffy, J. E., P. L. Reynolds, C. Boström, J. A. Coyer, M. Cusson, S. Donadi, et al. 2015. Biodiversity mediates top–down control in eelgrass ecosystems: A global comparative-experimental approach. Ecology Letters 18:696–705.

Dybas, C. 2011. NSF grants foster understanding of biological systems on regional to continental scales. National Science Foundation, Press Release 11-160. Washington, D.C.: National Science Foundation.

Ehleringer, J. R., and C. B. Field, eds. 1993. Scaling physiological processes: Leaf to globe. San Diego: Academic Press.

Ellis, E. C., K. K. Goldewijk, S. Siebert, D. Lightman, and N. Ramankutty. 2010. Anthropogenic transformation of the biomes, 1700 to 2000. Global Ecology and Biogeography 19:589–606.

Elser, J. J., W. F. Fagan, A. J. Kerkhoff, N. G. Swenson, and B. J. Enquist. 2010. Biological stoichiometry of plant production: metabolism, scaling and ecological response to global change. New Phytologist 186:593–608.

Fay, P. A., S. M. Prober, W. S. Harpole, J.M.H. Knops, J. D. Bakker, E. T. Borer, et al. 2015. Grassland productivity limited by multiple nutrients. Nature Plants 1:15080.

Fenchel, T., and B. J. Finlay. 2004. The ubiquity of small species: Patterns of local and global diversity. Bioscience 54:777–784.

Field, R., B. A. Hawkins, H. V. Cornell, D. J. Currie, J.A.F. Diniz-Filho, et al. 2009. Spatial species-richness gradients across scales: A meta-analysis. Journal of Biogeography 36:132–147.

Ford, M. 2008. AAAS: Large-scale collaborations in physics. and https://arstechnica.com/science/2008/02/aaas-large-scale-collaborations-in-physics/.

Fraune, S., and T.C.G. Bosch. 2010. Why bacteria matter in animal development and evolution. Bioessays 32:571–580.

Frey, S. D., J. Lee, J. M. Melillo, and J. Six. 2013. The temperature response of soil microbial efficiency and its feedback to climate. Nature Climate Change 3:395–398.

Fukami, T. 2004. Community assembly along a species pool gradient: Implications for multiple-scale patterns of species diversity. Population Ecology 46:137–147.

Gill, S. R., M. Pop, R. T. DeBoy, P. B. Eckburg, P. J. Turnbaugh, B. S. Samuel, J. I. Gordon, D. A. Relman, C. M. Fraser-Liggett, and K. E. Nelson. 2006. Metagenomic analysis of the human distal gut microbiome. Science 312:1355–1359.

Goring, S. J., K. C. Weathers, W. K. Dodds, P. A. Soranno, L. C. Sweet, K. S. Cheruvelil, J. S. Kominoski, J. Rüegg, A. M. Thorn, and R. M. Utz. 2014. Improving the culture of interdisciplinary collaboration in ecology by expanding measures of success. Frontiers in Ecology and the Environment 12:39–47.

Guimerà, R., B. Uzzi, J. Spiro, and L.A.N. Amaral. 2005. Team assembly mechanisms determine collaboration network structure and team performance. Science 308: 697–702.

Hampton, S. E., and J. N. Parker. 2011. Collaboration and productivity in scientific synthesis. Bioscience 61:900–910.

Heidorn, P. B. 2008. Shedding light on the dark data in the long tail of science. Library Trends 57:280–299.

Hooper, D. U., F. S. Chapin, J. J. Ewel, A. Hector, P. Inchausti, S. Lavorel, et al. 2005. Effects of biodiversity on ecosystem functioning: a consensus of current knowledge. Ecological Monographs 75:3–35.

Isbell, F., P. B. Reich, D. Tilman, S. E. Hobbie, S. Polasky, and S. Binder. 2013. Nutrient enrichment, biodiversity loss, and consequent declines in ecosystem productivity. Proceedings of the National Academy of Sciences 110:11911–11916.

Kareiva, P. M. 2011. Natural capital: Theory and practice of mapping ecosystem services. New York: Oxford University Press.

Keesing, F., R. D. Holt, and R. S. Ostfeld. 2006. Effects of species diversity on disease risk. Ecology Letters 9:485–498.

Kiers, E. T., and R. F. Denison. 2008. Sanctions, cooperation, and the stability of plant-rhizosphere mutualisms. Annual Review of Ecology Evolution and Systematics 39:215–236.

Lacroix, C., E. W. Seabloom, and E. T. Borer. 2014. Environmental nutrient supply alters prevalence and weakens competitive interactions among coinfecting viruses. New Phytologist 204:424–433.

Leibold, M. A., M. Holyoak, N. Mouquet, P. Amarasekare, J. M. Chase, M. F. Hoopes, et al. 2004. The metacommunity concept: A framework for multi-scale community ecology. Ecology Letters 7:601–613.

Levin, S. A. 1992. The problem of pattern and scale in ecology. Ecology 73:1943–1967.

Lombardo, M. P. 2008. Access to mutualistic endosymbiotic microbes: An underappreciated benefit of group living. Behavioral Ecology and Sociobiology 62:479–497.

Medina, M., and J. L. Sachs. 2010. Symbiont genomics, our new tangled bank. Genomics 95:129–137.

Melillo, J. M., P. A. Steudler, J. D. Aber, K. Newkirk, H. Lux, F. P. Bowles, C. Catricala, A. Magill, T. Ahrens, and S. Morrisseau. 2002. Soil warming and carbon-cycle feedbacks to the climate system. Science 298:2173–2176.

Milchunas, D. G., and W. K. Lauenroth. 1993. Quantitative effects of grazing on vegetation and soils over a global range of environments. Ecological Monographs 63:327–366.

Nabout, J. C., M. R. Parreira, F. B. Teresa, F. M. Carneiro, H. F. da Cunha, L. de Souza Ondei, S. S. Caramori, and T. N. Soares. 2015. Publish (in a group) or perish (alone): The trend from single- to multi-authorship in biological papers. Scientometrics 102:357–364.

Nash, K. L., C. R. Allen, D. G. Angeler, C. Barichievy, T. Eason, A. S. Garmestani, et al. 2014. Discontinuities, cross-scale patterns, and the organization of ecosystems. Ecology 95:654–667.

Nelson, K. E., G. M. Weinstock, S. K. Highlander, K. C. Worley, H. H. Creasy, J. R. Wortman, et al. 2010. A catalog of reference genomes from the human microbiome. Science 328:994–999.

Peay, K. G., M. I. Bidartondo, and A. E. Arnold. 2010. Not every fungus is everywhere: scaling to the biogeography of fungal-plant interactions across roots, shoots and ecosystems. New Phytologist 185:878–882.

Peters, D. P. C., B. T. Bestelmeyer, and M. G. Turner. 2007. Cross-scale interactions and changing pattern–process relationships: consequences for system dynamics. Ecosystems 10:790–796.

Polis, G. A. 1999. Why are parts of the world green? Multiple factors control productivity and the distribution of biomass. Oikos 86:3–15.

Rockström, J., W. Steffen, K. Noone, A. Persson, F. S. Chapin, E. F. Lambin, et al. 2009. A safe operating space for humanity. Nature 461:472–475.

Rodriguez, R. J., J. F. White, A. E. Arnold, and R. S. Redman. 2009. Fungal endophytes: diversity and functional roles. New Phytologist 182:314–330.

Running, S. W. 2012. A measurable planetary boundary for the biosphere. Science 337:1458–1459.

Schimel, D. S., B. H. Braswell, and W. J. Parton. 1997. Equilibration of the terrestrial water, nitrogen, and carbon cycles. Proceedings of the National Academy of Sciences of the United States of America 94:8280–8283.

Shively, J. M., R. S. English, S. H. Baker, and G. C. Cannon. 2001. Carbon cycling: The prokaryotic contribution. Current Opinion in Microbiology 4:301–306.

Shoo, L. P., and V. V. Ramirez. 2010. Global potential net primary production predicted from vegetation class, precipitation, and temperature: Comment. Ecology 91:921–923.

Smith, V. H., T. P. Jones, and M. S. Smith. 2005. Host nutrition and infectious disease. An ecological view. Frontiers in Ecology and the Environment 3:268–274.

Soranno, P. A., and D. S. Schimel. 2014. Macrosystems ecology: Big data, big ecology. Frontiers in Ecology and the Environment 12:3–3.

Tian, X., and P. Doerner. 2013. Root resource foraging. Does it matter? Frontiers in Plant Science 4:303.

U'Ren, J. M., F. Lutzoni, J. Miadlikowska, and A. E. Arnold. 2010. Community analysis reveals close affinities between endophytic and endolichenic fungi in mosses and lichens. Microbial Ecology 60:340–353.

U'Ren, J. M., F. Lutzoni, J. Miadlikowska, A. Laetsch, and A. E. Arnold. 2012. Host and geographic structure of endophytic and endolichenic fungi at a continental scale. Am. J. Bot., in press.

van der Heijden, M.G.A., J. N. Klironomos, M. Ursic, P. Moutoglis, R. Streitwolf-Engel, T. Boller, A. Wiemken, and I. R. Sanders. 1998. Mycorrhizal fungal diversity determines plant biodiversity, ecosystem variability and productivity. Nature 396:69–72.

Vitousek, P. M., J. D. Aber, R. W. Howarth, G. E. Likens, P. A. Matson, D. W. Schindler, W. H. Schlesinger, and D. G. Tilman. 1997. Human alteration of the global nitrogen cycle. Sources and consequences. Ecological Applications 7:737–750.

Wang, K., and R. E. Dickinson. 2012. A review of global terrestrial evapotranspiration: Observation, modeling, climatology, and climatic variability. Reviews of Geophysics 50.

Weathers, K., P. Hanson, P. Arzberger, J. Brentrup, J. Brookes, C. Carey, et al. 2013. The Global Lake Ecological Observatory Network (GLEON): The evolution of grassroots network science. Limnology and Oceanography Bulletin 22:71–73.

Wiens, J. A. 1989. Spatial scaling in ecology. Functional Ecology 3:385–397.

Williams, J. W., and S. T. Jackson. 2007. Novel climates, no-analog communities, and ecological surprises. Frontiers in Ecology and the Environment 5:475–482.

Williams, J. W., S. T. Jackson, and J. E. Kutzbacht. 2007. Projected distributions of novel and disappearing climates by 2100 AD. Proceedings of the National Academy of Sciences of the United States of America 104:5738–5742.

Worm, B., E. B. Barbier, N. Beaumont, J. E. Duffy, C. Folke, B. S. Halpern, et al. 2006. Impacts of biodiversity loss on ocean ecosystem services. Science 314:787–790.

Yatsunenko, T., F. E. Rey, M. J. Manary, I. Trehan, M. G. Dominguez-Bello, M. Contreras, et al. 2012. Human gut microbiome viewed across age and geography. Nature 486:222–227.

Zheng, S., H. Ren, W. Li, and Z. Lan. 2012. Scale-dependent effects of grazing on plant C: N: P stoichiometry and linkages to ecosystem functioning in the Inner Mongolia grassland. PLOS One 7:14.

How Will Organisms Respond to Complex, Novel Environments?

Emilie C. Snell-Rood and Megan E. Kobiela

The world is changing rapidly and organisms are experiencing environments drastically different from those in which they evolved. Biologists are increasingly interested in predicting how organisms will respond to such environmental change. Much of the initial work in this area has focused on species-level traits to make such predictions; for instance, using characteristics of a species' current distribution to predict range changes in the face of climate change or following an introduction to a new region (Peterson 2008, Arau and Peterson 2012). However, variation in individual responses to environmental change will likely play an important role in how populations and species respond. For instance, standing genetic variation will shape the evolutionary potential of a population and plastic, developmental responses will influence how individuals survive in novel conditions (Chevin et al. 2010, Hoffmann and Sgro 2011, Snell-Rood et al. 2018). Integrating evolutionary and ecological approaches has proven to be particularly successful with respect to predicting responses to climate change (Huey 2012, Foden 2008). However, environmental change is occurring simultaneously on many different axes, many of which are more challenging to quantify than changing climate, such as novel toxins, habitats, competitors, predators, or resources. Here we discuss some of the challenges of this question, and explore types of plasticity and individual-level traits that could result in adaptive responses to a wide range of complex environmental change.

Extending Reaction Norms into Novel Space

When will an organism have an adaptive plastic response to a novel environment? In other words, when will an individual be able to developmentally adjust their behavior, physiology, and morphology in a way that maintains performance in new conditions? Reaction norms capture the phenotypic response of genotypes to environmental variability (Schlichting and Pigliucci 1998), but do not necessarily predict whether such a response will be adaptive when an organism is presented with new conditions. What exactly is a "novel" environment? Here, we treat novelty on a continuum, as an environment that occurred infrequently in the recent evolutionary history of a species. It has been suggested that plastic responses to environmental change will likely only be adaptive if evolution has had an opportunity to shape that part of the reaction norm, meaning that adaptive plastic responses to completely novel environments will be rare (Ghalambor 2007). Although nonadaptive plasticity still has important implications for responses to novel environments, adaptive responses to novel environments are more likely to result in persistence of entire populations, which can impact the course of survival and diversification in the new environment. For instance, genetic bottlenecks are much less likely if the entire population survives in a new environment due to plasticity, thus fueling subsequent evolutionary change. We argue that there are several cases where adaptive plastic responses to novel environments may occur, and considering the developmental or physiological mechanisms underlying a reaction norm can give insight into the likelihood of an adaptive response to novel conditions.

In some cases, existing mechanisms of plasticity may be adaptive if they are "ramped up" in novel conditions. In other words, an underlying reaction norm may adaptively extend into novel environmental space. We know that organisms can cope with variable temperatures through physiological mechanisms, such as heat-shock proteins, and behavioral responses, such as microhabitat choice (Munoz and Losos 2018, Feder and Hoffmann 1999). For some range of temperatures, these plastic responses result in maintenance of performance, even for presumably novel temperature. At some point of temperature extremes, the adaptive plastic response breaks down, but there may be some window of novel temperature space where the plastic response is adaptive. What determines the breadth of this window of environmental space? It is hard to know, but it is possible that the degree of variation in their evolutionary history might matter. For instance, temperate and tropical species vary in their evolutionary history with variable temperatures, which shapes thermal tolerance and survival in novel climates (Janzen 1967, McCain 2009).

Reaction norms may adaptively extend into novel space for those traits that have evolved to fit a range of environments or respond to multiple stressors—traits that could be considered generalized or *multipurpose*. This idea of *multipurpose traits* has popped up in several disparate areas concerned with how organisms persist in the face of novel toxins. The term *cross-resistance* was first introduced by microbiologists who observed that bacteria that had evolved resistance to one type of antibiotic often gained protection against different classes of antibiotic to which they had never been exposed (Szybalski and Bryson 1952). Cross-resistance is also discussed in the insecticide resistance literature, where arthropods that have coevolved with certain plant defensive chemicals are usually the first to evolve resistance to novel pesticides (Dawkar et al. 2013). For instance, in a common crop pest, the spider mite *Tetranychus urticae*, similar suites of genes are important for coping with natural chemical defenses in tomato plants and for resistance to pesticides (Dermauw et al. 2013). In the ecological and ecotoxicology literature, the concept of multipurpose traits has been termed *cotolerance,* first used when it was observed that grass species adapted to one heavy-metal pollutant were often also tolerant of elevated levels of other metallic elements (Cox and Hutchinson 1979). Most intriguing for multidimensional anthropogenic change, cotolerance may also occur between seemingly disparate types of stressors, but those that share similar mechanisms of damage to organisms. For example, frogs locally adapted to the genotoxic effects of increased ultraviolet type B radiation at higher elevations exhibit increased resistance to polycyclic aromatic hydrocarbons, pollutants that also have genotoxic effects (Marquis et al. 2009). We argue that considering the natural stressors to which organisms are currently adapted may help us predict which species may tolerate novel, anthropogenic stressors in a way that current models may not.

Ramping up existing plastic responses and multipurpose traits represent two areas where extending reaction norms into novel space may result in an adaptive response. However, it is unclear to what extent these responses would result in adaptive plastic responses when considering complex, multidimensional environmental change. Ramping up heat-shock protein response may be adaptive given a simple linear change in temperature. But even for climate change, there are changes in extreme weather events, temperature variability, precipitation, and humidity. At the same time, there are changes on many axes that can be challenging to align in a linear fashion. Organisms are presented with novel toxins, predators, resources, habitat structure, and levels of available nutrients. For such complex landscapes of environmental change, we argue that "developmental selection" forms of plasticity are the most likely to produce adaptive, novel phenotypes.

Developmental Selection and Adaptive Plastic Responses to Novel Environments

A reaction norm approach can be useful for some applications, but it alone fails to capture the underlying developmental mechanism of plasticity, which can have major implications for the likelihood that an adaptive trait develops in a novel environment. Developmental selection is a mechanism of plasticity that involves both variability and refinement within an individual over developmental time (Snell-Rood 2012). In the case of learning and neural development, variability stems from different motor patterns produced in response to various environmental stimuli or on the basis of broad neural projections early in development; reinforcement stems from strengthening or weakening of neural connections that lead to positive or negative consequences, respectively. Similar processes occur in the development of adaptive immunity, from initial variability of B-cell antibodies to reinforcement based on interaction with antigens (Frank 1996). Although developmental selection is best characterized with respect to behavior and immunity, the basic components of the process are present throughout other levels of development (Snell-Rood et al. 2019, Snell-Rood 2012). The development of muscle and bone structure is responsive to mechanical load—in this case, there is spatial selection on cells that tend to experience more force, which may vary depending on the foraging or locomotor behavior of an individual (Adams et al. 2003, Menegaz 2010). Stochasticity in gene expression, coupled with epigenetic mechanisms that are responsive to environmental inputs suggests such mechanisms may even play out at the level of gene expression (Feinberg and Irizarry 2010). Finally, diversity in the microbiome may be selectable within an individual depending on their diet or environmental conditions (Gilbert 2010). It is possible that organisms that rely more on developmental selection at some levels (e.g., morphology), also rely on it at other levels (e.g., behavior, physiology), either because plasticity in one developmental system requires complementary plasticity in other systems, or due to common selective pressures (West-Eberhard 2003).

Forms of plasticity that rely on developmental selection are likely to result in an adaptive response to a novel environment because they involve sampling and selection within an individual. One can think of an individual exploring phenotypic space over developmental time, adopting traits that work particularly well in the local environment. However, the benefits of developmental selection scale directly with the costs, which stem from the sampling process (Frank 1996). Gathering information takes time, energy, and investment in structures to process such information, such as large, energetically expensive brains in the case of learning, or the production of many B cells in the case of acquired immunity. This "cost of being naïve" means that reliance on developmental selection should lead to lengthened developmental time, delays in reproduction, and dependence on greater parental investment to survive this sampling period, thus resulting in lower fecundity. In other words, developmental selection forms of plasticity should shift life histories from fast to slow life histories (Snell-Rood 2012). It is possible that quantifying aspects of an organism's life history might hint at their developmental selection potential, but this is an area in need of more research.

Developmental selection forms of plasticity should be particularly important in considering responses of organisms to novel environments that differ from ancestral environments in a range of different ways. Many of the classic examples of plasticity, such as social insect castes or seasonal polyphenisms, fall into a different category—that of evolved developmental programs, or plastic developmental switches (West-Eberhard 2003). However, these forms of plasticity tend to be specific to the environmental range in which they have evolved. Furthermore, switch mechanisms of plasticity are not conducive to complex landscapes of environmental variation because developmental programs specific to a wide range of environments are vulnerable to relaxed selection (Snell-Rood et al. 2010).

These ideas suggest that different mechanisms of plasticity should have different population-level consequences in the face of environmental change. Developmental selection forms of plasticity should be more likely to result in adaptive responses to novel environments for a significant portion of a population. Other forms of plasticity may be more likely to result in nonadaptive plastic responses in variable directions across a population (Ghalambor et al. 2007). These population-level consequences may generate interactions between different types of plasticity and evolutionary responses to new environments. Developmental selection forms of plasticity should be likely to shift an entire population to a novel selective peak. Models suggest such a shift will reduce selection intensity (Lande 2015), referred to as *behavioral inertia* with respect to plasticity in behavior (Huey et al. 2003). However, these forms of plasticity are extremely costly, which should select for loss of plasticity and the genetic assimilation of the novel phenotype if the novel environment is somewhat stable or can be stabilized through behavioral choices such as habitat preferences (Snell-Rood 2013, Stamps 1995). Because the entire population persists, underlying genetic variation may be maintained, facilitating evolutionary responses. At the same time, because developmental selection forms of plasticity are associated with life-history tradeoffs associated with a slower life history, it is possible that population growth rate may be slower, dampening the evolutionary response to rapid environmental change (Reznick and Ghalambor 2001). For nonadaptive forms of plasticity, where population-level responses are uncorrelated, it is more likely the population will go through a bottleneck, and the exact direction of phenotypic change may be less likely to predict. Regardless, it is clear there are multiple interesting and open questions about how plastic responses will interact with evolutionary responses to novel environments, paving the way for both theoretical and empirical work in the decades to come. Recent models of evolutionary rescue in novel environments have begun to incorporate plasticity, but we suggest that considering the mechanism of plasticity will substantially change the outcome (Ashander et al. 2016, Chevin et al. 2013).

Open Questions

We have suggested that certain forms of plasticity may be more likely to produce adaptive responses to novel environments. However, this is a wide-open area of inquiry. In considering reactions to environmental change along multiple axes, it is clear that adaptive responses will include plasticity in a range of traits (e.g., behavior, physiology), often coupled with traits that work well in a range of environments (multipurpose traits or generalized responses to stress). Can we predict which species are capable of enhanced developmental selection or generalized responses to stress? The costs associated with developmental selection forms of plasticity suggest that species with slower, K-selected life-history traits may be more likely to invest in such plasticity. Interestingly, it has also been suggested that long-lived species would be more likely to have multipurpose traits to cope with changing conditions over their lifetime, such as higher levels of enzymes to cope with oxidative stress (Beckman and Ames 1998). Future work is needed to clarify whether life-history traits would be enough to predict adaptive responses to novel conditions, or whether other traits, such as degree of specialization, might be more accurate.

In considering the costs and tradeoffs associated with plasticity, it's important to note that humans are drastically altering nutrient and resource availability. To what extent does anthropogenic change in nutrient availability interact with plasticity in responses to novel conditions? It is possible that increases in resource availability ameliorate tradeoffs between costly forms of plasticity and life history traits (Snell-Rood et al. 2015), which could explain patterns of plasticity in invasive species relative to non-invaders (Davidson et al. 2001). In this case, regions or species that see greater increases in resource availability may have not only adaptive plastic responses to novel environments, but also population-level evolutionary responses as tradeoffs with fecundity are minimized.

If plasticity plays an important role in survival in novel environments, to what extent will diversification in such environments be biased by existing developmental programs? Will the mechanisms of plasticity provide the axes upon which genetic divergence proceeds? Although some evidence suggests that diversification may be biased along axes of adaptive plasticity (Pfennig 2010, Parsons et al. 2016), other recent research calls into question the directionality of this change (Ghalambor et al. 2015). This underscores the importance of considering both adaptive and nonadaptive plasticity when evaluating the potential drivers of genetic change in novel environments.

There has been increasing interest over the last decade in understanding the role of plasticity in responses to rapid environmental change. However, the complexity of such environmental change demands a consideration of the types of plasticity that are likely to produce adaptive responses to novel environments. In the coming decades, advances in techniques that can be used to quantify and manipulate developmental mechanisms will no doubt help clarify the extent to which developmental selection, and associated life-history tradeoffs, are indeed important players in this equation.

References

Adams, C. E., C. Woltering, and G. Alexander. 2003. Epigenetic regulation of trophic morphology through feeding behaviour in Arctic charr, *Salvelinus alpinus*. Biological Journal of the Linnean Society 78(1):43–49.

Araujo, M. B., A. T. Peterson. 2012. Uses and misuses of bioclimatic envelope modeling. Ecology 93(7):1527–1539.

Ashander, J., L. M. Chevin, and M. L. Baskett. 2016. Predicting evolutionary rescue via evolving plasticity in stochastic environments. Proceedings of the Royal Society B: Biological Sciences 283(1839):20161690.

Beckman, K. B., and B. N. Ames. 1998. The free radical theory of aging matures. Physiological Reviews 78(2):547–581.

Chevin, L. M., R. Gallet, R Gomulkiewicz, R. D. Holt, and S. Fellous. 2013. Phenotypic plasticity in evolutionary rescue experiments. Philosophical Transactions of the Royal Society B: Biological Sciences 368(1610):20120089.

Chevin, L. M., R. Lande, and G. M. Mace. 2010. Adaptation, plasticity, and extinction in a changing environment: towards a predictive theory. Plos Biology 8(4):8.

Cox, R. M., and T. C. Hutchinson. 1979. Metal co-tolerances in the grass *Deschampsia cespitosa*. Nature 279(5710):231–233.

Davidson, A. M., M. Jennions, and A. B. Nicotra. 2011. Do invasive species show higher phenotypic plasticity than native species and, if so, is it adaptive? A meta-analysis. Ecology Letters 14(4):419–431.

Dawkar, V. V., Y. R. Chikate, P. R. Lomate, B. B. Dholakia, V. S. Gupta, and A. P. Giri. 2013. Molecular insights into resistance mechanisms of lepidopteran insect pests against toxicants. Journal of Proteome Research 12(11):4727–4737.

Dermauw, W., N. Wybouw, S. Rombauts, B. Menten, J. Vontas, M. Grbic, R. M. Clark, R. Feyereisen, and T. Van Leeuwen. 2013. A link between host plant adaptation and pesticide resistance in the polyphagous spider mite *Tetranychus urticae*. Proceedings of the National Academy of Sciences of the United States of America. 110(2):E113–E122.

Feder, M. E., and G. E. Hoffmann. 1999. Heat-shock proteins, molecular chaperones, and the stress response: Evolutionary and ecological physiology. Annual Review of Physiology 61:243–282.

Feinberg, A., and R. A. Irizarry. 2010. Stochastic epigenetic variation as a driving force of development, evolutionary adaptation, and disease. Proceedings of the National Academy of Sciences of the United States of America 107:1757–1764.

Foden, W. B., S. H. M. Butchart, S. N. Stuart, J.-C. Vie, H. R. Akcakaya, A. Angulo, L. M. DeVantier, A. Gutsche, E. Turak, L. Cao, et al. 2013. Identifying the world's most climate change vulnerable species: A systematic trait-based assessment of all birds, amphibians and corals. PLOS One 8(6).

Frank, S. 1996. The design of natural and artificial adaptive systems. In: M. Rose and G. Lauder, eds. Adaptation. New York: Academic Press, 451–505.

Ghalambor, C. K., J. K. McKay, S. P. Carroll, and D. N. Reznick. 2007. Adaptive versus non-adaptive phenotypic plasticity and the potential for contemporary adaptation in new environments. Functional Ecology 21:394–407.

Ghalambor, C. K., K. L. Hoke, E. W. Ruell, E. K. Fischer, D. N. Reznick, and K. A. Hughes. 2015. Non-adaptive plasticity potentiates rapid adaptive evolution of gene expression in nature. Nature 525(7569):372–375.

Gilbert, S. F., E. McDonald, N. Boyle, N. Buttino, L. Gyi, M. Mai, N. Prakash, and J. Robinson. 2010. Symbiosis as a source of selectable epigenetic variation: Taking the heat for the big guy. Philosophical Transactions of the Royal Society B: Biological Sciences 365(1540):671–678.

Hoffmann, A. A., and C. M. Sgro. 2011. Climate change and evolutionary adaptation. Nature 470(7335):479–485.

Huey, R. B., M. R. Kearney, A. Krockenberger, J. A. Holtum, M. Jess, and S. E. Willims. 2012. Predicting organismal vulnerability to climate warming: roles of behaviour, physiology and adaptation. Philosophical Transactions of the Royal Society B: Biological Sciences 367(1596):1665–1679.

Huey, R. B., P. E. Hertz, and B. Sinervo. 2003. Behavioral drive versus behavioral inertia in evolution: A null model approach. American Naturalist 161(3):357–366.

Janzen, D. H. 1967. Why mountain passes are higher in the tropics. American Naturalist 101(919):233–249.

Lande, R. 2015. Evolution of phenotypic plasticity in colonizing species. Molecular Ecology 24(9):2038–2045.

Marquis, O., C. Miaud, G. F. Ficetola, A. Boscher, F. Mouchet, S. Guittonneau, and A. Devaux.2009. Variation in genotoxic stress tolerance among frog populations exposed to UV and pollutant gradients. Aquatic Toxicology 95(2):152–161.

McCain, C. M. 2009. Vertebrate range sizes indicate that mountains may be "higher" in the tropics. Ecology Letters 12(6):550–560.

Menegaz, R. A., S. V. Sublett, S. D. Figueroa, T. J. Hoffman, M. J. Ravosa, and K. Aldridge.2010. Evidence for the influence of diet on cranial form and robusticity. Anatomical Record: Advances in Integrative Anatomy and Evolutionary Biology 293(4):630–641.

Munoz, M. M., and J. B. Losos. 2018. Thermoregulatory behavior simultaneously promotes and forestalls evolution in a tropical lizard. American Naturalist 191(1):E15–E26.

Parsons, K. J., M. Concannon, D. Navon, J. Wang, I. Ea, K. Groveas, C. Campbell, M. Rose, and G. Lauder, and R. C. Albertson. 2016. Foraging environment determines the genetic architecture and evolutionary potential of trophic morphology in cichlid fishes. Molecular Ecology 25(24):6012–6023.

Peterson, A. T. 2003. Predicting the geography of species' invasions via ecological niche modeling. Quarterly Review of Biology 78(4):419-433.

Pfennig, D. W., M. A. Wund, E. C. Snell-Rood, T. Cruickshank, C. D. Schlichting, and A. P. Moczek. 2010. Phenotypic plasticity's impacts on diversification and speciation. Trends in Ecology & Evolution 25(8):459–467.

Reznick, D. N., and C. K. Ghalambor. 2001. The population ecology of contemporary adaptations: What empirical studies reveal about the conditions that promote adaptive evolution. Genetica 112:183–198.

Schlichting, C. D., and M. Pigliucci. 1998. Phenotypic evolution: a reaction norm perspective. Sunderland, Mass.: Sinauer Associates, 387.

Snell-Rood, E. 2012. Selective processes in development: implications for the costs and benefits of phenotypic plasticity. Integrative and Comparative Biology 52:31–42.

Snell-Rood, E. 2013. An overview of the evolutionary causes and consequences of behavioural plasticity. Animal Behaviour 85(5):1004–1011.

Snell-Rood, E., R. Cothran, A. Espeset, P. Jeyasingh, S. Hobbie, and N. I. Morehouse. 2015. Life history evolution in the anthropocene: Effects of increasing nutrients on traits and tradeoffs. Evolutionary Applications 8(7):635–649.

Snell-Rood, E. C., M. E. Kobiela, K. L. Sikkink, A. M. Shephard. 2018. Mechanisms of plastic rescue in novel environments. Annual Review of Ecology, Evolution, and Systematics 49:331–354.

Snell-Rood, E., J. D. Van Dyken, T. Cruickshank, M. J. Wade, and A. P Moczek. 2010. Toward a population genetic framework of developmental evolution: costs, limits, and consequences of phenotypic plasticity. BioEssays 32:71–81.

Stamps, J. 1995. Motor learning and the value of familiar space. American Naturalist 146(1):41–58.

Szybalski, W., and V. Bryson. 1952. Genetic studies on microbial cross resistance to toxic agents. I. Cross resistance of Escherichia coli to fifteen antibiotics. Jour Bact 64(4):489–499.

West-Eberhard, M. J. 2003. Developmental plasticity and evolution. New York: Oxford University Press.

Variance-Explicit Ecology

A Call for Holistic Study of the Consequences of Variability at Multiple Scales

Marcel Holyoak and William C. Wetzel

Variability or heterogeneity is everywhere in ecology and evolution. For instance, Levins (1968) introduces his classic work "Evolution in Changing Environments: Some Theoretical Explorations" as "a series of explorations . . . around the common theme of the consequences of environmental heterogeneity." We have many reasons for studying variability at different levels, including within-individual variation through time or in modular organisms (e.g., tree branches); among or across individual variation in genetics or traits (including behaviors) within a population, guild, community, or ecosystem; or as environmental (e.g., meteorological, hydrological, limnological, oceanographic) drivers of processes of interest. Yet, ecologists most frequently manipulate the mean value of a driving variable of interest and look at its ecological effects and ignore variation in that driving variable or process of interest. For instance, we might rear an insect or plant at three average temperatures and then use analysis of variance to compare individual growth rates at these temperatures but overlook variation in temperature over time and its effects on growth rate. Most frequently, patterns of spatial or temporal variation in either biotic or abiotic factors are used to make inferences about underlying mechanisms (e.g., using geostatistical techniques, Rossi et al. 1992 or using power law plots, Taylor 1961). Variation may even be treated as an annoyance, requiring larger sample sizes to achieve statistical power or as unexplained variation for things that are stochastic or where mechanisms are not understood. As we aim to illustrate, although we have frameworks and sometimes good knowledge about direct effects of single forms of variation, we frequently miss important questions and opportunities about the mechanisms involved, how multiple forms and scales of variation combine, and effects across organizational levels.

A major unsolved problem in ecology is resolving the relative importance between different types and scales of variability to ecological processes.

Organisms experience and respond to variation at many different biological and ecological levels, ranging from physiological to behavioral to populations and communities, and eventually to metapopulations, metacommunities, and geographic ranges. We will argue that certain forms of variation have been quite well studied, but that we lack research programs that might provide information about the relative importance of different mechanisms and interactions among them. Investigating these gaps might provide a mechanistic framework for how to understand how different forms of variation combine to affect ecological problems. A variety of general questions follow from our line of reasoning. What is the relative importance of different mechanisms by which variability influences ecology and what is the relative importance of variability at different scales? At what scales is variability averaged over so that it does not matter? At what scales does variability most influence ecology and how does it do so?

There have been several calls for more explicit consideration of the consequences of variability in ecology, either limited to particular mechanisms by which variation acts (e.g., through nonlinear averaging, Ruel and Ayres 1999), or effects on particular levels of ecological organization (e.g., Bolnick et al. 2011). There are a growing number of studies that do just that, but most are restricted to one scale and type of variability. Moreover, most studies either focus solely on one mechanism or ignore mechanisms altogether and instead just measure the net effect of variation. An example of a kind of problem that ecologists have worked extensively on and for which we have a relatively good understanding of the role of variation is the literature linking plant diversity with plant yield (biomass production). Most of this literature indicates that increasing functional trait diversity (variability) in plant communities leads to increased plant biomass and greater overall resource utilization (e.g., Cardinale et al. 2006). This positive effect on biomass comes partly from a sampling effect and more substantially through niche complementarity and/or positive interspecific interactions (van Ruijven and Berendse 2005). Moving beyond plants, there is less understanding of the effects of plant trait diversity on higher trophic levels (e.g., Ruel and Ayres 1999, Benedetti-Cecchi 2000). A recent meta-analysis that we took part in suggests there could be relatively consistent negative effects of (within-species) variation in plant nutritional quality traits on average herbivore performance through Jensen's Inequality (Wetzel et al. 2016). Considering predators and herbivores, Mason et al. (2014) suggested that some generalist herbivores perform better feeding on a diversity of resources and that this may affect higher trophic levels; Arctiid caterpillars (*Grammia incorrupta*) were well defended against predators when they sequestered secondary metabolites from several different plant species but poorly defended when they sequestered compounds from only one plant species. Overall, such studies show that for certain problems we understand some mechanisms by which a particular form of variation acts on processes of interest. Equally well, the effects of certain forms of individual variation in altering population dynamics have been widely studied (e.g., Grimm 1999), as have several other problems relating variation to processes within a single species or trophic level in ecology and evolution. There has been less work exploring how variation among multiple trophic levels combines to affect herbivore performance. For instance, how does variation in herbivore traits relate to variation in plant traits to affect herbivore performance (Moreira et al. 2016)?

To describe the background and elements required to proceed towards an integration of scales and types of variation and mechanisms by which variation acts, we present the following: (1) An overview of scales and types of biotic and abiotic variation by describing three frameworks for classifying them. (2) A summary of common mechanisms by which variation influences ecological dynamics. (3) A description of what might be gained by integrating different types and scales of variation. (4) We conclude by highlighting some next steps that could move us towards a conceptual framework for how organisms integrate multiple types and scales of variation.

Ways of Classifying Scales and Types of Variation

The literature describes a range of ways of classifying variation or that can be borrowed from classifications of other ecological patterns. We present three such classifications, one based on the structure of environmental variation, a second recognizing the hierarchical nature of biological or ecological organization, and a general scheme that might be applied to any type of variation.

Environmental Variation

Environmental variation is most commonly viewed as the physical, chemical, and geological factors that are largely independent of biotic factors at least over the time scales of most concern to ecologists; such factors were termed *scenopoetic* in an ecological niche context by Soberón and Arroyo-Peña (2017). Such environmental variation merits separate consideration from biotic variation because it has its own scaling and structure, occurring continuously from microscopic to global scales. For example, temperature varies temporally at a scale of minutes as clouds pass in front of the sun, at a scale of hours as the sun rises, peaks, and sets; at a scale of months as the seasons progress; at scales of years to decades (sunspot cycles, el Niño cycles etc.); and at geologic timescales through glacial cycles. Environmental variation may have stochastic and predictable components. Some work also separates recurrent stochastic components from extreme events, including hurricanes, floods, and fires (e.g., Shaffer 1981, Yang et al. 2008, Yang et al. 2010). Temporal environmental variation is often somewhat cyclical and predictable, as exemplified by daily temperature cycles, seasonal variation, and sunspot cycles. Spatial variation often increases with distance. For instance, Bell et al. (1993) studied variation in physical variables in lakes or soil nutrients from a variety of geographic areas and found that, in general, environmental variation continued to increase with spatial scale of study (distance). The scaling of different environmental factors with distance or with time has been used to identify relevant processes in studies of scaling (Levin 1992, Storch et al. 2007). Denny (2015) describes how to use principles from engineering and physics to understand both physical environment interactions and subsequent species interactions through what he terms "ecological mechanics."

The extent to which environmental (and biotic) variation is encountered by an organism depends on its scale of movement, longevity, and life cycle. Within life cycles, periods of dormancy versus intense resource use are particularly relevant. Spatial and temporal variation are both potentially relevant in several ways. McPeek and Kalisz (1998) modeled the effect of spatial, temporal, and spatiotemporal variation on the evolution of dormancy versus dispersal, finding that pure temporal variation promotes dormancy and that spatial and spatiotemporal variation promote dispersal. Cyclical seasonal migration of North American and European passerine birds is known to be a response to extreme temperatures (Newton and Dale 1996a, 1996b), whereas Australian butterflies respond to extreme dry conditions (Dingle et al. 2001).

In some cases, biotic factors may interact with abiotic factors, and even then, it may be a valid simplification to separately consider abiotic environmental factors if we are studying processes that operate at very different timescales relative to the rate of change of environmental factors through biotic–abiotic coupling (e.g., many ecosystem processes). However, if we were studying long-term tree growth, then a feedback between habitat fragmentation and microclimate might be relevant (e.g., Laurance and Williamson 2001); for long-lived perennial grasses mineral nutrients in soils may depend on grazing history (e.g., McNaughton et al. 1997). For such processes it would make sense to instead think about how to combine different forms of biotic and abiotic variation into analyses.

Biotic Variation

Biotic variation in traits relevant to ecological interactions occurs from subindividual to between individuals within a species or across species. Raw genetic and somatic variation within individuals (or part of them), expressed as traits including behaviors that vary in timing and sequence, may relate to subspecific (e.g., races, morphs) variation, other taxonomic levels, and to higher organizational levels within ecology, paleobiology, biogeography, and other biological sciences. A brief tour of relevant levels of biotic organization helps to identify some of the things that each level contributes or emphasizes. Of course, lower-level variation is included in higher organizational levels but may or may not have effects on higher-level processes. For instance, there is a growing literature on community and ecosystem genetics that investigates the effects of genotype on processes from communities to ecosystems (Whitham et al. 2003).

At the level of within-individual variation, individual organisms frequently respond in plastic ways to ambient environmental and biotic conditions, including behaviors, physiological acclimation, developmental flexibility, life-historical changes in timing, and as ecological engineers (Jones et al. 1997). Critically, such plasticity changes both variability encountered and the relationship between this variability and emergent or higher-level processes performed by the organism. Although there is a great deal of literature on behavioral plasticity, developmental plasticity, life histories, and related subjects, it is unusual for studies to make links to emergent higher-level processes of interest. Beyond plasticity, individual history may produce changes in organisms. A plant phenotype might vary through time depending on the history of herbivory and plant responses to herbivory through inducible defenses (Adler and Karban 1994, Karban and Baldwin 1997). Individual history of infection may alter the susceptibility to the same or new diseases in the future in ways that are either positive or negative. Carryover effects from one habitat to another may produce a relevance of spatial history (e.g., Talley et al. 2006), and there are several named temporal carryover effects (e.g., maternal effects) that produce time-lagged responses (e.g., Ratikainen et al. 2008). Organisms with repeating structures, such as plants with multiple leaves and reproductive organs, may produce especially high variation among organs within individuals (Herrera 2009).

Variability among individuals within a population is recognized as intraspecific trait variation (e.g., Bolnick et al. 2011), arising through phenotypic plasticity, genetic diversity (Hughes et al. 2008), and ontogeny, including life histories and history more generally. Just as species may have different population dynamics, or serve different roles in communities or ecosystems, the same is true of individuals with different traits within a species. Intraspecific variation has been a recent focus of study in ecology (e.g., reviews by Hughes et al. 2008, Bolnick et al. 2011), yet as far back as the 1970s Lomnicki (1978) pointed out that population regulation could not occur if all individuals within a population were identical. Recent synthetic analyses indicate that approximately 30% to 50% of the total variation in plant functional traits in plant communities occurs at the intraspecific level, with intraspecific variation being especially large for chemical traits and smaller for physical traits (Albert et al. 2010, Messier et al. 2010, Siefert et al. 2015).

A population is not necessarily the appropriate scale at which to study variation. At a higher level, sections or subpopulations within a population may sometimes be identified, or analogously populations with a metapopulation. They are described using variables such as phenotype frequencies, population densities (e.g. aggregations, congregations), or sex ratios within populations. Population cohorts may be identifiable based on time of birth, leading to a temporal structure, and such temporal variation is known to produce cyclical population dynamics (e.g. Kendall et al. 1999). Alaska sockeye salmon provide a good example of population segments, with stream- versus lake-spawning individuals varying in morphology (e.g. Blair et al. 1993).

Another form of variation is created by species diversity, making guilds, communities, and ecosystems relevant. Ecologists are familiar in population, community, and ecosystem ecology with studying the effects of species diversity or interspecific differences (both forms of variability) on processes of interest. Interspecific variation may have effects through direct or indirect species interactions within a guild, or more diffuse community or ecosystem-level effects. Species richness, multivariate dispersion of communities (e.g., principal components analysis (PCA) of species' abundances), functional diversity, phylogenetic diversity, and variation in interaction strength within a food web all capture elements of across species variation. Such variation may also be the complex outcome of the action of biological and environmental factors, and emergent effects of such variation, which is the sum of what we describe in this chapter. At some level, diversity begets diversity, in that the variation experienced by an organism may be a response to variation within a community. Hence there may be a rapid scaling up of the potential for complex effects of species diversity on organisms, just as the potential for higher-order species interactions increases rapidly with the number of species in a community.

Pattern and Structure of Variation

Irrespective of whether variation is biotic or abiotic we can consider whether variation is essentially unstructured within the scope of the process under exploration, or whether there is a pattern or structure involved. In a more specific form of this, Adler et al. (2001) pointed out that spatial heterogeneity is composed of spatial variance and spatial pattern (structure). Although spatial variance is necessary for spatial heterogeneity, spatial variation may or may not be organized into a spatial pattern.

Unstructured snapshots may be typical of a foraging herbivore if plants of different quality are essentially randomly distributed within the area within which it can forage and during the relevant time period. On the other hand, Tobler's first law of geography reminds us that near things tend to be more similar than far things, which lends structure and predictability to spatial variance (Tobler 1970). Such spatial autocorrelation typically has characteristic spatial scales. For example, a species of herbivorous beetle tended to occur in clumps of its host plant of 25–50 m in diameter and separated from neighboring clumps by 200–300 m (Talley 2007). Temporal variation reflects daily, lunar, solar, and longer-term processes such as El Niño, Pacific Decadal Oscillation, and sunspot and glacial cycles, but also the less-predictable components of weather and seemingly random and often extreme events. Temporal autocorrelation and the unidirectionality of history structure such variation.

Some Existing Mechanisms by Which Variation Influences Ecological Processes

Mechanisms by which variation alters a process of interest include general mechanisms that can apply to any ecological level of organization and some that are specific to particular organizational levels. Consequently, it is a large topic and we aim to be illustrative rather than encyclopedic in our descriptions but encourage readers to think beyond the mechanisms we include. We first describe some more general mechanisms and then describe those that relate more closely to biology and ecology.

General Mechanisms: Mathematical Functions as Filters, and Effects of Nonlinearity of Functions

Considering mathematical functions (e.g., $y = f(x)$) as filters provides a broad view of their role in changing variation between the input (x) and output (y) variables (Denny and Benedetti-Cecchi 2012). Trait variation interacts with responses to biotic and abiotic variation to determine the inputs to filters. The outputs are the ecological processes of interest and either the average outputs or variation in outputs may be of interest. Such a filtering view is frequently expressed in the literature about scaling in ecology, asking if variation at one level is present at another (e.g., Storch et al. 2007). Peter Chesson's scale-transition theory provides a mathematical framework for formally analyzing systems of equations to investigate such changes (Chesson 2012). Feedback processes, such as density and frequency dependence, can either amplify or cancel out variation from the input to the output. More generally, different forms of variation may act additively, synergistically, or antagonistically. Ideas about resonance emphasize that processes acting at different temporal or spatial frequencies may amplify or cancel out variation (e.g., Blarer and Doebeli 1999). Such ideas are interesting and poorly explored given that both individual growth and population growth have associated timescales, and that density-dependent functions produce characteristic return times for populations returning to an equilibrium (e.g., Luckinbill and Fenton 1978). Nonlinear equations have the ability to amplify variation, as is emphasized in the literature on chaotic dynamics (e.g., Hastings et al. 1993); viz., small amounts of variation in initial conditions can lead to large differences in the emergent (population) dynamics. The approach led Hastings et al. (1993) to ask questions about nonlinear dynamics, such as what are the respective roles of endogenous and exogenous factors, and do they interact? More generally, determining the role of variation in an input variable on a process of interest requires us to determine if the dynamics are nonlinear or not.

Another important effect of nonlinearity of functions is how variation in an input variable affects the average value of the output variable, our process of interest. Jensen's Inequality describes the role of nonlinearity in altering the output from a mathematical function (reviewed by Ruel and Ayres 1999). Variation in an input variable to a function that is concave down will reduce the average value that is given by the function relative to a linear function, and a convex function does the opposite. Sibly et al. (2005) found that most population time series produced nonlinear and concave curves for per-capita growth as a function of population size (or density); consequently, variation in population density reduces average population size below the equilibrium abundance (carrying capacity). (The statistics of Sibly et al. were criticized in several published comments but the general point about the shape of functions and effect of variation is well illustrated by the example.) Nonlinear or non-monotonic functions are common in ecology and arise through a variety of mechanisms, as reviewed by Zhang et al. (2015). Mechanisms leading to nonlinearity include the law of tolerance, whereby species underperform with either too little or too much of a required ecological factor (Shelford 1931), through the action of adaptive behaviors or physiological adaptation altering relationships between environmental factors and organismal responses, or by sequentially combining multiple synergistic (or antagonistic) factors so as to produce nonlinear outcomes (Zhang et al. 2015). The strong role of nonlinearity leads us to question whether we should be using general mechanistic functional forms for particular problems (e.g., functional responses of predators to prey, or allometric equations), or whether we should use more flexible functional forms to represent arbitrary forms of nonlinearity (e.g., cubic splines (Schluter 1988), or response surface methodologies (Inouye 2005)). Nonmechanistic statistical equations can still be used to infer things like the size of a Jensen effect or whether environmental variation as an input is amplified or damped down in the output from the mathematical function of interest. In some cases, nonlinear averaging may serve as a null model to predict the expected effect (Koussoroplis et al. 2017). For instance, Pearse et al. (2018) looked at how experimental variance in the concentration of a plant toxin in artificial diet altered herbivore performance and found that nonlinear averaging predicted toxin variance would enhance performance, whereas the observed effect was negative. The authors hypothesized that the costs of physiological acclimation in the face of trait variance (Wetzel and Thaler 2016) explained the difference between the predicted and observed results.

Mechanisms Involving Biology and Ecology

Physiological Responses and Consequences

When individuals directly encounter biotic or abiotic variability within their lifetime and are unable to use behavioral mechanisms to avoid it, it is likely to have important physiological consequences. Variability is especially important for organismal physiology because when it is high it encompasses extreme values, which is when physiological stress is expected to be greatest and the consequences of not dealing with conditions may be most harmful. This occurs because relationships between environmental variables and organismal performance tend to be concave-down over large environmental ranges as expressed by Shelford's law of tolerance (1932); the general mechanism behind this is Jensen's Inequality (or nonlinear averaging), discussed previously.

The physiological responses of consumers to diet species diversity–trait diversity at the guild or community level—are especially well studied. It was long believed that diverse diets helped consumers achieve balanced nutrient intake and diluted the effects of toxic defenses associated with any one prey species (Bernays et al. 1994). A recent meta-analysis, however, indicates that mixed-species diets tend to be no better for consumers than the best single-species diet, and they are typically worse than the best single-species diet when diet species possess chemical defenses (Lefcheck et al. 2013). This suggests that consumers facing greater diet variability may experience reduced physiological performance (Wetzel and Thaler 2018). It is often not clear how to view heterogeneity within diets. For instance, Marzetz et al. (2017) show that the chemical composition of algal species as food are more important to growth rates of *Daphnia* than are the algal species' identities or diversity. One general way forward may be to use colimitation theory to integrate several physical and/or biotic factors into a single unified conceptual framework that incorporates potential nonlinearities that arise in a multivariate context, but which are not apparent when factors are considered unidimensionally (Koussoroplis et al. 2017).

Organisms can have important physiological adaptations that help them cope with variability. These take the form of physiological plasticity, which allows organisms to change their physiology to maximize performance under current conditions, or fixed phenotypes that are useful for coping with variable environments. Examples of plastic responses to variability include insect herbivores that reshape their digestive chemistry in response to changing plant conditions (e.g., Bolter and Jongsma 1995), and tadpoles, which change gut size in response to predation risk and food availability (Relyea and Auld 2004). If phenotypic alterations of this nature are costly, which they certainly are for insect herbivores acclimating to plant conditions, then high variability could lead to costly repeated acclimation (Wetzel and Thaler 2016). Rather than changing physiology to match current biotic and abiotic conditions, some organisms pay a permanent cost to be constantly ready for changing conditions. For example, 38 predatory fish species maintained gut sizes two- to three-fold larger than necessary for the average amount of prey they encountered; this allowed them to be ready to process rare pulses of high food abundance (Armstrong and Schindler 2011).

Behavioral Responses and Consequences

Movement, activity patterns and resource selection are major ways that organisms modulate the amount and type of abiotic and biotic variability that they experience. We often think about resource selection as having the goal of getting an organism to resources of a certain quality or quantity, but resource selection is likely to be vital for coping with variability in resource quality and quantity. Optimality theory suggests a wide range of ways organisms reduce costs, such as decisions when to leave patches in response to declining food quality (from the marginal value theorem) (MacArthur and Pianka 1966; Charnov 1976), or when to consume less-profitable food items (from optimal diet theory) (e.g. Emlen 1966). Game theory shows how such decisions can be contingent on other individuals if an organism is maximizing resource intake (or some other currency). Similarly, habitat selection behavior modifies the environmental variation that an organism experiences (e.g., Morris 2003). These central ideas in behavioral ecology alter the relationship between variability encountered and a fitness-related output. Of course, for real organisms the ability of such behaviors to reduce variability between input and output has its limits. For instance, Sih and Christensen (2003) identified conditions such as prey mobility that prevented predators from foraging optimally, and which may therefore lead to a more direct relationship between prey variation and variation in food intake.

Extreme variation may also be coupled with unusual and interesting behaviors. Nomadism is thought to arise in response to extremes of spatiotemporal variation in resource availability or environmental conditions. For instance, desert locust outbreaks track spatiotemporally variable rainfall and subsequent periods of plant germination and growth (Jonzén et al. 2011). Environmental variation that is novel to an organism may also produce different ecological effects to that which is routinely encountered. Hence, Sih and colleagues coined the term human-induced rapid environmental change (HIREC) to draw attention to anthropogenic changes that place organisms under conditions (e.g., population) the species has not experienced before and may produce either individual- or population-level responses (Robertson et al. 2013).

Population and Community Responses and Consequences

Population and community responses to variation are numerous. Various mechanisms for the consequences of and responses to trait variation were reviewed by Bolnick et al. (2011) and serve as a starting point: (1) diversification of species interactions, such as increased generalism, through traits affecting the kinds of interactions and with which other individuals or species focal individuals interact; (2) a portfolio effect produced by covariation among individuals with different traits; (3) phenotypic subsidy whereby genetic variation or plasticity decouple phenotype and fitness; (4) trait variation as a source of adaptive variation in rapid evolution; and (5) sampling effects whereby small populations contain only a small number of traits. Population ecologists often relate variation to extinction risk through population viability analyses and decompose mechanisms into those involving demographic and environmental stochasticity, or more extreme catastrophes (Shaffer 1981). At a multispecies level there is a large body of research on indirect interactions that are trait- or density-mediated (e.g. Bolker et al. 2003), indirect effects (e.g., Menge 1995), and positive versus negative species interactions (e.g., Tylianakis et al. 2008). All the above serve as potential mechanisms for how interspecific variation modifies processes of interest. Surprisingly little empirical work has evaluated the mechanisms described by Bolker et al. (2003) and their relative importance.

Metapopulation and Metacommunity Responses and Consequences

Source–sink dynamics (Pulliam 1988) and habitat-specific demography recognize that habitat areas have heterogeneous effects on population dynamics. Similarly, species sorting and mass effects ideas from metacommunity theory do the same for whole communities (Leibold et al. 2004). Rescue and mass effects across space alter the ability of species to cope with low-quality habitats (Pulliam 1988; Leibold et al. 2004). Most frequently, such ideas are applied to constant habitat heterogeneity, but we can also view habitats as changing, as captured by ideas about source–sink inversions (e.g., Boughton 1999) or temporally autocorrelated environmental conditions (Gonzalez and Holt 2002). Local adaptation may drive whether populations are sources or sinks and modify source–sink dynamics (e.g. Dias 1999), or metacommunity dynamics (Urban et al. 2008).

The Motivation for Integrating Different Forms of Variation and Processes of Interest

As outlined in the introduction, although the effects of variation in some processes have been quite well studied, three related problems have been poorly explored: (1) consideration of multiple forms of variation, (2) consideration of variation at multiple scales, and (3) the relative importance of different mechanisms by which variation influences ecology. Nor are interactions among types of variability usually a subject of study. Consider a focal herbivore species feeding on a single species of plant and which is fed on by a specialist predator. This scenario includes variation in the physical environment and that which arises from variation within, and emergence of, variation across three interacting species or trophic levels. Variation in the physical environment might (for example) include spatial variation in mineral nutrients, water availability, and climatic variation at multiple space and time scales that can act on any species. A natural question is what forms of physical variation affect each species? To what extent does plasticity of any kind reduce the relationship between variation in a physical variable and processes of interest in each species? Or conversely do some forms of physical variation actually lead to increased variation in the process of interest? Does individual variation produce different outcomes of the process of interest, and is the net effect to dampen or enhance variation in the process of interest? The answers to such questions mean that we should also be interested in what the biological and ecological mechanisms are, and what kinds of mathematical functions can be used to represent them. Species interactions could also dampen (filter) or amplify variation in traits of one species in their population dynamics or other processes in which the species participates. Again the ecological and biological mechanisms and functional forms are of interest.

Individuals experience variance in abiotic and biotic conditions within their lifetime, and simultaneously the population encompassing those individuals' experiences interindividual variability, and again simultaneously populations within a metapopulation experience landscape-level variability. It is well established that variability on one scale matters for adjacent scales. An unresolved question is if variability matters for more distant scales. For example, we know fine-scale abiotic variability matters for individual physiology, but how does it influence population or meta-population dynamics? This question harks back to Levin's (1992) MacArthur Award lecture, in which he argues, "The key to understanding how information is transmitted across scales is to determine what information is preserved and what information is lost as one moves from one scale to the other." In the quarter-century since Levin pointed out this gap in our understanding, surprisingly little empirical work has been done on the topic.

For the species involved, their encounter with variation and responses to it will depend on their movements, activity periods, and degree of selection of resources (or physical conditions). Although we frequently study foraging, we rarely view it from a variation perspective to understand how foraging behaviors affect quality and quantity of resources acquired relative to variability in these things. The pattern (structure) and scale of resources are relevant (as well as resource-specific aspects of depletability and substitutability). Other questions are: Does the species adjust its foraging movements in relation to resource quality? What is the net effect of any resource selection behavior? Does selection (and other plasticity) reduce variation in individual growth or survival of the consumer? How is the population growth rate of the consumer species affected? There may also be feedbacks such that resource quality or quantity is affected by previous species interactions, and spatiotemporal patterns of variation in resources may alter competition among individual consumers. The consumer interacts with its predator through a variety of top-down and bottom-up forces. Are there detectable signals from predators of variation in the physical environment that affect plant quality and subsequently herbivores? If it is a top-down process, how do predators integrate spatiotemporal variation into processes of interest?

Questions also arise at more general levels. For instance, thinking about tritrophic interactions in general we might ask about the importance of different mechanisms in producing variation in a response variable of interest. Further, are there characteristic nonlinear shapes of processes (functions) of interest? How does the scale of movement of different kinds of species relate to variation in resource quality in the environment?

Conclusion: What Are the Next Steps in Moving toward an Understanding of How Multiple Scales and Types of Variation Interact to Influence Ecological Processes?

As stated above, a major unsolved problem in ecology is resolving the relative importance of different types and scales of variance, and the relative importance of the different mechanisms by which variance can influence ecological dynamics. With the exception of variation through species

composition or trait variation (e.g., biodiversity and ecosystem functioning studies), studies that actually manipulate variation to look at the effects on ecological dynamics are rare (e.g., Underwood 2004, 2009,
Pearse et al. 2018). And studies that do manipulate variation tend to manipulate genetic diversity but ignore the traits and mechanisms underlying the ecological effects of variability (Crutsinger 2015). We need more
studies that examine and manipulate variance at multiple scales and compare the consequences; for instance, studies that manipulate trait diversity
at both intra- and interspecific scales (Cook-Patton et al. 2011). Model species such as crop plants in which species have been bred to exhibit particular traits may be a good starting point for the often-difficult job of
creating trait variation under carefully controlled conditions. However,
investigating many forms of variation require us simply to recognize and
quantify existing patterns and use them in new manipulations of variation.
We also need more studies that experimentally isolate different mechanisms by which variance influences ecology (section 3). Does variation at
certain scales have typical mechanisms of action or types of consequences?
Under what circumstances are different types of mechanisms (e.g., Jensen's Inequality versus effects via phenotypic plasticity) more important?

Making progress in answering these questions is likely to involve collaborations among different types of biologists, from physiologists to behavioral ecologists to community and ecosystem ecologists. We already
possess a formidable array of mathematical tools for investigating the effects of variation. Certain forms of environmental variation have been
the targets of research in population ecology, such as reddening spectra
to produce autocorrelated temporal variation (Gonzalez and Holt 2002),
or the effects of timescales of variation in producing resonance through
interaction with intrinsic population dynamics (Orland 2003). There is no
reason why we cannot look at the effects of multiple scales of environmental variation on ecological dynamics. In such investigations, like any
investigations of variation, careful parameterization of mathematical functional forms and analysis of any nonlinearities can show whether there is
an effect on the average or variance of the output variable. Collaborations
between mathematical ecologists and empiricists are likely to be especially fruitful. Ultimately, we need to move towards research programs
aimed at investigating the role of variation in ecological dynamics.

References

Adler, F. R., and R. Karban. 1994. Defended fortresses or moving targets? Another
model of inducible defenses inspired by military metaphors. American Naturalist
144:813–832.

Adler, P., D. Raff, and W. Lauenroth. 2001. The effect of grazing on the spatial heterogeneity of vegetation. Oecologia 128:465–479.

Albert, C. H., W. Thuiller, N. G. Yoccoz, R. Douzet, S. Aubert, and S. Lavorel. 2010. A multi-trait approach reveals the structure and the relative importance of intra- vs. interspecific variability in plant traits. Functional Ecology 24:1192–1201.

Armstrong, J. B., and D. E. Schindler. 2011. Excess digestive capacity in predators reflects a life of feast and famine. Nature 476:84–87.

Bell, G., M. J. Lechowicz, A. Appenzeller, M. Chandler, E. DeBlois, L. Jackson, B. Mackenzie, R. Perziosi, M. Schallenberg, and N. Tinker. 1993. The spatial structure of the physical environment. Oecologia 96:114–121.

Benedetti-Cecchi, L. 2000. Variance in ecological consumer–resource interactions. Nature 407:370–374.

Bernays, E. A., K. L. Bright, N. Gonzalez, and J. Angel. 1994. Dietary mixing in a generalist herbivore: Tests of two hypotheses. Ecology 75:1997–2006.

Blair, G. R., D. E. Rogers, and T. P. Quinn. 1993. Variation in life history characteristics and morphology of sockeye salmon in the Kvichak River system, Bristol Bay, Alaska. Transactions of the American Fisheries Society 122:550–559.

Blarer, A., and M. Doebeli. 1999. Resonance effects and outbreaks in ecological time series. Ecology Letters 2:167–177.

Bolker, B., M. Holyoak, V. Krivan, L. Rowe, and O. Schmitz. 2003. Connecting theoretical and empirical studies of trait-mediated interactions. Ecology 84:1101–1114.

Bolnick, D. I., P. Amarasekare, M. S. Araújo, R. Bürger, J. M. Levine, M. Novak, V.H.W. Rudolf, S. J. Schreiber, M. C. Urban, and D. A. Vasseur. 2011. Why intraspecific trait variation matters in community ecology. Trends in Ecology & Evolution 26:183–192.

Bolter, C. J., and M. A. Jongsma. 1995. Colorado potato beetles (*Leptinotarsa decemlineata*) adapt to proteinase inhibitors induced in potato leaves by methyl jasmonate. Journal of Insect Physiology 41:1071–1078.

Cardinale, B. J., D. S. Srivastava, J. Emmett Duffy, J. P. Wright, A. L. Downing, M. Sankaran, and C. Jouseau. 2006. Effects of biodiversity on the functioning of trophic groups and ecosystems. Nature 443:989–992.

Charnov, E. L. 1976. Optimal foraging, the marginal value theorem. Theoretical Population Biology 9:129–136.

Chesson, P. 2012. Scale transition theory: Its aims, motivations and predictions. Ecological Complexity 10:52–68.

Cook-Patton, S.C., S. H. McArt, A. L. Parachnowitsch, J. S. Thaler, and A. A. Agrawal. 2011. A direct comparison of the consequences of plant genotypic and species diversity on communities and ecosystem function. Ecology 92:915–923.

Crutsinger, G. M. 2015. A community genetics perspective: Opportunities for the coming decade. New Phytologist 210:65–70.

Denny, M. 2015. Ecological mechanics: Principles of life's physical interactions. Princeton, N.J.: Princeton University Press.

Denny, M., L. Benedetti-Cecchi. 2012. Scaling up in ecology: mechanistic approaches. Annual Review of Ecology, Evolution, and Systematics 43:1–22.

Dias, P. C. 1996. Sources and sinks in population biology. Trends in Ecology and Evolution 11:326–330.

Dingle, H., W. A. Rochester, and M. P. Zalucki. 2000. Relationships among climate, latitude and migration: Australian butterflies are not temperate-zone birds. Oecologia 124:196–207.

Emlen, J. M. 1966. The role of time and energy in food preferences. American Naturalist, 100:611–617.

Gonzalez, A., and R. D. Holt. 2002. The inflationary effects of environmental fluctuations in source-sink systems. Proceedings of the National Academy of Sciences of the United States of America 99:14872.

Grimm, V. 1999. Ten years of individual-based modelling in ecology: What have we learned and what could we learn in the future? Ecological Modelling 115:129–148.

Hastings, A., C. L. Hom, S. Ellner, P. Turchin, and H.C.J. Godfray. 1993. Chaos in ecology: Is Mother Nature a strange attractor? In: D. G. Fautin, eds. Annual Review of Ecology and Systematics, Vol. 24. Palo Alto: Annual Reviews, 1–33.

Herrera, C. M. 2009. Multiplicity in unity: Plant subindividual variation & interactions with animals. Chicago: University of Chicago Press.

Hughes, A. R., B. D. Inouye, M. T. J. Johnson, N. Underwood, and M. Vellend. 2008. Ecological consequences of genetic diversity. Ecology Letters 11:609–623.

Inouye, B. D. 2005. The importance of the variance around the mean effect size of ecological processes: comment. Ecology 86:262–265.

Jones, C. G., J. H. Lawton, and M. Shachak. 1997. Positive and negative effects of organisms as physical ecosystem engineers. Ecology 78:1946–1957.

Jonzén, N., E. Knudsen, R. D. Holt, and B. E. Sæther. 2011. Uncertainty and predictability: The niches of migrants and nomads. In: E. J. Milner-Gulland, J. M. Fryxell, A. and R. E. Sinclair, eds. Animal migration: A synthesis. Oxford: Oxford University Press.

Karban, R., and I. T. Baldwin. 1997. Induced responses to herbivory. Chicago: University of Chicago Press.

Kendall, B. E., C. J. Briggs, W. W. Murdoch, P. Turchin, S. P. Ellner, E. McCauley, R. M. Nisbet, and S. N. Wood. 1999. Why do populations cycle? A synthesis of statistical and mechanistic modeling approaches. Ecology 80:1789–1805.

Koussoroplis, A.-M., S. Pincebourde, and A. Wacker. 2017. Understanding and predicting physiological performance of organisms in fluctuating and multifactorial environments. Ecological Monographs 87:178–197.

Laurance, W. F., and G. B. Williamson. 2001. Positive feedbacks among forest fragmentation, drought, and climate change in the amazon. Conservation Biology 15:1529–1535.

Lefcheck, J. S., M. A. Whalen, T. M. Davenport, J. P. Stone, and J. E. Duffy. 2013. Physiological effects of diet mixing on consumer fitness: A meta-analysis. Ecology 94:565–572.

Leibold, M. A., M. Holyoak, N. Mouquet, P. Amarasekare, J. M. Chase, M. F. Hoopes, et al. 2004. The metacommunity concept: A framework for multi-scale community ecology. Ecology Letters 7:601–613.

Levin, S. A. 1992. The problem of pattern and scale in ecology: The MacArthur award lecture. Ecology 73:1943–1967.

Levins, R. 1968. Evolution in changing environments: Some theoretical explorations. Princeton, N.J.: Princeton University Press.

Lomnicki, A. 1978. Individual differences between animals and the natural regulation of their numbers. Journal of Animal Ecology 47:461–475.

Luckinbill, L. S., and M. M. Fenton. 1978. Regulation and environmental variability in experimental populations of protozoa. Ecology 59:1271–1276.

MacArthur, R. H., and E. R. Pianka. 1966. On optimal use of a patchy environment. American Naturalist 100:603–609.

Marzetz, V., A.-M. Koussoroplis, D. Martin-Creuzburg, M. Striebel, and A. Wacker. 2017. Linking primary producer diversity and food quality effects on herbivores: A biochemical perspective. Scientific Reports 7:11035. http://doi.org/10.1038/s41598-017-11183-3.

Mason, P., M. Bernardo, and M. Singer. 2014. A mixed diet of toxic plants enables increased feeding and anti-predator defense by an insect herbivore. Oecologia 176:477–486.

McNaughton, S. J., F. F. Banyikwa, and M. M. McNaughton. 1997. Promotion of the cycling of diet-enhancing nutrients by African grazers. Science 278:1798–1800.

McPeek, M. A., and S. Kalisz. 1998. On the joint evolution of dispersal and dormancy in metapopulations. Advances in Limnology 52:33–51.

Menge, B. A. 1995. Indirect effects in marine rocky intertidal interaction webs: patterns and importance. Ecological Monographs 65:21–74.

Messier, J., B. J. McGill, M. J. Lechowicz. 2010. How do traits vary across ecological scales? A case for trait-based ecology. Ecology Letters 13:838–848.

Moreira, X., L. Abdala-Roberts, B. Castagneyrol, K. A. Mooney. 2016. Plant diversity effects on insect herbivores and their natural enemies: current thinking, recent findings, and future directions. Current Opinion in Insect Science 14:1–7.

Morris, D. W. 2003. Toward an ecological synthesis: a case for habitat selection. Oecologia 136:1–13.

Newton, I., and L. C. Dale. 1996a. Bird migration at different latitudes in eastern North America. Auk 113:626–635.

Newton, I., and L. Dale. 1996b. Relationship between migration and latitude among west European Birds. Journal of Animal Ecology 65:137–146.

Orland, M. C. 2003. Scale-dependent interactions between intrinsic and extrinsic processes reduce variability in protist populations. Ecology Letters 6:716–720.

Pearse, I. S., R. Paul, P. J. Ode. 2018. Variation in plant defense suppresses herbivore performance. Current Biology 28:1981–1986.

Pulliam, H. R. 1988. Sources, sinks, and population regulation. American Naturalist 132:652–661.

Ratikainen, I. I., J. A. Gill, T. G. Gunnarsson, W. J. Sutherland, H. Kokko. 2008. When density dependence is not instantaneous: theoretical developments and management implications. Ecology Letters 11:184–198.

Relyea, R. A., and J. R. Auld. 2004. Having the guts to compete: how intestinal plasticity explains costs of inducible defences. Ecology Letters 7:869–875.

Robertson, B. A., J. S. Rehage, and A. Sih. 2013. Ecological novelty and the emergence of evolutionary traps. Trends in Ecology and Evolution 28:552–560.

Rossi, R. E., D. J. Mulla, A. G. Journel, and E. H. Franz. 1992. Geostatistical tools for modeling and interpreting ecological spatial dependences. Ecological Monographs 62:277–314.

Ruel, J. J., and M. P. Ayres. 1999. Jensen's inequality predicts effects of environmental variation. Trends in Ecology and Evolution 14:361–366.

Schluter, D. 1988. Estimating the form of natural selection on a quantitative trait. Evolution 42:849–861.

Shaffer, M. L. 1981. Minimum population sizes for species conservation. Bioscience 31:131–134.

Shelford, V. E. 1931. Some concepts of bioecology. Ecology 12:455–467.

Sibly, R. M., D. Barker, M. C. Denham, J. Hone, and M. Pagel. 2005. On the regulation of populations of mammals, birds, fish, and insects. Science 309:607–610.

Siefert, A., C. Violle, L. Chalmandrier, C. H. Albert, A. Taudiere, A. Fajardo, et al. 2015. A global meta-analysis of the relative extent of intraspecific trait variation in plant communities. Ecology Letters 18:1406–1419.

Sih, A., and B. Christensen. 2001. Optimal diet theory: When does it work, and when and why does it fail? Animal Behaviour 61:379–390.

Soberón, J., and B. Arroyo-Peña, B. 2017. Are fundamental niches larger than the realized? Testing a 50-year-old prediction by Hutchinson. PLOS One 12:e0175138. doi: https://doi.org/10.1371/journal.pone.0175138.

Storch, D., P. A. Marquet, and J. H. Brown. 2007. Introduction: Scaling biodiversity—what is the problem? In: D. Storch, P. A. Marquet, and J. H. Brown, eds. Scaling biodiversity. Cambridge: Cambridge University Press.

Talley, T. S. 2007. Which spatial heterogeneity framework? Consequences for conclusions about patchy population distributions. Ecology 88:1476–1489.

Talley, D. M., G. R. Huxel, and M. Holyoak. 2006. Connectivity at the land-water interface. In: K. Crooks and M. Sanjayan, eds. Connectivity conservation. Cambridge: Cambridge University Press.

Taylor, L. R. 1961. Aggregation, variance and the mean. Nature 189:732–735.

Tobler, W. R. 1970. A computer movie simulating urban growth in the Detroit region. Economic Geography 46:234–240.

Tylianakis, J. M., R. K. Didham, J. Bascompte, and D. A. Wardle. 2008. Global change and species interactions in terrestrial ecosystems. Ecology Letters 11:1351–1363.

Underwood, N. 2004. Variance and skew of the distribution of plant quality influence herbivore population dynamics. Ecology 85:686–693.

Underwood, N. 2009. Effect of genetic variance in plant quality on the population dynamics of a herbivorous insect. Journal of Animal Ecology 78:839–847.

Urban, M. C., M. A. Leibold, P. Amarasekare, L. De Meester, R. Gomulkiewicz, M. E. Hochberg, et al. 2008. The evolutionary ecology of metacommunities. Trends in Ecology & Evolution 23:311–317.

van Ruijven, J., and F. Berendse. 2005. Diversity–productivity relationships: initial effects, long-term patterns, and underlying mechanisms. Proceedings of the National Academy of Sciences of the United States of America 102:695–700.

Wetzel, W.C., H. M. Kharouba, M. Robinson, M. Holyoak, and R. Karban. 2016. Variability in plant nutrients reduces insect herbivore performance. Nature 539:425–427.

Wetzel, W. C., and J. S. Thaler. 2016. Does plant trait diversity reduce the ability of herbivores to defend against predators? The plant variability-gut acclimation hypothesis. Current Opinion in Insect Science 14:25–31.

Wetzel, W. C., and J. S. Thaler. 2018. Host-choice reduces, but does not eliminate, the negative effects of a multi-species diet for an herbivorous beetle. Oecologia 186: 483–493.

Whitham, T. G., W. P. Young, G. D. Martinsen, C. A. Gehring, J. A. Schweitzer, S. M. Shuster, et al. 2015. Community and ecosystem genetics: a consequence of the extended phenotype. Ecology 84:559–573.

Whittle, P. 1956. On the variation of yield variance with plot size. Biometrika 43:337–343.

Yang, L. H., J. L. Bastow, K. O. Spence, and A. N. Wright. 2008. What can we learn from resource pulses? Ecology 89:621–634.

Yang, L. H., K. F. Edwards, J. E. Byrnes, J. L. Bastow, A. N. Wright, and K. O. Spence. 2010. A meta-analysis of resource pulse-consumer interactions. Ecological Monographs 80:125–151.

Zhang, Z., C. Yan, C. J. Krebs, and N. C. Stenseth. 2015. Ecological non-monotonicity and its effects on complexity and stability of populations, communities and ecosystems. Ecological Modelling 312:374–384.

Why Does Intragenotypic Variance Persist?

C. Jessica E. Metcalf and Julien F. Ayroles

Evidence for Intragenotypic Variability

Phenotypes vary across traits and species, and it has long been understood that such differences emerge from genetic and environmental variation. However, striking phenotypic differences are also observed among individuals with identical genotypes and that experience identical and constant environments. Such variation is known as intragenotypic variability (*Bradshaw 1965*) and refers to the nongenetic component of phenotypic variance. Note that our focus is specifically on variation evident even in organisms reared in nominally identical environments, and therefore we are explicitly not considering phenotypic plasticity.

Evidence for intragenotypic variability is fairly ubiquitous. In plants, variability in flowering time and germination provide well-studied examples (Bradford 2002). In insects, the fruit fly Drosophila spp. has been a useful model system for titrating the magnitude of intragenotypic variance. For example, by inbred lines, a suite of phenotypic traits can be measured for a large number of individuals within a single genotype in rigorously controlled environmental conditions (Kain et al. 2015). With these techniques, intragenotypic variability has been characterized in bristle number (Mackay and Lyman 2005), sleep (Harbison et al. 2013), thermal or phototactic preference (Kain et al. 2015), and locomotor behavior (Ayroles 2015), to list only a few. One of the key advances that has allowed this progress in characterizing variability is the development of automated instruments that can capture large numbers of phenotypes with high precision and low statistical error (Kain et al. 2015, Ayroles 2015). Using similar techniques, studies of wild isolates of the worm model organism *Caenorhabditis elegans* have, for example, uncovered high intragenotypic variability in lifetime fecundity (Diaz and Viney 2014). Interestingly, this variance is negatively correlated with the mean lifetime fecundity, suggesting a trade-off between these two levels of variation (mean and variance). Beyond model organisms and academic investigations, a large body of work on this topic has been driven by agricultural breeding programs, where homogeneity is a key goal, and variability a source of considerable frustration. Traits ranging from litter size in rabbits (Garreau et al. 2008), teat numbers in pigs (Felleki and Lundeheim 2015), or tassel length in corn (Ordas et al. 2008) all show evidence of intragenotypic variability. Moving to less-tangible traits, the adaptive immune system in vertebrates is a rich source of intragenotypic variability. For example, B- and T-cell receptor proteins vary substantially across genetically identical individuals; the underlying mechanism (VDJ recombination) and its consequences for variation are increasingly well characterized (e.g., Murugan (2012)). The underlying proximal mechanisms determining the magnitude of intragenotypic variability are also beginning to come to light in other systems: for example, subtle differences in sensitivity to microenvironmental perturbations between individuals (e.g., seeds developing on microenvironmental gradients in the soil (Simons and Johnston 2006); messenger RNA gradients in insect eggs (Yucel and Small 2006)), state–behavior feedback loops (Sih et al. 2015), developmental stochasticity (Raser and O'Shea, Topalidou and Chalfie 2011) and cell-to-cell transcriptional noise/variation (Kim and Marioni 2013, Levy et al. 2012), have all been suggested.

To date, the heritability of intragenotypic variability, a key requirement for its evolutionary significance has been difficult to estimate (Simons and Johnston 2006, Wagmann et al. 2010), but work on model organisms (Kain et al. 2015) and from agriculture (Hill and Mulder 2010) provide empirical support for its existence. A recent review of available data (Hill and Mulder 2010) (dominated by estimates from agricultural species) suggests that heritability estimates are themselves highly variable, ranging from 0.0 to 0.2. Interestingly, although generally low, these magnitudes of heritability are often on par with the additive contribution to trait means. Our ability to extend this evidence to natural populations is about to experience a step-change: the field is on the cusp of statistical and technological innovation. The low cost of genome sequencing now allows us to efficiently search for additive genetic effects on variability, using well-defined genome-wide association mapping techniques, but focusing on variance instead of the mean as would traditionally be the approach (Ayroles et al. 2015, Yang et al. 2012, Shen et al. 2012). Simultaneously, new statistical approaches are being developed to studying variance heterogeneity, such as Bayesian frameworks to identify variance-controlling quantitative trait loci (QTLs) (Dumitrascu et al. 2015) or double–generalized linear modeling (Rönnegård and Valdar 2012, Mulder 2013).

Having established that the intragenomic variation is widespread, and often appears to be heritable, the next question is establishing the degree to which this trait might be adaptive.

Adaptive Explanations

Interestingly, although studies of development, morphology, and animal breeding have long demonstrated the heterogeneity of variance among genotypes, and this evidence is increasingly augmented by data from field systems, and paired with a deepening understanding of underlying mechanisms, this axis of variation has received relatively little attention in evolutionary genetics. As a result, the evolutionary forces that shape and maintain the magnitude of such variability remain poorly characterized (Hill 2007). Here, we lay out some of the main hypotheses.

No Selection against It: Flat Fitness Landscapes

Perhaps the simplest, and most boring, explanation for persistent intragenomic variation is simply that there is no selection against this variation—i.e., deviations from the phenotypic mean within a certain range have no effect on survival or fecundity, and thus do not translate into fitness differences. Although this might sometimes be the case, phenotypes whose intragenomic variance have attracted research focus have tended to be ones likely to have a considerable impact on life-history evolution; e.g., traits such as germination, or phototaxis (Kain et al. 2015). We note this explanation but do not consider it further.

Facing an Unpredictable Environment: Bet-Hedging

Variability in a trait might be selected for where environments fluctuate unpredictably over space (e.g., latitude or altitude) or time (e.g., season), and different phenotypes are optimal under different conditions. Under this scenario, developmental stochasticity is maintained to produce a distribution of phenotypes ensuring that at least some individuals within the population will be well suited to a range of ecological conditions (Levy et al 2012, Simons 2011, Hopper 1999). Formally, bet-hedging theory requires that phenotypic fluctuations reduce mean fitness, as well as reducing variance in fitness, thereby increasing the population stochastic growth rate (Childs et al. 2011). The strongest tests of this theory to date have perhaps been in the area of seed germination (Gremer and Venable 2014) or timing of flowering in monocarpic plants (Metcalf et al. 2008, Rees et al. 2006) as well as work on phototaxis in *Drosophila* (Kain et al. 2015). For such life-history traits, the benefit of variance lies in either risk avoidance at the individual level (termed *conservative bet-hedging*) or risk spreading among individuals of the same genotype (termed *diversifying bet-hedging*) (Rees et al. 2004). Intragenotypic variability has the potential to fall within this second category.

Determining that diversifying bet-hedging is occurring may be non-trivial (Simons 2011). Variance in fitness is highly likely to increase within a generation as a result of intragenotypic variability—rather than the reduction expected under bet-hedging. Consequently, multigenerational tests of the effects of phenotypic variation on fitness variance will be necessary to establish that bet-hedging is the driving selective force, as the key point is reduction in the variance *across* generations (Simons 2011). Combining empirical data with mathematical models and simulation has been a core feature of grappling with this challenge (Kain et al. 2015, Gremer and Venable 2014, Metcalf et al. 2008, Rees et al. 2006).

Facing Competitors: Game Theory, Adaptive Dynamics

A key feature of the environment of most individuals is the presence of other individuals in the population, their phenotypes, and their absolute or relative abundances. Where frequency or density dependence is operating, the success of a phenotype will be predicated on what other individuals are doing. If similarity of phenotypes results in increased competition, and thus reduced fitness, this can result in selection for intragenotypic variability among offspring (Metcalf et al. 2015), an effect that might be further amplified by variation in the environment. In this case, variable phenotypes represent unexploited strategies (von Neumann and Morgenstern 1944). An example might be phenological traits, such as timing of germination within the year. If all individuals germinate at once, this will maximize competition, and selection is expected to favor individuals who produce offspring with a distribution of germination times. Although some of the offspring will be germinating in suboptimal conditions in terms of the abiotic environment (e.g., too dry or too wet), these will also be conditions that feature fewer competitors, and thus might be associated with higher fitness. Under a simple model of phenology in an annual plant under competition for recruitment microsites, the convergent stable strategy (i.e., one that can invade all other strategies and cannot be invaded by any strategy) can be shown to reflect a degree of variance under a range of different contexts (Metcalf et al. 2015), indicative of selection for intragenotypic variance.

Understanding the fitness consequences of phenotypic variability is generally challenging. In a comprehensive set of simulations, Bruijning et al. (2020) investigated such consequences under a range of commonly encountered relationships between traits, fitness, and environmental conditions and showed that, especially under fluctuating environments, there can be clear fitness advantages of maintaining intragenotypic variability.

As for bet-hedging, theoretical approaches are key to characterizing the conditions in which competition may lead to the expression of alternative phenotypes, in this case building on evolutionary game theory. However, unlike bet-hedging, generation of hypotheses about the role of competition in maintaining phenotypic variability linked to specific systems, and formal tests with empirical data, remain rare.

Probing the Adaptive Basis of Intragenotypic Variance

Tests Using Experimental Evolution

A major barrier to investigation into the fitness consequences of intragenotypic variability has long been the necessity of characterizing the phenotypes of many individuals across a population, to appropriately characterize variance, often a nontrivial task. Further, for bet-hedging scenarios, experiments must feature multiple generations under different scales of environmental fluctuations. Given these challenges, experimental evolution (Fuller et al. 2005) can be a powerful approach to testing hypotheses about the nature of the forces maintaining variable levels of phenotypic variability, offering both experimental flexibility and the opportunity to pair experiments with modeling. Although measuring the ultimate phenotype of interest (i.e., fitness) remains challenging, as it requires characterizing both survival and fertility of each individual, a number of life-history traits known to be highly correlated with fitness can be measured and manipulated. Examples of such experiments aimed at testing bet-hedging strategies remain rare, but the emergence of novel technologies (Ayroles 2015) allowing automation of phenotype characterization opens the way to a series of experimental tests of the adaptive basis of intragenotypic variation. This has been particularly true for microbial systems (Kawecki et al. 2012), where, for example, *de novo* evolution of bet-hedging traits has been demonstrated in bacteria, and the nuances of the context specificity of this evolution has been described (Beaumont et al. 2009).

An experimental evolutionary assay of the role of competition in selecting for intragenotypic variance will necessarily feature some form of resource limitation. For example, the outcome of competition between individuals and populations featuring differential levels of heritable variance could be characterized. If intragenotypic variability increases fitness in competitive settings, we expect that genotypes encoding higher variability will spread within experimental populations. More complex outcomes, including evolutionary branching, are also possible, and, as noted previously, experiments paired with theoretical models are likely to be an important driver of progress in understanding the underlying selective pressures. Again, recent technical innovations provide the scope of automatic measurement required to build the necessary empirical evidence. Advances in genomic technology may be particularly powerful: The fitness advantage of alternative strategies, tested in a competitive setup, can be monitored over time indirectly using molecular markers tagging individuals harboring alleles favoring different strategies. In such an experiment, one could, for example, coculture genetic strains with high- and low-intragenotypic variability for body size in environments with both constant and fluctuating nutritional resources, to see if the theoretical prediction that fluctuating environments favor variability is born out experimentally. Genomic technology helps bypass the difficulties of phenotyping each individual or of counting a large number of progeny in each generation. Rather, one can simply genotype individuals at a number of characteristic loci, where the high and low genetic variability backgrounds have been differentially tagged. This would not only help to determine the trajectory of these alleles over time but also provide one of the most intuitive and direct estimates of fitness: the survival and change in frequency of alleles associated with each of these strategies. This approach has the potential to offer a clear picture of which genome dominates the population, a strong signature of a strategy associated with a fitness advantage.

Leveraging Environmental Variation

Environmental variation offers a unique opportunity to ask whether the degree of intragenotypic variability in a population tracks uncertainty in environmental conditions. For example, the climate varies regionally for a variety of parameters ranging from temperature to rainfall to luminance. Some regions are relatively stable, with predictable conditions, others experience seasonal variation and may have a more unpredictable climate.

In an elegant study, Kain et. al. (2015) noticed striking intragenotypic variability for the light- and temperature-preference behaviors of *D. melanogaster* individuals. Intrigued by this pattern, they hypothesized that this variability may be maintained as a bet-hedging strategy driven by variable and unpredictable regional weather. Using simulations based on real-world weather and climate data, they characterized the temperature preference–dependent survival and reproduction under such variable conditions compared with under stable (predictable) conditions. They found that a bet-hedging strategy favors the maintenance of interindividual behavioral variability for these traits in fluctuating conditions. Further, they were able to show experimentally that the mean light- and temperature-preference behaviors of individual flies were not heritable. Parents with strong light preference or avoidance behavior do not produce F1s with a

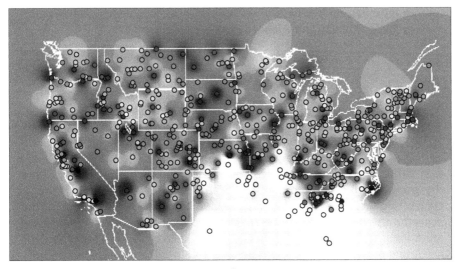

0

−16 [ ] 16

bet-hedging
advantage (%)

FIGURE 1. Predicted geographic variation in the relative growth advantage of bet-hedging. Warm colors indicate conditions conducive to bet-hedging, and generally reflect strong seasonality. Blue indicates conditions unfavorable to bet-hedging, generally associated with lower seasonal variation, such as the southeast and coastal California. (Reproduced with permission from the authors.)

biased preference; rather, these F1s go back to the mean preference of the population, with a distribution reflecting the degree of phenotypic variability of the parental population. This result matched the prediction of their model, suggesting that intragenotypic variability provides an advantage under unpredictable environmental conditions, and that a bet-hedging strategy outcompetes one in which individuals would simply evolve a mean preference for a given set of environmental parameters. This example highlights how, using environmental variation, by combining simulations and experimental validation, can be a powerful approach to study the forces driving and maintaining intragenotypic variability, generating testable predictions across broad scales (Fig. 1).

In the same vein, a study conducted in Madagascar suggested that regional variation in the predictability of precipitation and its effects on timing of fruiting and flowering of the local vegetation has shaped the life histories of animals that depend upon these resources, resulting in unusual patterns of selection on the variability in life-history traits, such as reproductive variance or basal metabolic rate, across a wide range of mammalian fauna (Dewar and Richard 2007).

Open Questions

As detailed previously, there is now clear evidence that the degree of intragenotypic variability may be itself under genetic control (Kain et al. 2015). However, only a handful of loci affecting intragenotypic variability have been mapped in higher eukaryotes (e.g., Kain et al. 2015, Yang et al. 2012, Shen et al. 2012). As a result, we know virtually nothing about the population genetic dynamics driving the evolution of variance-controlling loci. Are these loci maintained by balancing selection? What prevents the fixation of alleles favoring adaptive tracking? What is the mutational target size of variance alleles? Given that, at the mechanistic level, one of the favored hypotheses to explain intragenotypic variability is the maintenance of transcriptional noise (Yvert 2014, Willmore et al. 2007), it seems that the mutational target size for phenotypic variability has the potential to be substantially larger than that for mutations affecting trait means. How might this affect evolutionary trajectories of traits? What mechanisms mediate trade-offs between an organism's need for variability and the need to be locally adapted to its environment (Fig. 2)? Are particular timescales of environmental fluctuations relative to core life-history features such as lifespan likely to lead to the evolution of bet-hedging? Can the sampling of the distribution of phenotypes be biased to increase fitness in situations where information becomes available about the environmental context? What underlying mechanisms would be involved, and what are the necessary tests to identify such factors? Last, the nature of the relationship between intragenotypic variability (resulting from microenvironmental perturbation) and changes in variance (induced by macroenvironmental perturbation; e.g., changes in temperature or nutrient availability) is largely unexplored. Are these two levels of variability genetically correlated, such that a high level of microenvironmental sensitivity (i.e., high level of intragenotypic variability) would be predictive of macroenvironmental sensitivity (e.g., plasticity)? In other words, do the mechanisms of macroenvironmental robustness differ from mechanisms of microenvironmental robustness? Are there settings where selection might be expected to favor such coupling? Settings where the selection might be in different directions at different scales? Effectively addressing these questions will require the mapping of more mutations affecting phenotypic variability, in constant environments and across environments. Fortunately, the popularization of genomic tools in molecular ecology now allows such endeavors in natural populations.

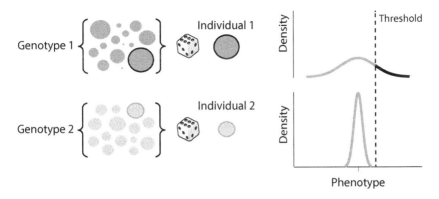

FIGURE 2. Genotypes 1 and 2 differ in their degree of phenotypic variability. The sets of circles represent the range of possible outcomes for each genotype. An individual drawn at random from genotype 1 (high variance) may land in the tail of the distribution, potentially in adaptive or maladaptive phenotypic space (e.g., disease threshold). On the other hand, an individual drawn randomly from genotype 2 never gets a chance to explore the phenotypic space explored by genotype 1.

Conclusions

Although intragenotypic variability appears to be ubiquitous in biology, under genetic control, and with important consequences for life history, very little is known about this phenomenon, both at the proximate and ultimate level, particularly in multicellular or other higher eukaryotes. A traditional focus on phenotypic means and relative neglect of phenotypic variance (sometimes inappropriately dismissed as experimental noise) is perhaps partly responsible. With a greater diversity of fields in biology realizing what breeders have long noticed, namely that variance itself is a trait under genetic control, the characterization of the proximate mechanisms of intragenotypic variability is in a phase of considerable expansion. This progress is driven by a combination of analysis of model systems (Ayroles et al. 2015), availability of ever more detailed genetic data from natural systems, and the development of statistical methods to characterize genotypes through to variances in phenotypes (Dumitrascu et al. 2015). Linking patterns of diversity in this trait through to its ultimate ecological and evolutionary drivers remains an open research area in the field of evolutionary ecology (despite some notable progress). Tackling this question will require careful evaluation of alternative hypotheses (e.g., careful distinction between adaptive tracking and bet-hedging). Matching models to empirical data (Kain et al. 2015, Childs et al. 2010, Gremer and Venable 2014) is likely to be key.

Why should we care? Besides being a deeply intriguing phenomenon, so much effort in biology is dedicated to understanding the factors that drive differences between cells, individuals, populations, or even species. By focusing primarily on the effects of genetic variation on trait averages and ignoring its effects on variance, we are missing an important axis contributing to phenotypic variation at each of these levels, especially if genetic control of mean and variance are decoupled. This underappreciated connection between genotype and phenotype may be a missing link allowing us to improve our ability to make better predictions from genotype to phenotype, to predict short term evolutionary responses based on allele frequencies and effect sizes; or in genetic epidemiology, to understand disease emergence for non-communicable diseases (Yang et al. 2012, Gibson 2009).

Lastly, from a more anthropocentric perspective, variation among individuals of the same genotype speaks directly to our sense of individuality and refutes misconceptions about genetic determinism. Our genetic background does not determine a phenotypic outcome; rather, it determines a statistical distribution and the probability of a given outcome given that distribution. Formally recognizing this has implications for inference into what shapes phenotypes across scales.

References

Ayroles, J. F., S. M. Buchanan, C. O'Leary, K. Skutt-Kakaria, J. K. Grenier, A. G. Clark, D. L. Hartl, and B. L. de Bivort. 2015. Behavioral idiosyncrasy reveals genetic control of phenotypic variability. Proceedings of the National Academy of Sciences of the United States of America 112:6706–6711.

Beaumont, H. J., J. Gallie, C. Kost, G. C. Ferguson, and P. B. Rainey. 2009. Experimental evolution of bet hedging. Nature 462:90–93.

Bradford, K. J. 2002. Applications of hydrothermal time to quantifying and modeling seed germination and dormancy. Weed Science 50:248–260.

Bradshaw, A. D. 1965. Evolutionary significance of phenotypic plasticity in plants. Advances in Genetics 13:115–155.

Bruijning, M., C.J.E.Metcalf, E. Jongejans, and J. F. Ayroles. 2020. The evolution of variance control. Trends in Ecology & Evolution 35:22–33.

Childs, D. Z., C. Metcalf, and M. Rees. 2010. Evolutionary bet-hedging in the real world: empirical evidence and challenges revealed by plants. Proceedings of the Royal Society B: Biological Sciences 277:3055–3064.

Dewar, R. E., and A. F. Richard. 2007. Evolution in the hypervariable environment of Madagascar. Proceedings of the National Academy of Sciences of the United States of America 104:13723–13727.

Diaz, S. A., and M. Viney. 2014. Genotypic-specific variance in *Caenorhabditis elegans* lifetime fecundity. Ecology and Evolution 4:2058–2069.

Dumitrascu, B., G. Darnell, J. Ayroles, and B. E. Engelhardt. 2015. A Bayesian test to identify variance effects. arXiv:1512.01616.

Felleki, M., and N. Lundeheim. 2015. Genetic heteroscedasticity of teat count in pigs. Journal of Animal Breeding and Genetics 132:392–398.

Fuller, R. C., C. F. Baer, and J. Travis. 2005. How and when selection experiments might actually be useful. Integrative and Comparative Biology 45:391–404.

Garreau, H., G. Bolet, C. Larzul, C. Robert-Granié, G. Saleil, M. SanCristobal, and L. Bodin. 2008. Results of four generations of a canalising selection for rabbit birth weight. Livestock Science 119:55–62.

Gibson, G. 2009. Decanalization and the origin of complex disease. Nature Reviews Genetics 10:134–140.

Gremer, J. R., and D. L. Venable. 2014. Bet hedging in desert winter annual plants: Optimal germination strategies in a variable environment. Ecology Letters 17:380–387.

Harbison, S. T., L. J. McCoy, and T. F. Mackay. 2013. Genome-wide association study of sleep in *Drosophila melanogaster*. BMC Genomics 14:281.

Hill, W. G. 2007. The variance of the variance-genetic analysis of environmental variability. Journal of Animal Breeding and Genetics 124:49–49.

Hill, W. G., and H. A. Mulder. 2010. Genetic analysis of environmental variation. Genetics Research 92:381–395.

Hopper, K. R. 1999. Risk-spreading and bet-hedging in insect population biology 1. Annual Review of Entomology 44:535–560.

Kain, J. S., S. Zhang, J. Akhund-Zade, A. D. T. Samuel, M. Klein, and B. L. de Bivort. 2015. Variability in thermal and phototactic preferences in *Drosophila* may reflect an adaptive bet-hedging strategy. Evolution 69:3171–3185.

Kawecki, T. J., R. E. Lenski, D. Ebert, B. Hollis, I. Olivieri, and M. C. Whitlock. 2012. Experimental evolution. Trends in Ecology & Evolution 27:547–560.

Kim, J. K., and J. C. Marioni. 2013. Inferring the kinetics of stochastic gene expression from single-cell RNA-sequencing data. Genome Biology14:R7.

Levy, S. F., N. Ziv, and M. L. Siegal. 2012. Bet hedging in yeast by heterogeneous, age-correlated expression of a stress protectant. PLOS Biology 10:e1001325.

Mackay, T. F., and R. F. Lyman. 2005. *Drosophila* bristles and the nature of quantitative genetic variation. Philosophical Transactions of the Royal Society of London B: Biological Sciences 360:1513–1527.

Metcalf, C.J.E., L. T. Burghardt, and D. N. Koons. 2015. Avoiding the crowds: The evolution of plastic responses to seasonal cues in a density-dependent world. Journal of Ecology 103:819–828.

Metcalf, C., K. E. Rose, D. Z. Childs, A. W. Sheppard, P. J. Grubb, and M. Rees. 2008. Evolution of flowering decisions in a stochastic, density-dependent environment. Proceedings of the National Academy of Sciences of the United States of America 105:10466–10470.

Mulder, H. A., L. Rönnegård, W. F. Fikse, R. F. Veerkamp, and E. Strandberg. 2013. Estimation of genetic variance for macro-and micro-environmental sensitivity using double hierarchical generalized linear models. Genetics Selection Evolution 45:23.

Murugan, A., T. Mora, A. M. Walczak, and C. G. Callan. 2012. Statistical inference of the generation probability of T-cell receptors from sequence repertoires. Proceedings of the National Academy of Sciences of the United States of America 109:16161–16166.

Ordas, B., R. A. Malvar, and W. G. Hill. 2008. Genetic variation and quantitative trait loci associated with developmental stability and the environmental correlation between traits in maize. Genetics Research 90:385–395.

Raser, J. M., and E. K. O'Shea. 2005. Noise in gene expression: Origins, consequences, and control. Science 309:2010–2013.

Rees, M., D. Z. Childs, K. E. Rose, and P. J. Grubb. 2004. Evolution of size-dependent flowering in a variable environment: partitioning the effects of fluctuating selection. Proceedings of the Royal Society of London. Series B: Biological Sciences 271:471–475.

Rees, M., et al. 2006. Seed dormancy and delayed flowering in monocarpic plants: Selective interactions in a stochastic environment. American Naturalist 168:E53–E71.

Rönnegård, L., and W. Valdar. 2012. Recent developments in statistical methods for detecting genetic loci affecting phenotypic variability. BMC Genetics 13:63.

Shen, X., M. Pettersson, L. Rönnegård, and Ö. Carlborg. 2012. Inheritance beyond plain heritability: Variance-controlling genes in *Arabidopsis thaliana*. PLOS Genetics 8:e1002839.

Sih A., K. J. Mathot, M. Moirón, P. O. Montiglio, M. Wolf, and N. J. Dingemanse. 2015. Animal personality and state-behaviour feedbacks: A review and guide for empiricists. Trends in Ecology & Evolution 30:50–60.

Simons, A. M. 2011. Modes of response to environmental change and the elusive empirical evidence for bet hedging. Proceedings of the Royal Society of London B: Biological Sciences:rspb20110176.

Simons, A. M., and M. O. Johnston. 2006. Environmental and genetic sources of diversification in the timing of seed germination: Implications for the evolution of bet hedging. Evolution 60:2280–2292.

Topalidou, I., I., and M. Chalfie. 2011. Shared gene expression in distinct neurons expressing common selector genes. Proceedings of the National Academy of Sciences of the United States of America 108:19258–19263.

von Neumann, J., and O. Morgenstern. 1944. The theory of games and economic behavior. Princeton, N.J.: Princeton University Press.

Wagmann, K., N.-C. Hautekèete, Y. Piquot, and H. Van Dijk. 2010. Potential for evolutionary change in the seasonal timing of germination in sea beet (*Beta vulgaris* ssp. *maritima*) mediated by seed dormancy. Genetica 138:763–773.

Willmore, K. E., N. M. Young, and J. T. Richtsmeier. 2007. Phenotypic variability: Its components, measurement and underlying developmental processes. Evolutionary Biology 34:99–120.

Yang, J., R. J. F. Loos, J. E. Powell, S. E. Medland, E. K. Speliotes, D. I. Chasman, et al. 2012. FTO genotype is associated with phenotypic variability of body mass index. Nature 490:267–272.

Yucel, G., and S. Small. 2006. Morphogens: Precise outputs from a variable gradient. Current Biology 16:R29–R31.

Yvert, G. 2014. "Particle genetics": Treating every cell as unique. Trends in Genetics 30:49–56.

Population Dynamics of Species with Complex Life Cycles

Andrew Dobson

Many species have complex life cycles, particularly parasitic helminths, but also insects and fungi that sequentially use different host species at different stages of their life cycle. Indeed, complex life cycles may ultimately prove to be more common than the relatively simple life-histories that are characteristic of mammals, birds, and most higher plants. At present, most of our understanding of population dynamics is based on models for species with the simple life cycle of the form: birth, reproduce, care for offspring, repeat until senescence and/or death. I suspect this reflects the interaction between the actuarial origins of population biology and ecology's origins in wildlife management (Kingsland 1982, 1991, 1995). Although there have been occasional explorations of the dynamics and evolution of complex life cycles (Caswell 1986, 1989; Dobson et al. 1985; Dobson 1988; Parker et al. 2015a; Parker et al. 2003), it is still an area that would profit from much more study.

My own interest in complex life cycles stems from a lifelong interest in the parasitic helminths, most of which have life cycles that involve two or more hosts that must be gone through sequentially to complete the life cycle. Sometimes the parasite exhibits highly specific host choice at one stage of its life cycle, in other cases a large diversity of hosts can be used. I will mainly use parasitic helminths to guide the structure of this essay, while noting that many of the results that I outline will also apply to many insects, fungi, and marine invertebrates where "alternation of generations" between sexual and asexual reproduction, and dispersal at different stages in the same life cycle, lead to a spectrum of life-cycle complexities whose dynamics and evolution can only reward further study.

Vector-Borne Disease as a Complex Life Cycle

Initially, let me lay out the nexus of the problem, using malaria as an example; after the measles virus, the malaria pathogens are arguably the world's most intensively studied pathogens from a population perspective. These studies are distorted by the logical perception that malaria is an important disease of humans; this is certainly true from a social and medical perspective, but from an evolutionary and population dynamic perspective, *Plasmodium* is a parasite of mosquitoes that has been using humans and other primates as a sort of "parking lot" for keeping infections ticking over, when mosquitos are in low abundance. Sexual reproduction of *Plasmodium* spp. occurs in mosquitoes, and the different species of this protozoan parasite have evolved to reproduce asexually in the blood and liver cells of the large variety of vertebrate hosts that they use. The five *Plasmodium* species that use humans are a small subset of those that use primates, which in turn are a small subset of those that use birds, lizards, and some other mammal groups (Garnham 1966, Faust and Dobson 2015). The human *Plasmodium* species are members of a very ancient lineage of pathogens that have used mosquitoes as hosts since their earliest blood meals.

To understand the evolutionary and population dynamic constraints on species with complex life cycles like that of *Plasmodium* it is useful to derive an expression for the basic reproductive ratio, R_0, the number of offspring, or new infections produced during the lifetime of the pathogen when it first arrives in a new host population, a keystone concept in population dynamics. Intriguingly, one of the first uses of this concept was in the derivation of an expression for R_0 for malaria by Ronald Ross, the British army polymath who won the second Nobel Prize for Physiology or Medicine for his work describing the life cycle of malaria (Ross 1916). His contemporary derivation of an expression for the "*a priori pathometry for malaria*" was essentially a derivation of R_0 (Fine 1975, Smith et al. 2012) (predating Fisher's later derivation for age-structured populations by nearly 20 years). It led Ross to emphasize that the most effective way to control malaria was by reducing the rate at which mosquitoes bite human hosts, an insight that is still central to any form of control for vector-borne disease (Yoeli 1973).

Ross's original formulation was formalized by MacDonald (1957, 1952) and further explored by Aron and May (1982) and Dietz (1980, 1992). All of these formulations assume that the dynamics of *Plasmodium* can be characterized by two equations, one for the proportion of mosquitoes infected, Y, and one for the proportion of humans infected, X. The essential dynamics of this complex life cycle may be captured by two differential equations:

$$\frac{dx}{dt} = \frac{abMy(1-x)}{H} - rx \tag{0.1}$$

$$\frac{dy}{dt} = ax(1-y) - \mu y \tag{0.2}$$

Here a is the biting rate of female mosquitoes, b is the proportion of bites that give rise to infections in humans, $m = M/H$ is the mean number of mosquitoes per human, r is the recovery rate of infected humans, and μ is the death rate of mosquitoes. Algebraic examination of the zero-growth isoclines for these two equations provide an expression for the threshold condition when $R_0 = 1$, based upon the slope of the two isoclines and the condition that they just intersect, by definition a condition that the pathogen is just able to establish in the host population (Aron and May 1982).

$$R_0 = \frac{ma^2b}{r\mu} \tag{0.3}$$

As Ross noted, this means that anything that reduces the mosquito biting rate will be the most efficient way to reduce the spread of malaria. A key point to notice here is that this definition of R_0 is always true at the threshold when $R_0 = 1$, but is it always true at other combinations of host and vector abundance? To understand this, we need a more general approach to estimating R_0, this is based on a rather wonderful piece of theory developed by Dieckmann et al. suggesting that this might not be the case (Diekmann, Heesterbeek, and Metz 1990).

Next-Generation Matrices

Their approach uses next-generation matrices to derive an expression for R_0, the order of these matrices are determined by the number of stages in the life cycle of the pathogen, so they will be central to developing an understanding of the dynamics of species with more complicated life cycles than that of *Plasmodium*. Each element of the matrix corresponds to the rate at which infections at this stage of the life cycle are passed to the next. In the case of *Plasmodium* we reduce the two equations (Equations 1.1 and 1.2) that the transient dynamics of infection to a 2×2 next-generation matrix takes the form

$$\begin{bmatrix} 0 & \dfrac{abm}{r} \\ \dfrac{a}{\mu} & 0 \end{bmatrix} \tag{0.4}$$

An expression of R_0 is given by the trace, or dominant eigenvalue of this matrix:

$$R_0 = \sqrt{\frac{ab^2m}{\mu r}} \qquad\qquad (0.5)$$

This expression is very similar to the one derived previously and is identical at the threshold condition of $R_0 = 1$. Importantly, as R_0 exceeds this linear approximation it increasingly overestimates the true value of R_0, similarly when $R_0 < 1$ the threshold approximation overestimates R_0. There are important consequences to this as many estimates of R_0 based on the threshold equation will egregiously overestimate the magnitude of R_0 and whence the level of control needed to achieve control. Thus, estimates of R_0 suggesting it might be close to 100, actually reflect values of 10. An unsolved problem in the control of vector-borne diseases is broad scale failure to recognize this.

While emphasizing that important caveats always need to be underlined when models are used to create risk maps and drive broad-scale control strategies for pests and pathogens, let us now use this machinery to return to the central question of this essay: What determines the dynamics of parasitic species with complex life cycles?

Cestodes of the Serengeti

Although the Serengeti is characterized in the eyes of tourists and couch-bound tourists watching wildlife videos by its huge diversity and abundance or by the herbivores and the carnivores that feed upon them, these species are also intimately coupled together by the parasitic cestodes whose life cycles sequentially utilize these predator and prey species.

Parasitic Trematodes, Complexity, and Specificity: Model Structure for Trophically Transmitted Parasites

The population dynamics of parasites with two sequential hosts coupled by free-living egg stages and trophic transmission from intermediate to definitive host may be described by expanding the standard Anderson and May macroparasite equations to include the details of parasite stages in the intermediate host. We assume intermediate hosts are either uninfected, C, or infected, C_I, and ignore the distribution of the numbers of parasite infective stages they contain. We assume transmission to the definitive host takes place via a predator–prey relationship and this can most simply be characterized by a type II (Michaelis–Menten) functional response, characterized by a half-saturation constant that is determined by the prey abundance at which predators can locate and eat prey at 50% of their maximum consumption rate. We also assume that the predator birth rate is dependent upon the abundance of the intermediate hosts as a food resource. The equation for each of the populations linked by the parasite life cycle follow (Equations 1.6–1.11):

Definitive hosts, H:

$$\frac{dH}{dt} = abH\left(\frac{C+C_I}{C_{50}+C+C_I}\right) - dH - (\alpha + \delta)H \tag{1.6}$$

Free-living parasite infective stages, W:

$$\frac{dW}{dt} = \lambda P - \sigma W - \beta_C(C+C_I)W \tag{1.7}$$

Uninfected intermediate hosts, C:

$$\frac{dC}{dt} = fC - d_cC - \frac{aHC^2}{(C_{50}+C+CI)} - \beta_C CW \tag{1.8}$$

Infected intermediate hosts, C_I:

$$\frac{dC_i}{dt} = \beta_C CW - \frac{aH(C_I)^2}{(C_{50}+C+C_I)} - d_cC_I \tag{1.9}$$

Parasites in intermediate host, P_C:

$$\frac{dP_C}{dt} = \beta_C(C+CI)W - (\mu_c+d_c)Pc - \frac{aHCIPc}{(C_{50}+C+CI)} \tag{1.10}$$

Parasites in definitive hosts, P:

$$\frac{dP}{dt} = \frac{aHC_IP_C}{(C_{50}+C+C_I)} - (\mu + d + \alpha)P - \alpha\frac{P^2}{H} \tag{1.11}$$

Table 1. Descriptions of Model's Parameters

PARAMETER	DEFINITION
a	Predation attack rate by definitive host, H, on intermediate host, C
b	Birth rate of definitive host, H
d	Death rate of definitive host, H
C_{50}	Half-saturation constant of definitive host predation rate on intermediate host
A	Per capita impact of parasite on definitive host survival
δ	Per capita impact of parasite on definitive host fecundity
λ	Birth rate of adult parasites, O, in definitive host, H
σ	Death rate of free-living parasite infective stages, W
β_C	Consumption rate of free-living parasites, W, by intermediate host, C
f	Birth rate of intermediate hosts, C
d_C	Death rate of intermediate hosts, C
μ_C	Mortality rate of parasites, P_C, in intermediate hosts, C
μ	Mortality rate of adult parasites, P, in definitive hosts, H.

Descriptions of the model's parameters are given in Table 1.

We can gain an understanding for the dynamics of this more complex life cycle by obtaining an expression for the basic reproductive number of the life cycle, R_0. This can be done in the same way as for the malaria example by writing down the next-generation matrix for the rate of spread of the pathogen between each stage of the life cycle when first introduced into the host populations. This matrix has the form

$$
\begin{bmatrix}
0 & \dfrac{\lambda H}{(\mu + d + \alpha)} & 0 \\[2ex]
0 & 0 & \dfrac{\beta C}{\sigma} \\[2ex]
\dfrac{aCH}{(C_{50} + C)(\mu_C + d_C)} & 0 & 0
\end{bmatrix}
\tag{1.12}
$$

Notice the element in the top row, second column, is the rate at which infective stages are produced in the definitive hosts, λH, multiplied by the average life expectancy of adult parasites $1/(\mu + d + \alpha)$; the expression in the second row, third column, is the life expectancy of free-living stages $1/\sigma$, times the rate they at which they are consumed by intermediate hosts,

$\beta_C C$. The bottom row, first column, is the rate of consumption of infected intermediate hosts by definitive hosts, times the life expectancy of parasites in intermediate hosts. The expression for R_0 is again given by the dominant eigenvalue of this matrix, whence

$$R_0 = \sqrt[3]{\left(\frac{\lambda H}{(\mu + d + \alpha)} \cdot \right) \left(\frac{aCH}{(C_{50} + C)(\mu_c + d_c)} \right) \left(\frac{\beta C}{\sigma} \right)} \tag{1.13}$$

Notice that R_0 is now the cube root of the product of the three transmission stages in the life cycle. Obviously, this has to be greater than unity if the parasite is to persist. This can only be achieved if each of the stages in the parasite always increase the abundance of the parasite, or if losses at one stage are compensated for by significant amplification in others. In general, increases in host abundance will increase R_0, but this is offset slightly by the trophic transmission stage from intermediate to definitive host, where the functional response of the predator–prey interaction saturate as prey and intermediate host abundances increase. Note also that taking a cube root, rather than a square root, will likely reduce the overall magnitude of R_0; this suggests that complex life cycles may be less likely to give rise to large epidemics as occur for species with simpler life cycles and large values of R_0.

It would be very intriguing to integrate this next-generation approach to R_0 for complex life cycles with the optimization methods developed by Parker et al. (2015b) and Parker et al. (2003). It would allow us to explore trade-offs between different life-history parameters at different stages of the parasite's life cycle.

This approach can be expanded to include situations in which more than one host species is used at different stages in the life cycle. This strategy is employed commonly by the cestodes that live in the muscle tissues and viscera of herbivores and the alimentary canals of carnivores. Cestodes have wonderfully complex life cycles that always include an arthropod host, usually a beetle in the case of most terrestrial species, but an arthropod in the case of most aquatic species. When the beetle, or the parasite's eggs excreted from the beetle, are accidentally ingested by a herbivore they migrate through the tissue to either muscle or nervous system tissue where they may cause additional stress that increases susceptibility to predation by the carnivore species that prey on the herbivore host. Trophic transmission then leads to further development of the cestode in the alimentary tract of the carnivore hosts, where sexual reproduction produces the eggs that are excreted and eventually consumed by arthropods.

Here we consider two cases with either two possible hosts at the second intermediate stage, I_2, or two possible hosts at the definitive host stage, D_2. Rather than write down the many equations that correspond to Equations 1.6 through 1.11 above, let us simplify the exercise and simply use a shorthand for elements of the matrix that reflects that we are dealing with production of infective stages (or predation events) that transmit the parasite from one host to the next stage in its life cycles (as we saw in the matrix, in Equation 1.12, above). In matrix I2 (for the I_2 stage) following, WD and WJ are rates of production of eggs by adult parasites in dogs and jackals, BW is the rate at which eggs are eaten by beetles, MB is the rate at which mice eat beetles, and DM and JM are the rate at which dogs and jackals eat infected mice. In the second matrix, we have ignored beetle stage and assumed that mice and voles eat the eggs directly and that wolves and jackals both prey on mice and voles, so there are multiple ways around the life cycle.

The transmission matrices for these two different systems may be written as:

$$
I2 = \begin{bmatrix}
0 & 0 & 0 & 0 & DM \\
0 & 0 & 0 & 0 & JM \\
WD & WJ & 0 & 0 & 0 \\
0 & 0 & BW & 0 & 0 \\
0 & 0 & 0 & MB & 0
\end{bmatrix}
\tag{1.14}
$$

$$
D2 = \begin{bmatrix}
0 & 0 & 0 & WJ & WD \\
MW & 0 & 0 & 0 & 0 \\
VW & 0 & 0 & 0 & 0 \\
0 & JM & JV & 0 & 0 \\
0 & DM & DV & 0 & 0
\end{bmatrix}
\tag{1.15}.
$$

We can readily derive expressions for R_0 for each of these two matrices. In the case of I2 this takes the form

$$
R_0 = \sqrt[4]{(BW.DM.MB.WD + BW.JM.MB.WJ)}
\tag{1.16}.
$$

In the case of the second life cycle matrix D2 (for the D_2 stage), then R_0 has the form

$$
R_0 = \sqrt[3]{(DM.MW.WD + JM.MW.WJ + DV.VW.WD + JV.VW.WJ)} \tag{1.17}.
$$

Although these expressions look at first sight slightly incongruous, they have similar properties to those derived above for *Plasmodium* and the trophically transmitted trematode. In every case R_0 is defined as the *n*th root of the sum of all the possible transmission routes around the life cycle; notice that *n* is the number of trophic levels that the parasite passes through in the course of its life cycle. All of which suggests that the next-generation matrix is providing a way of calculating R_0 as the geometric mean of R_0 from all possible starting points in the life cycle. This is why it was demonstrative to drop the beetle stage in the D2 example, we have lost a trophic level and end up with an expression in the form of a cube root; if we returned this stage back into D2, we would again obtain an expression with a route to the fourth power. If we added two beetle species, we would now have eight different ways for the parasite to go round its life cycles and the expression for R_0 would look increasingly complex, as every time we add in two or more possible hosts on the same trophic level we double the number of possible routes the parasite can take around its life cycle. *This creates a beautiful link to the need to study complex life cycles parasites within a food-web context. Initial explorations of life cycles with this structure and more hosts at either the definitive, or intermediate host level, consistently produce equations of the same form.*

Stability in Complex Life Cycles

The results previously described provide a couple of glimpses into the factors that determine the dynamics of species with complex life cycles. Although I have wrestled with their formulation for several years, I can get no closer to an exact solution that those adumbrated above. So for myself at least this remains an unsolved problem. Discussions with Mick Roberts (who first introduced me to these methods) and Hans Heesterbeek suggested that deeper analytical solutions are attainable; their fine papers on this provide a map for those who wish to pursue this journey further (Roberts and Heesterbeek 1998, Roberts and Heesterbeek 2003, Roberts and Heesterbeek 2013). I will leave it at this time as a conjecture and hope that younger, sharper minds find ways to explore these problems by using either these approaches, or ones that might provide sharper insights, such as those developed by Parker et al. (2015a, 2015b) and Parker et al. (2003).

Nonetheless, I see several insights in the results above that deserve further comment. First, there is considerable debate in the host–parasite literature about the dilution effect (Lafferty 2013, Ostfeld 2013); any attempt to resolve this debate analytically has to involve a formulation based on this next-generation-matrix approach. A recent paper by Faust et al. (2017) provides some important insights here. They examine transmission of mosquito and tick-borne pathogens in increasingly diverse communities of hosts. As host diversity increases some species are likely to be better or worse hosts for the parasites or their vectors. The addition of these additional routes around the life cycle creates a dilution effect that reduces the magnitude of the R_0.

Second, a curious effect that emerges, that I confess to only partly understand, is that as life cycles become more complex, then the root term needed to describe R_0 will have a higher and higher term; e.g., 2 for *Plasmodium*, 3 for trematode, 4 for the cestodes. This means that it is very hard for species with complex life cycles to have explosive birth rates; instead, they are more likely to have R_0 values increasingly close to unity and thus be characterized by rather stable dynamics. While I half-joked earlier that *Plasmodium* uses humans as a parking lot, when mosquito abundance is transiently reduced, this use of multiple hosts may provide an intrinsic form of stability that allows persistence of species that would otherwise be susceptible to extinction during the boom–bust cycles of individual host species in their life cycles.

Evolution in Complex Life Cycles

Our traditional perspective is that complex life cycles have taken millions of years to evolve; a coarse underlying generality is that parasitic species have had had association with their invertebrate hosts for millions of years before their mammalian and avian hosts appeared on the scene. There are two things wrong with this simplification: Evolution is a constant and ongoing process, so although there may be a general pattern of older evolutionary hosts, being the ones that define the basal host in a complex life cycle, there are likely to be many examples when parasites of vertebrates evolved to use a biting insect as an efficient vector for transfer from one patch of host to the next. Similarly, the efficiency of transfer of free-living infective stages may be considerably enhanced when these either attach to, or are ingested by, an invertebrate that occurs in the diets of the definitive host. Fascinating insights into the evolution of complex life cycles have come from the recent work of Parker et al. (2003, 2015a, 2015b) as well as Benesh et al. (2012) and Michaud et al. (2006).

Secondly, the idea that it takes millions of life cycles for complex life cycles to evolve is eloquently falsified by recent work from the Panama Canal where the mass of invasive host and parasite species introduced by global trade passing through the canal have allowed new complex life cycles to evolve in entirely novel sequences of hosts on a decadal time scale (Frankel 2016, Frankel et al. 2015).

All of this raises intriguing questions about how the genomes of species with complex life cycles are organized. How are different genetic processes switched on or suppressed when the same individual lives in a very different environment at different stages of its life cycle? How much of its genome is expressed throughout all of its life cycle? Or is their total differentiation with vital processes replicated in the genome and differentially expressed at different stages? Plainly we still have a lot to learn about the dynamics and evolution of species with complex life cycles. The methods I have described above can fairly readily be applied to the many non-parasitic invertebrate species with complex life cycles.

References

Aron, J. L., and R. M. May. 1982. The population dynamics of malaria. In: R. M. Anderson, ed. Population dynamics of infectious diseases: Theory and applications. London: Chapman & Hall, 139–179.

Benesh, D. P., J. C. Chubb, and G. A. Parker. 2012. Complex life cycles: Why refrain from growth before reproduction in the adult niche? American Naturalist 181(1):39–51.

Caswell, H. 1986. Life cycle models for plants. Lectures on Mathematics in the Life Sciences 18:171–233.

Caswell, H. 1989. Matrix population models. Sunderland, Mass.: Sinauer Associates Inc.

Diekmann, O., J.A.P. Heesterbeek, and J.A.J. Metz. 1990. On the definition and the computation of the basic reproduction ratio R_0 in models for infectious diseases in heterogeneous populations. Journal of Mathematical Biology 28:365–382.

Dietz, K. 1980. Models for vector-borne parasitic diseases. Lecture Notes in Biomathematics 39:264–277.

Dietz, K. 1992. The estimation of the basic reproduction number for infectious diseases. Statistical Methods in Medical Research 2:23–41.

Dobson, A. P. 1988. The population biology of parasite-induced changes in host behavior. Quarterly Review of Biology 63:139–165.

Dobson, A. P., A. E. Keymer, D.W.T. Crompton, and B. B. Nickol. 1985. Biology of the Acanthocephala. In D.W.T. Crompton and B. B. Nickol, eds. Biology of the Acanthocephala. Cambridge: Cambridge University Press.

Faust, C., and A. P. Dobson. 2015. Primate malarias: Diversity, distribution and insights for zoonotic *Plasmodium*. One Health 1:66–75.

Faust, C. L., A. P. Dobson, N. Gottdenker, L. S. P. Bloomfield, H. I. McCallum, T. R. Gillespie, M. Diuk-Wasser, and R. K. Plowright. 2017. Null expectations for disease dynamics in shrinking habitat: dilution or amplification? Philosophical Transactions of the Royal Society B 372(1722):20160173.

Fine, P.E.M. 1975. Tropical disease—a challenge for epidemiology: Ross's a priori pathometry—a perspective. Washington, D.C.: SAGE Publications.

Frankel, V. M. 2016. Natural history of host-parasite interactions in an invaded community. Thesis, McGill University.

Frankel, V. M., A. P. Hendry, G. Rolshausen, and M. E. Torchin. 2015. Host preference of an introduced "generalist" parasite for a non-native host. International Journal for Parasitology 45(11):703–709.

Garnham, P.C.C. 1966. Malaria parasites and other haemosporidia. In: G. Valkiūnas, ed. Avian malaria parasites and other haemosporidia. Washington, D.C.: CRC Press.

Kingsland, S. 1982. The refractory model: the logistic curve and the history of population ecology. Quarterly Review of Biology 57(1):29–52.

Kingsland, S. E. 1991. Defining ecology as a science. In: L. A. Real and J. H. Brown, eds. Foundations of ecology. Chicago: University of Chicago Press, 1–13.

Kingsland, S. E. 1995. Modeling nature. Chicago: University of Chicago Press.

MacDonald, G. 1952. The analysis of equilibrium in malaria. Tropical Diseases Bulletin 49:813–829.

MacDonald, G. 1957. The epidemiology and control of malaria. Oxford: Oxford University Press.

Michaud, M., M. Milinski, G. A. Parker, and J. C. Chubb. 2006. Competitive growth strategies in intermediate hosts: experimental tests of a parasite life-history model using the cestode, *Schistocephalus solidus*. Evolutionary Ecology 20(1):39–57.

Parker, G. A., J. C. Chubb, M. A. Ball, and G. N. Roberts. 2003. Evolution of complex life cycles in helminth parasites. Nature 425(6957):480.

Parker, G. A., M. A. Ball, and J. C. Chubb. 2015. Evolution of complex life cycles in trophically transmitted helminths. I. Host incorporation and trophic ascent. Journal of Evolutionary Biology 28(2):267–291.

Parker, G. A., M. A. Ball, and J. C. Chubb. 2015. Evolution of complex life cycles in trophically transmitted helminths. II. How do life-history stages adapt to their hosts? Journal of Evolutionary Biology 28(2):292–304.

Roberts, M. G., and J.A.P. Heesterbeek. 2003. A new method for estimating the effort required to control an infectious disease. Proceedings of the Royal Society London B 270:1359–1364.

Roberts, M. G., and J.A.P. Heesterbeek. 1998. A simple parasite model with complicated dynamics. Journal of Mathematical Biology 37:272–290.

Roberts, M. G., and J.A.P. Heesterbeek. 2013. Characterizing the next-generation matrix and basic reproduction number in ecological epidemiology. Journal of Mathematical Biology 66(4–5):1045–1064. doi: https://doi.org/10.1007/s00285-012-0602-1.

Ross, R. 1916. An application of the theory of probabilities to the study of *a priori* pathometry. I. Proceedings of the Royal Society London A 92(638):204–230.

Smith, D. L., K. E. Battle, S. I. Hay, C. M. Barker, T. W. Scott, and F. E. McKenzie. 2012. Ross, Macdonald, and a theory for the dynamics and control of mosquito-transmitted pathogens. PLOS Pathogens 8(4):e1002588.

Yoeli, M. 1973. Sir Ronald Ross and the evolution of malaria research. Bulletin of the New York Academy of Medicine 49(8):722.

What Determines Population Density?

Robert M. May

This chapter focuses on "What determines population density?"[1]

Ecology is a young science. Arguably the first ecological text is Gilbert White's *Natural History of Selborne*, published in 1789. This work goes beyond earlier fascination with descriptive natural history to begin to frame analytical questions about, for instance, what governs the abundance—and vastly different fluctuations in abundance—of swifts and wasps in Selborne, in the United Kingdom. The nineteenth century saw what I believe to be the most important advance in humanity's intellectual history, with the advent of Darwin's and Wallace's understanding of the basic ideas of evolution by natural selection. But this nineteenth-century advance in our comprehension of the evolutionary drama was not matched by analytical advances in exploration of the ecological stage on which it is played. In describing the "struggle for existence" that underlies evolution, Darwin reached, for instance, for metaphors of wedges in a barrel to illuminate a discussion of what we might call "competition for niche space among species." But there was, at that time, no attempt—either quantitative or qualitative—to measure the shape and size of the wedges or niches, much less to see if or how they fit together in the barrel. In short, ecological studies lagged behind evolutionary ones in the nineteenth century.

All biological populations fluctuate in abundance over time. The degree of fluctuation varies greatly among populations, affected both by their external environment and by other factors, whose effects may or may not depend on population density. The long-term average about which any given population fluctuates (either a little, as with Gilbert White's swifts, or a lot, as with his wasps) will ultimately be set by the density-dependent factors. In the absence of such nonlinear regulatory mechanisms, a population will either pursue a fluctuating trajectory to extinction, or balloon toward infinity. Perhaps the most striking illustration of ecology's comparative lack of an analytical tradition—thinking of populations as dynamical systems—is the long-running verbal debate, reminiscent of medieval scholasticism, about the relative roles of density-dependent and density-independent mechanisms in setting population sizes (for an excellent review, see Sinclair (1989)).[2]

This being said, the task of elucidating the density-dependent and in-dependent regulatory mechanisms for specific populations continues to prove very difficult. White observed that there were eight breeding pairs of swifts in Selborne, year after year. Given their capacity to double their population each year by producing typically two offspring, he asked what governed their steady abundance. When I ask my ornithologist friends, I get a range of guesses, but no agreed answer. Maybe nesting sites. But in White's day, the swifts nested in the church tower or in the thatched roofs of cottages, whereas for today's steady population of roughly 12 pairs (which, to ecological accuracy, is equal to eight, representing 200 years of constancy!) the church tower is sealed against squirrels and thatched roofs are essentially gone. Maybe food supplies. But, although field patterns have changed little, it strains credulity to suggest that the abundance of swifts' insect food has remained constant under great changes in farming practice. In short, there is no predictive understanding of why the population density of swifts in Selborne is what it is, and has been for more than 200 years.

On the one hand, we could give a huge list of density-dependent factors that are involved in setting the long-term abundances of plant and animal populations. I could, however, multiply endlessly the Selborne story, giving example upon example of well-studied populations where we have no confident understanding as to what ultimately sets their abundance. On the other hand, for practical problems—e.g., fisheries management,[3] whaling quotas,[4] harvesting terrestrial animals,[5] predicting outbursts of insect pests[6]—we do have effective methods for short-term prediction. They are mainly phenomenological, based to varying degrees of sophistication on projecting trends. In this sense, I could make a fairly confident projection that the swift population of Selborne will be 12 pairs in ten years' time. But lacking fundamental ecological understanding of the underlying regulatory mechanisms, we can only guess at the long-term changes in the same populations that will be caused by various kinds of environmental and other changes. What level of harvesting will cause fisheries to collapse? What, quantitatively, will be the effects on particular bird populations of further changes in agricultural practice? Phenomenology has its uses, but until we really understand why there were and are about 10 pairs of swifts in Selborne, rather than 1 or 100, we cannot claim to predict the effects on them of present and future changes in climate, landscape, or whatever. To put it another way, I think ecologists have their analogue of the chemists' heuristic periodic table, which is adequate for many practical purposes, but we lack the analogue of the underpinning atomic chemistry.

It is to be hoped that the twenty-first century will see advances in fundamental understanding of the average magnitude, and differing degrees of fluctuation, in plant and animal populations. The difficulty has been that most density-dependent regulatory effects acting on a population[7]— food supplies, predators, disease, competing species, and so on—are themselves affected by the populations they are interacting with.[8] The nonlinearities inherent in such dynamical systems make it essentially impossible to take the system apart, studying it piece by piece in a controlled and comparative way,[9] in the manner that has been so effective for the linear problems that dominate so much of classical physics. But recent advances in understanding nonlinear dynamical systems give cause for optimism.

Having described at length what is arguably one of the main problems of twentieth-century ecology, we now give a telegraphic sketch of some of the developments that hold hope.

The simplest, purely deterministic, models of density-dependent population change were shown in the 1970s to be capable of amazingly complicated dynamics, ranging from stable points, through stable cycles, to apparently random or chaotic fluctuations (May 1974, 1976). These simplest models were first-order difference equations (such as $N_{t+1} = N_t (1 - aN_t)$), originally propounded as crude first approximations to the dynamics of particular insect and fish populations. This ecological work was one of the two strands that brought chaos to center stage,[10] the other being Lorentz's work on three-dimensional continuous systems in meteorology (Gleick 1987). The earlier, and to my mind silly, debates between proponents of density-independent versus density-dependent regulation had implicitly assumed that strong density independence led to erratic fluctuations and density dependence to population constancy. The advent of the concept of deterministic chaos stood all this on its head, with strong density dependence causing population fluctuations as erratic, and—as a result of sensitivity to initial conditions—as unpredictable as anything density-independent external noise could produce.[11] In this light, the agenda for understanding the dynamical behavior of populations becomes one of unravelling density-dependent signals from density-independent noise, in complex nonlinear systems where even a purely deterministic signal may be apparently random, and long-term unpredictable, chaos.[12] For an early overview, see Zimmer (1999) in the special issue of *Science* devoted to complexity.

New techniques[13] for thus distinguishing the apparent randomness of chaotic signals from the "real" randomness of density-independent noise are being developed (e.g., Sugihara and May 1990; Hastings et al. 1993[14]), with applications ranging well beyond ecology.[15]

The most clear-cut illustration of the complexities inherent in nonlinear population dynamics have been demonstrated in the laboratory.[12] In a series of beautiful experiments,[16] Constantino et al. (1995) have shown how, essentially by changing development rates, laboratory populations of *Drosophila*[17] in constant environments can move, in a way that is predictable, from steady cycles to chaotic fluctuations. Some other, less definitive,[18] examples are reviewed by Murdoch and McCaughley (1985).

Modern data-inductive techniques[9,10,13] for thus distinguishing the apparent randomness of chaotic signals from the "real" randomness of density-independent noise, and for understanding how this problem plays out in field data are being developed (e.g., Sugihara & May 1990; Hastings & Higgins 1994; Sugihara 1994; Dixon et al. 1999; Stenseth et al. 2004),[3,13] with applications ranging well beyond ecology.[14] For the most part, these datacentric methods broadly involve building attractors from time series and then validating the attractor model by how well it forecasts out of sample.

Field studies are obviously much trickier, owing partly to environmental noise and partly to nonlinear interactions with other populations. But these are the problems we need to deal with. Using the methods mentioned previously to distinguish chaos (density-dependent signals) from noise (density-independent effects), Stenseth (1995) has, I think, gone a long way towards resolving the long-standing question of what causes the celebrated cycles (with a roughly 11-year period) seen over the past 150 or so years in lynx and snowshoe hares in Canada. He finds the hare dynamics to show the signature of almost-periodic chaos, with two interactive variables in the dynamical system. The methods used in decoding such time-series do not tell you what these active variables are; they only tell you the dimensionality of the system.[19] For the lynx, in contrast, the time-series has the signature of almost-periodic chaos with one interactive variable. We could conjecture that the lynx dynamics are driven by its interaction with hares, whereas hare dynamics involve both lynx and food supply. But whatever the biological factors actually are, the dimensionality of the system can be fairly confidently ascribed. Interestingly, this dimensionality agrees with earlier suggestions, based on ecological arguments (May 1973).

Dixon et al. (1999) have applied the modern time series techniques to obtain insights into the factors controlling the larval supply of pomacentrid reef fish populations whose dynamics look suspiciously chaotic with irregular episodic spikes. Their analysis found that the dimensionality of the problem—the number of active variables—was roughly three to four; they then used multivariate forecasting to try to identify them. As it turned out, spikes in larval abundance involved a perfect storm of three controlling factors: wind direction (blowing larvae toward the reef), lunar phase (phototropic spawning on the full moon) and wind speed (intermediate levels of turbulence required for larval fish to feed). Pulses of larvae occurred nonlinearly with the alignment of stars in a way that quite interestingly could not be ascribed to simple correlations among variables. It is described as a case of physics convolved with biology to produce what could be called "stochastic chaos" (Sugihara 1994), where instability in an otherwise stable simple model is provoked by process error (where the simple model represents an incomplete specification of the actual phenomenon).[15] Thus, although the early time series methods for assessing dimensionality and for distinguishing chaos from noise were primarily focused on short-term forecasting and not on mechanisms, this has changed, and with it our understanding of the problem has deepened, providing a more realistic path to Selbourne's puzzle.

To this end, two modern developments for studying this question from field data that appear to be particularly intriguing (and which I have unabashedly participated in), include a method, convergent cross-mapping, for detecting causation (to identify the active variables and how they relate to each other in a fully-fledged nonlinear network) (Sugihara et al. 2012), and a way of measuring interactions that are not constant (as we typically imagine them) but have the difficult nonlinear tendency to change as the system state changes (Deyle et al. 2017, Ushio et al. 2019) (also see https://www.youtube.com/watch?v=fevurdpiRYg).

Bishop Berkeley's (1710) warning notwithstanding, correlation has been the default for finding active variables in nearly all of the natural sciences, but this fails both because of the obvious—correlation does not imply causation—but also because of the less obvious (and highly confounding) converse: Variables that are causative may be totally uncorrelated with their effects (as in the above larval fish example). The latter is common in ecosystems, and to be able to identify the active variables or the relevant ecosystem, as it were, is key to extracting mechanism. The new method for measuring causation, mentioned above, does not rely on correlation, and, for example, shows that although Pacific sardines and anchovies are negatively correlated in their abundances (once thought to be evidence for competition) they show no mutual causal effect on each other, but appear to be driven by sea surface temperature, which, interestingly, through the twentieth century is not cross-correlated with the abundances of either species. Understanding how populations are situated within their networks of causal influence advances the problem.

Equally promising, from time-series data taken in the field we are beginning to be able to quantify interactions among populations as they occur on the fly, instantaneously. Whereas much of the work from the last century, particularly with regard to models, assumes interactions that are constant, what is potentially game-changing from this work is emerging evidence from the application of these attractor-based empirical dynamic methods showing that the interactions among species in nature are highly variable through time and can be volatile and episodic, something never captured in the lab or in simple models.

Finally, a potential problem with all such analysis of time-series data (e.g., seeking to identify nonlinear signals versus noise, unraveling causal mechanisms, population forecasting, quantifying interactions) is that they need long runs of data (long compared with the time interval on which significant changes can occur in the population size). Such long series were once considered rare in ecology, but as data itself becomes more of a focus and the need to monitor our environment escalates, this is beginning to change.[16] Thus, as data constraints become less of a concern, I think that progress in developing techniques initially aimed at disentangling deterministic signals from external noise in nonlinear ecological systems[17] is at last clearing the ground for a more mature and complete approach to the fundamental problem of ecology that we have been circumambulating here: namely, what determines the density and controls change of populations that are ineluctably linked to the ecosystems surrounding them. I think there is good reason for optimism.

Notes

1. At the very beginning of this project, Lord Robert May, upon request from Andy Dobson, rapidly wrote a short essay. With his characteristically graceful and incisive style, he articulated a long-standing puzzle in the ecological sciences, and highlighted some key papers from the 1970s through the 1990s that contribute towards addressing this puzzle. Due to his declining health, he has not been able to edit his contribution, and it would appear, left his essay incomplete. George Sugihara believes that Bob would have ended his essay with a deft discussion of current exciting work about teasing out the mechanistic causality of noisy nonlinear systems, extracting causes inductively from patterns in the data. It is beginning to be possible to model interactions on the fly, which are proving to be highly labile, so the interaction structure of communities can change even over short time periods (as alluded to in the essay). We decided, after consultation with his wife Judith May to publish what Bob wrote, verbatim (with the exception of trivial corrections).

     But as a service to readers, we provide pointers to more recent literature on the theme of population regulation. These numbered superscripts in the text correspond to the list that follows. We thank Michael Bonsall, George Sugihara, and Jake Ferguson for providing suggestions towards this end.

2. See the special issue: Sibly, R. M., J. Hone, and T. H. Clutton-Brock, eds. 2002. Population growth rate: Determining factors and role in population regulation. Proceedings of the Royal Society of London B: Biological Sciences 357:1147–1320.

     Workers in the metabolic theory of ecology have uncovering intriguing allometric relationships in population abundance, e.g., Figure 2 in Allen, A. P., J. H. Brown, and J. F. Gillooly. 2002. Global biodiversity, biochemical kinetics, and the energetic-equivalence rule. Science 297:1545–1548 and in Yeakel, J. D., C. P. Kempes, and S. Redner. 2018. Dynamics of starvation and recovery predict extinction risk and both Damuth's law and Cope's rule. Nature Communications 9: 657; doi: https://doi.org/10.1038/s41467-018-02822-y. These regularities hint that explanations for density might be in part sought in fundamental organismal traits, such as body size and metabolic rate.

3. Deyle, E. R., M. Fogarty, C-h. Hsieh, L. Kaufman, A. D. MacCall, S. B. Munch, C. T. Perretti, H. Ye, and G. Sugihara. 2013. Predicting climate effects on Pacific sardine. Proceedings of the National Academy of Sciences of the United States of America 110:6430–6435.
4. Punt, A.E. and G.P. Donovan. 2007. Developing management procedures that are robust to uncertainty: lessons from the International Whaling Commission. ICES Journal of Marine Science 64:603–612.
5. Brook, C. E., J. P. Herrera, C. Borgerson, E. Fuller, P. Andriamahazoarivosoa, B. J. R. Rasolofoniaina, J.L.R. Ravoavy Randrianasolo, Z R.E. Rakotondrafarasata, H. J. Randriamady, A. P. Dobson, and C. D. Golden. 2018. Population viability and harvest sustainability for Madagascar lemurs. Conservation Biology 33:99–111.
6. Hudgins, E. J., A. M. Liebhold, and B. Leung. 2017. Predicting the spread of all invasive forest pests in the United States. Ecology Letters 20:426–435.
7. Royama, T. 1992. Analytical population dynamics. London: Chapman and Hall.
8. Berryman, A. A., M. L. Arce, and B. A. Hawkins. 2002. Population regulation, emergent properties, and a requiem for density dependence. Oikos 99:600–606.
9. Stenseth, N. C., K. S. Chan, G. Tavecchia, T. Coulson, A. Mysterud, T. Clutton-Brock, and B. Grenfell. 2004. Modelling non-additive and nonlinear signals from climatic noise in ecological time series: Soay sheep as an example. Proceedings of the Royal Society of London B: Biological Sciences 271:1985–1993; Hsieh, C. H., C. S. Reiss, J. R. Hunter, J. R. Beddington, R. M. May, and G. Sugihara. 2006. Fishing elevates variability in the abundance of exploited species. Nature 443:859–862.
10. Hassell, M. P., J. H. Lawton, and R. M. May. 1976. Patterns of dynamical behavior in single-species populations. Journal of Animal Ecology 45:471–486; Schaffer, W. M., and M. Kot. 1986. Chaos in ecological systems: the coals that Newcastle forgot. Trends in Ecology and Evolution 1:58–63; Turchin, P., and S. P. Ellner. 2000. Living on the edge of chaos: population dynamics of Fennoscandian voles. *Ecology* 81:3099–3116.
11. Hastings, A., C. L. Hom, S. Ellner, P. Turchin, and H.C.J. Godfray. 1993. Chaos in ecology: is Mother Nature a strange attractor? Annual Review of Ecology and Systematics 24:1–33.
12. Boettiger, C. 2018. From noise to knowledge: how randomness generates novel phenomena and reveals information. Ecology Letters 21:1255–1267.
13. Bjornstad, O. N., and B. T. Grenfell. 2001. Noisy clockwork: time series analysis of population fluctuations in animals. *Science* 293:638–643; de Valpine, P., and A. Hastings. 2002. Fitting population models incorporating process noise and observation error. Ecological Monographs 72:57–76; Sugihara, M., R. May, H. Ye, C-h. Hsieh, E. Deyle, M. Fogarty, and S. Munch. 2012. Detecting causality in complex ecosystems. *Science* 338:496–500.
14. Ye, Hao, R. J. Beamish, S. M. Glaser, S.C.H Grant, C-h. Hsieh, L. J. Richards, J. T. Schnute, and G. Sugihara. 2015. Equation-free mechanistic ecosystem forecasting using empirical dynamic modeling. Proceedings of the National Academy of Sciences of the United States of America 112:E1569–E1576; Perretti, C. T., S. B. Munch, and G. Sugihara. 2013. Model-free forecasting outperforms the correct mechanistic model for simulated and experimental data. Proceedings of the National Academy of Sciences of the United States of America 110:5253–5257.
15. Finance: Soofi, A.S., L. Cao. 2012. Modelling and forecasting financial data: techniques of nonlinear dynamics. In: Studies in Computational Finance, vol. 2. New York: Springer. Hydrology: Nayak, P. C., K. P. Sudheer, D. M. Rangan, and K. S. Ramasastri. 2004. A neuro-fuzzy computing technique for modeling hydrological time series. Journal of Hydrology 291:52–66. Physics: Hamilton, F., T. Berr, and T. Sauer. 2016. Ensemble Kalman filtering without a model. Physical Review X 6:011021. Statistics: Fan, J. and

Q. Yao. 2008. Nonlinear time series: Nonparametric and parametric methods. New York: Springer.

16. Benincà, E., J. Huisman, R. Heerkloss, K. D. Jöhnk, P. Branco, E. H. Van Nes, M. Scheffer, and S. P. Ellner. 2008. Chaos in a long-term experiment with a plankton community. *Nature* 451:822–826.

17. Muller, L. D. and A. Joshi. 2001. Stability in Model Populations. Monographs in Population Biology, Vol. 31. Princeton, N.J.: Princeton University Press.

18. Benincà, E., B. Ballantine, S. P. Ellner, and J. Huisman. 2015. Species fluctuations sustained by a cyclic succession at the edge of chaos. Proceedings of the National Academy of Sciences of the United States of America 112:6389–6394.

19. Abbott, K. C., J. Ripa, and A. R. Ives. 2009. Environmental variation in ecological communities and inferences from single-species data. *Ecology* 90:1268–1278.

20. Earn, D.J.D, P. Rohani, B. M. Bolker, and B. T. Grenfell. 2000. A simple model for complex dynamical transitions in epidemics. Science 287:667–670; Ferrari, M. J., R. F Grais, N. Bharti, A.J.K. Conlan, O. N. Bjornstad, L. J. Wolfson, P. J. Guerin, A. Djibo, and B. T. Grenfell. 2008. The dynamics of measles in sub-Saharan Africa. Nature 451:679–684.

21. Ferguson, J. M. and J. M. Ponciano. 2015. Evidence and implications of higher-order scaling in the environmental variation of animal population growth. Proceedings of the National Academy of Sciences of the United States of America 112:2782–2787; Ferguson, J. M., F. Carvalho, O. Murillo-Garcia, M. L. Taper, and J. M. Ponciano. 2016. An updated perspective on the role of environmental autocorrelation in animal populations. Theoretical Ecology 9:129–148; Huffaker, R., M. Bittelli, and R. Rosa. 2017. Nonlinear time series analysis with R. Oxford: Oxford University.

References

Berkeley, G. 1710. A treatise concerning the principles of human knowledge. D. R. Wilkins, ed. 2002. Journal of Nervous and Mental Disease 85:468.

Constantino, R. F., J. M. Cushing, B. Dennis, and R. A. Desharnais. 1995. Experimentally induced transitions in the dynamic behaviour of insect populations. Nature 375:227–230.

Deyle, E. R., R. M. May, S. B. Munch, and G. Sugihara. 2016. Tracking and forecasting ecosystem interactions in real time. Proceedings of the Royal Society B: Biological Sciences:283.

Dixon, P. A., M. J. Milicich, and G. Sugihara. 1999. Episodic fluctuations in larval supply. Science 283:1528–1530.

Gleick, J. 1987. Chaos: Making a new science. New York: Viking.

Hastings, A., and K. Higgins. 1994. Persistence of transients in spatially structured models. Science 263:1133–1136.

Levin, S. A., B. Grenfell, A. Hastings, and A. S. Perelson. 1997. Mathematical and computational challenges in population biology and ecosystems science. Science 275:334–343.

May, R. M. 1973. Stability and complexity in model ecosystems. Princeton, N.J.: Princeton University Press.

May, R. M. 1974. Biological populations with nonoverlapping generations: stable points, stable cycles, and chaos. Science 186:645–647.

May, R. M. 1976. Simple mathematical models with very complicated dynamics. Nature 261:459–467.

Murdoch, W. W., and E. McCaughley. 1985. Three distinct types of dynamic behaviour shown by a single planktonic system. Nature 316:628–630.

Sinclair, A.R.E. 1989. The regulation of animal populations. In: J. M. Cherrett, ed. Ecological concepts. Oxford: Blackwell, 197–241.

Stenseth, N. C. 1995. Snowshoe hare populations: Squeezed from below and above. Science 269:1061–1062.

Stenseth, N. C., K. S. Chan, G. Tavecchia, T. Coulson, A. Mysterud, T. Clutton-Brock, and B. Grenfell. 2004. Modelling non–additive and nonlinear signals from climatic noise in ecological time series: Soay sheep as an example. Proceedings of the Royal Society of London B: Biological Sciences 271:1985–1993.

Sugihara, G., and R. M. May. 1990. Nonlinear forecasting as a way of distinguishing chaos from measurement error in time series. Nature 344:734–741.

Sugihara, M., R. May, H. Ye, C. H. Hsieh, E. Deyle, M. Fogarty, and S. Munch. 2012. Detecting causality in complex ecosystems. Science 338:496–500.

Ushio, M., C. H. Hsieh, H. Matsuda, E. R. Deyle, H. Ye, C. W. Chang, G. Sugihara, and M. Kondoh. 2019. Fluctuating interaction network and time-varying stability of a natural fish community. Nature 554:360–363.

White, G. F. 1789. The natural history of Selborne. Harmondsworth, UK: Penguin. (Reprint, R. Mabey, ed. 1977.)

Zimmer, C. 1999. Life after chaos. Science 284(5411):83–86.

PART II

POPULATION BIOLOGY AND THE ECOLOGY OF INDIVIDUALS

Neglected Problems in Ecology

Interdependence and Mutualism

Egbert Giles Leigh Jr.

The Nature of This Neglect

From 1950 on, community ecologists collectively reckoned most inadequately with the role of interdependence—among species within a habitat or among habitats—and mutualism (cooperation between species, as between plants and their pollinators) in ecosystem function and evolution. Odum's (1959) *Fundamentals of Ecology*, the era's dominant ecology text, defines mutualism and protocooperation (nonobligate mutualism) (p. 225) and gives examples (pp. 242–244), but the topic is otherwise ignored. This neglect is puzzling. Mutualism is a familiar story (Smith and Douglas 1987). So is interdependence: migratory birds that depend on both tropical and temperate-zone habitats (Keast and Morton 1980), heaths that depend on cattle to exclude trees and shrubs (Darwin 1859, pp. 71–72), coralline algae that depend on sea urchins to exclude kelps (Paine and Vadas 1969), and the like. This neglect is ending: Polis et al. (1997) wrote an influential paper that made many community ecologists aware of the pervasive importance of interdependence among habitats, and Bronstein (2015) edited a volume to do the same for mutualism. Yet a synoptic view of the crucial role in ecosystem function of mutualism and interdependence, and the factors that limit their evolution, is still lacking. This circumstance is remarkable, considering that the easiest way to disturb a natural ecosystem is to disrupt a crucial mutualism (as in overfrequent bleaching of corals), or a relationship of interdependence (as in degrading a forest by eliminating its top carnivores).

When ecosystems were considered to be superorganisms whose parts were designed to serve the whole and each stage of plant succession designed to prepare the site for the next (Clements 1936, p. 245), interdependence attracted more attention. Ecologists of the superorganism school, however, were unable to refute Gleason's (1926) assertion that tree species were distributed independently of each other, nor could they show how the presumed functional unity of this superorganism was actually manifested in natural phenomena. Elton (1927) suggested instead that the relationships of organisms with competitors, predators, parasites, and the physical environment were what organized ecological communities. His view eventually prevailed.

Later, ecology split into three disciplines—ecosystem, population and community ecology—a split apparent by 1959 (Odum 1959, p. 8). Population ecologists ask what factors limit a population's numbers and its distribution and seek to relate these limits to how the population's members live and reproduce. Employing mutualists such as mycorrhizae, nitrogen-fixing bacteria, and pollinators, often helps individuals live and reproduce, as does behavior entailing interdependence. Therefore, population ecologists often study these phenomena.

Community ecology focuses on two questions (Elton 1927, Hutchinson 1959). How do interactions of different species with each other and their environment govern community organization, species composition and species diversity? What conditions allow species to coexist? Answers center on mechanistic processes. Thanks to the influence of Lotka's (1925) and Volterra's (1931) equations, ecologists focused on competition and predation. Theoretical (Skellam 1951) and empirical (Denslow 1980) work forced ecologists to reckon with disturbance as well, thereby enthroning the trinity of competition, predation, and disturbance as the three basic processes of ecology. Trade-offs, whereby improving one of an organism's abilities necessarily imposes sacrifices on another, drives diversification (Fisher 1930, pp. 127–129). Most trade-offs considered by ecologists involve coping with competition, predation, and disturbance. The disturbance-related trade-off between growing fast in bright light versus surviving in shade (Wright et al. 2003) enhances tree diversity in a forest disturbed by tree falls. The competition-related trade-off between being able to defend and exploit rich resources and surviving on scarce resources (Robertson 1996) enhances alpha diversity; the competition-related trade-off between growing fast on good soil and surviving on poor soil (Russo et al. 2008) enhances beta diversity. The trade-off between outgrowing competitors and resisting herbivores or predators (Coley et al. 1985, Wulff 2005) also fits within this trinitarian scheme. This scheme, however, does not force neglect of mutualism or interdependence. Mutualism among sponges helps them survive predation, disease, and disturbance (Wulff 1997). Mutualism between plants and their pollinators and seed dispersers helps plants survive intense pest pressure while investing less in antiherbivore defense, thereby increasing their competitiveness and their forest's primary production (Leigh 2010).

By 1970, the idea was revived that ecosystems were functional units where nutrient-releasing decomposers, nutrient-requiring primary producers, and herbivores and carnivores that enhance the recycling of resources and maintain ecosystem health were designed to serve their ecosystem by fulfilling these roles (Odum 1971). H. T. Odum provided an appealing justification for this idea, based on an analogy between natural ecosystems and free-market economies. He argued that selection on a population would favor harming other populations that injured it and benefitting other populations that benefitted it, as much as possible. Although recent work suggests the correctness of Odum's argument (Vermeij 2013, p. 9), the ecological theory of the day was unable to evaluate this proposition (see later), and it was ignored. Moreover, many ecosystem ecologists (Odum 1971) focused on energy flows and energy stocks, whereas natural selection acts on the populations that generate them. Treating ecosystems as superorganisms, emphasizing energy stocks and energy flows, and advocating a form of overly coordinated big science that often generated shaky data, brought this new superorganism school of ecosystem ecology into disrepute among other ecologists.

Empirical ecosystem ecology fared better. In forestry schools, some ecosystem biologists studied mutualisms with genuine care. For example, Vogt et al. (1982) assessed the costs and benefits to a forest of supporting mycorrhizae. Moreover, an ecosystem-wide study of the Serengeti grassland (McNaughton 1985) revealed the mutualistic interdependence between the grassland and its grazers, and the dependence of migratory grazers on several different habitat types within the Serengeti. Ecologists respected these studies and recommended their publication, but never integrated them into a bigger picture.

A new ecosystem approach—on an earthwide scale—is the Gaia hypothesis (Lovelock 1988). Whereas a traditional ecosystem study focused on how its species served their ecosystem's function, Gaia focuses on how the activities of the earth's organisms collectively make its environment, especially the earth's temperature, atmospheric composition, and its ocean, river, and soil chemistry more hospitable to life. Automatic, active but unconscious feedback processes maintain this homeostasis. However, "the conditions are only constant in the short term and evolve in synchrony with the changing needs of the biota as it evolves." (Lovelock 1988, p. 19). Neither an ecosystem nor the ensemble of biogeochemical processes affecting life on earth are true superorganisms because, unlike honeybee colonies, these entities are not organized to reproduce. Nonetheless, natural ecosystems are organized to maintain high productivity and diversity, as evidenced by the circumstance that disturbance outside an ecosystem's evolutionary experience reduces its productivity and/or diversity (Leigh and Vermeij 2002). Moreover, living things have collectively modified the earth's surface, making it more suitable for life. Most geochemical cycles have come increasingly under biotic control as life evolved (Vermeij 2013, p. 13; Vermeij 2011, pp. 199–202), so the world is a far better home for life now than 3.7 billion years ago. Ecosystems, in-dividually and collectively, have these features because "by establishing feedbacks between species and enabling factors [factors enhancing access of species to essential resources], effective competitors regulate and en-hance resource supply (Vermeij 2013, p. 1).

Nonetheless, community ecologists rejected the hypothesis of ecosys-tems as adaptively organized and the Gaia hypothesis that, thanks to their relationships of interdependence, living things collectively make the earth better for life. They did not see how natural selection within popula-tions could favor these developments. Similarly, geologists initially re-jected the idea of continental drift even though it fit the facts, because they knew no mechanistic cause for it—but they had to discover new phenomena to explain continental drift, whereas Odum had already pro-posed the mechanism that shaped ecosystem adaptation and life's impact on its environment.

Although how mutualism helps make organisms better competitors or predators has been studied for decades (Smith and Douglas 1987), as has interdependence among ecosystems (Keast and Morton 1980), William-son (1972, p. 95) remarked that mutualism "is a fascinating biological topic, but its importance in populations in general is small." In their book on symbiosis (a form of mutualism where one partner lives in the other), Smith and Douglas (1987) assembled evidence that mutualism—between eukaryotes and their mitochondria, plants and their chloroplasts and my-corrhizae, wood-decomposing termites and the cellulose-consuming pro-tists in their guts, corals and their zooxanthellae—is as vital to the pro-ductivity of many, if not most, natural ecosystems as cooperative enterprise is to modern human economies, However, they found that, intracellular organelles excepted, "most biologists consider that the evolutionary im-portance of symbiosis has been trivial compared to the other mechanisms by which novel and heritable characteristics are produced" (Smith and Douglas 1987, p. 237). Pastor (2008, pp. 169–173) briefly discussed mod-els of mutualism, but this discussion had no impact on the rest of his book. How could this happen?

Why the Neglect?

Theoretical Biases

Several factors tended to marginalize mutualism and interdependence. First, the equations of Lotka (1925) and Volterra (1931), which Pastor (2008) still tested as a possible foundation for mathematical theory in ecol-ogy, were not suited to analyzing interdependence and mutualism. These equations assume a homogeneous community, where the population density of each species is everywhere uniform. Skellam (1951) and Muller-Landau (2010) needed a different approach that explicitly incorporated the heterogeneity in habitat due to light gaps opened by fallen trees (in Muller-Landau's case, gaps of different sizes), to show how the trade-off between exploiting light gaps and surviving in shade allows different species to coexist. No approach as elegant as Muller-Landau's springs to mind for modelling interdependence between widely separated habitats.

Although May (1976) recognized the importance of mutualism in the tropics (though not in the temperate zone), he suggested that theorists had ignored mutualism because reversing the signs of the coefficients of competitive interaction to represent mutualism in Lotka's equations for two competing species often caused their populations to explode. Representing mutualism more realistically, moreover, would destroy these equations' simplicity.

A more cryptic and therefore more disastrous aspect of these equations is their implicit assumption that each individual encounters a random sample from the same homogeneous boundaryless community. Consequently, if an animal somehow, at no cost to itself, benefits a species on which it feeds, all its conspecifics benefit equally from the resulting increase of food, so the benefactor reaps no differential advantage from its "good deed." These equations thus cannot represent the evolution in consumers of any feature that benefits a food species, such as a grazer replacing deer pellets with grassland-nourishing cowpies (McNaughton 1985), for cooperators and noncooperators in the consumer species would benefit equally from this change's improvement of grass production. Wilson (1980) devised a theory confining the immediate effects of interactions to near neighborhoods whereas young dispersed more widely, that resolved this anomaly. Mainstream theoretical ecologists, however, ignored his work.

Ignoring Adam Smith

Lotka-Volterra equations were not used widely enough to be an impassable roadblock to reckoning with mutualism and interdependence. Although they kept theoreticians from showing how a species could evolve to benefit species that benefit it, and to harm those that harm it, it is still puzzling why Odum's insight (1971), based on an analogy between natural ecosystems and human free-market economies, had so little impact. Adam Smith's *Wealth of Nations* (1776) shows how, given adequate transport and other technology, competition among individuals, families, and larger groups for the means to live and reproduce favors innovation, diversification of occupations (ways of making a living), cooperation among individuals to compete better with others, and interdependence at many different scales. These are just the ingredients needed to understand how ecological communities evolve higher productivity and diversity. Few biologists, however, read Adam Smith. Politicians depicted him as an advocate of unbridled competition—an infallible disincentive for reading him to learn how cooperation can evolve. Von Neumann and Morgenstern (1944) were unable to show how economic competition could serve society's common good, so Smith's approach seemed invalid. Smith (1759, Part II, Section ii, Chapter 2, paragraph 1) refutes the first claim: if competition is to benefit society, society must enforce the fairness of competition. The second is answered by the circumstance that in eukaryotes, natural selection on autosomal loci enforces the fairness of meiosis, thereby ensuring that an allele spreads only if it serves the autosomal genome's common interest (Leigh 2010).

The Tunnel Vision of Overspecialization and Overgeneralization

Community ecologists usually focus on a single guild or taxonomic group, as did many justly celebrated classics of ecology, such as Lack (1947), MacArthur (1958), and Losos (2009). These studies emphasized the Lotka-Volterra inspired themes of competition, niche differentiation and coexistence. Moreover, as mutualism usually, but not always (Wulff 1997), involves members of different kingdoms, such studies rarely unearth mutualism or interdependence.

Programmatic comparisons of different communities, and studies shaped by a single overarching approach, such as representing an ecosystem as a network of energy flows between different energy stocks (populations) (Odum 1971) leave too little place for individual natural history studies, driven by a student's own questions, which often produce surprises unwelcome to such visionaries. Seemingly frivolous individual studies of how figs enforce pollination by their pollinator wasps, and how the number of wasps pollinating each fig fruit affects the virulence of these wasps' parasitic nematodes, taught us much about how mutualism is maintained (Herre et al. 2008, Herre 1993).

What Is to Be Done?

How can the role of mutualism and interdependence in ecological communities be brought into clear focus? A first step is for biologists to read Adam Smith's (1776) *An Inquiry into the Nature and Causes of the Wealth of Nations.* Unlike many biologists, Smith understood how competition can favor cooperation, and how cooperative enterprise and interdependence within and among economies enhances their productivity and diversity of occupations. Smith (1776) also realized the importance of transport technology in making possible the interdependence and cooperative enterprise so essential to complex human economies. I will argue that movement ecology (Nathan et al. 2008) can be a source of analogous understanding for natural ecosystems. Finally, I will show how the concentration of diverse studies at one site (the ecology of place) often reveal relationships of mutualism and interdependence.

The Economic Analogy

Evolution . . . consists in raising the upper level of organization reached by living matter, while still permitting the lower types of organization to survive. This gradual rise in the upper level of control and independence . . . . in living things . . . may be called *evolutionary* or *biological progress*. It is . . . another necessary consequence of the struggle for existence.
(Haldane and Huxley 1927, p. 234) (emphasis in original)

There are striking parallels between natural and human economies, and in the processes that organize them (Vermeij and Leigh 2011). Both are driven by competition among individuals, families, and larger groups for the means to live and raise young. In both, two or more individuals may cooperate to compete more effectively with a third. Insofar as technological capacity, human, other animal, or plant allow, cooperation among individuals with different abilities, and interdependence among different groups, maintain productivity, and diversity of occupations or ways of life, alike in human economies and natural ecosystems. In both, diversity and productivity tend to increase over historical or evolutionary time. On the other hand, cultural transmission of ideas about how to make a living and handle social life drives evolution in human economies, whereas natural selection on genomes drives evolution in natural ecosystems. Cultural evolution can be much faster, especially in economies with rapid long-distance mass communication, but it can be dysfunctional, as when the internet propagates appealing but dangerous delusions. Genetic evolution is much slower, but genetic systems are organized to make evolution adaptive (Vermeij and Leigh 2011). Fair meiosis ensures that alleles spread only if they serve the common good of their autosomal genome. Sexual reproduction and recombination ensure that a new allele spreads more nearly according to its own contribution to fitness than according to the merits of the genome in which it first occurs. The parallel between natural ecosystems and human economies is far from exact, but each study can offer useful pointers for the other (Vermeij 2009).

The Uses of Movement Ecology

Developing effective transport paced the development of productive, diverse human economies. Likewise, the increasing range and sophistication of animal movements, and the increasing variety of ways plants take advantage of these movements, paced increase in the scale and intensity of interdependence and mutualism in natural ecosystems. Capacity for these movements, however, is limited by the technological capacities of the movers. Nathan et al. (2008) accordingly set out to found a new ecological subdiscipline, movement ecology, concerned with these questions. Movement ecology can promote a more unified vision of ecology by shedding light on how mutualism and interdependence work, and the technologies upon which they depend. Indeed, many movements reflect interdependence between regions, nearby or widely distant, ranging from leaving one's home range in search of food to seasonal migration between the tropics and the temperate zone. Other movements, such of those of pollinators among flowers, or animals dispersing seeds far from their parents, express mutualism.

These movements involve many animal technologies—vision (or echolocation), flight (or swimming), navigation, often learning and memory as well. Consider, for example, the technologies behind the mutualism enabling the luxuriance and diversity of tropical forest—animal pollination. Plants kept rare by specialized pests must attract faithful pollinators that will travel far to seek out other members of its species (Leigh 2010). Most tropical plants attract pollinators. Some domesticate their own, species-specific pollinators. Fig trees domesticated seed-eating wasps (Herre et al. 2008). A fig flowerhead, turned outside-in so the flowers line the inside of a ball with an entry hole, attracts fertilized female wasps carrying pollen from their natal fig, which pollinate its flowers and lay eggs in half of them. Each larva matures within a single fig seed. When adult wasps emerge, they mate with each other, and fertilized females fly off in search of new trees to pollinate. Fig trees often attract these wasps from 10–20 km away, so even rare species maintain extraordinary genetic diversity (Nason et al. 2008). The technology underlying this mutualism, including how the wasps' cooperation is enforced, comprise a remarkable chapter in natural history (Herre et al. 2008). Other pollinators face different problems. How can they locate other flowers of a rare and scattered species, and then find their way home? To answer we must learn why they move, how they do so, and how they find their way around—standard questions of movement ecology (Nathan et al. 2008). Honeybees show how complex pollinators' technology can be (Seeley 1995). A honeybee hive has one queen and 30,000 worker daughters that help her reproduce. The colony has special mechanisms to ensure that workers help their queen rather than reproducing on their own and promote appropriate division of tasks among workers. Foraging workers gather pollen and nectar for the hive: They have special dances to tell others how far away the food is, in what direction, and how good it is. To interpret the dance, a forager must infer from the angle a bee's waggle run makes with the vertical, the angle which the direction to the food makes with the horizontal component of the direction to the sun. Many other technologies, including biological clocks, help honeybees to be major pollinators in many parts of the world.

Migration enables animals to spend most of their time where food is abundant. Many tropical birds migrate to the temperate zone to breed in spring and summer, when insects and fruit are superabundant there, and return to the tropics to avoid winter's scarcity. These migrants play a crucial role in controlling the insect populations of some temperate-zone forests (Holmes and Sturges 1975). These birds must know when to migrate, remember where to go, and know how to find their way. Birds that navigate by the sun must be able to adjust for the sun's change in position as the day progresses. Migrants that save energy by gliding, such as vultures (Mandel et al. 2008), must be able to detect and take advantage of updrafts. Migration poses many technological problems, rather different from those posed by pollination.

The "Ecology of Place"

Studies of particular communities, such as the Serengeti grasslands in East Africa (Sinclair and Norton-Griffiths 1979), the tropical forest of Barro Colorado Island (Leigh 1999) in central Panama, or the rocky shores of Tatoosh in the northeastern Pacific (Paine et al. 2010) are likely to reveal mutualism and interdependence. At the most fruitful of these sites, "investigator independence, set within a supportive academic environment, enhances research creativity and the development of novel ideas and techniques" (Paine et al. 2010, p. 230). At these sites, "unity of place" makes different projects more relevant to each other. Such projects contribute to understanding the community as a whole, especially if the researchers feel themselves to be full members of the community devoted to that goal. Moreover, long acquaintance with a site ripens interest in and understanding of the natural history of its inhabitants (Paine et al. 2010). For example, work in the Serengeti grasslands reveals the mutualistic interdependence between these grasslands and their grazers, and the dependence of migratory grazers on access to several different habitats (Sinclair and Norton-Griffiths 1979). Larvae of the starfish, sea urchins, and mussels that play such crucial roles in the intertidal at Tatoosh, drift over the sea from where they hatched, who knows how far away (Paine 1977).

The importance of movements expressing interdependence and mutualism are best expressed in terms of the ecology of place. Here, I sketch how movement of animals, pollen, and seeds between the 1500-ha Barro Colorado Island and nearby mainland (200 m distant at its closest point) maintain the island ecosystem's integrity. Then I will sketch technology underlying some crucial mutualisms and relationships of interdependence.

Students on Barro Colorado were the first to show conclusively that each tree species is kept rare by its specialist pests or sets of pests (Comita et al. 2010, Mangan et al. 2010). They often need animals to disperse seeds out of reach of parental pests and always need pollinators to maintain genetic diversity. Thus, each tree species depends on other tree species, sometimes other forests, to maintain "their" animals when they themselves are not flowering or fruiting (Leigh 2010).

This island's ecosystem also depends on movements between it and the mainland of many animals, ranging from bats seeking figs on nearby mainland when they find none on the island to migratory birds, primary dispersers of the seeds of some island trees that fruit as the birds are making their way northward (Greenberg 1981), and other birds that must migrate to other, perhaps distant, tropical forests during the island's seasonal fruit shortage (Morton 1977). Toucans and parrots left Barro Colorado during a fruit famine in 1970, returning after the famine ceased (Foster 1982): They might have died out if they had no place to go. If mainland immigrants did not replenish genetic diversity of essential but rare animals, such as the island's 30 ocelots, now its top predators (Ziegler and Leigh 2002, p. 175), these populations would die out, as have 35 species of understory forest bird, whose members will not cross 100 m of open water (Moore et al. 2008). The island's strangler figs—a set of rare species that collectively provide a reliable source of fruit all year long—depend on pollen from mainland figs to maintain genetic diversity (Nason et al. 2008). This is only a sample of the many movements between island and mainland on which the island's ecosystem depends.

Conclusions

The high productivity and diversity of occupations in modern human economies depends largely on relationships of interdependence made possible by cheaper, more effective transport (Smith 1776), which also enhances the advantages of the cooperative enterprise so essential to modern economies. The importance of such relationships to human economies suggests that we must learn what relationships of interdependence and mutualism underlie the diversity and productivity of natural ecosystems. Many, if not most, such relationships involve animal movement. Movement ecology shows what technologies these movements require. Thus, movement ecology provides a way to learn how, and to what extent, the progressive improvement from the Cambrian onward of some animals' awareness of their surroundings (vision, hearing, smell, taste), locomotory and navigational capacities, and aptness for social life, paced the development of interdependence and mutualism. Thus, assessing the importance of mutualism and interdependence, and investigating the "organismic technologies" that make such relationships possible, would greatly widen the perspectives of ecological thought.

References

Bronstein, J. L., ed. 2015. Mutualism. Oxford: Oxford University.

Clements, F. E. 1936. Nature and structure of the climax. Journal of Ecology 24:252–284.

Coley, P. D., J. P. Bryant, and F. S Chapin III. 1985. Resource availability and plant anti-herbivore defense. Science 230:895–899.

Comita, L. S., H. C. Muller-Landau, S. Aguilar, and S. P. Hubbell. 2010. Asymmetric density dependence shapes species abundance in a tropical tree community. Science 329:330–332.

Darwin, C. 1859. On the origin of species. London: John Murray.

Denslow, J. S. 1980. Gap partitioning among tropical rain forest trees. Biotropica 12 supplement:47–55.

Elton, C, 1927. Animal ecology. London: Sidgwick and Jackson.

Fisher, R. A. 1930. The genetical theory of natural selection. Oxford: Clarendon.

Foster, R. B. 1982. Famine on Barro Colorado Island. In E. G. Leigh Jr., A. S. Rand, and D. M. Windsor, eds. The ecology of a tropical forest: Seasonal rhythms and long-term changes. Washington, DC: Smithsonian Institution, pp. 201–212.

Gleason, H. A. 1926. The individualistic concept of the plant association. Bulletin of the Torrey Botanical Club 53:7–26.

Greenberg, R. 1981. Frugivory in some migrant tropical forest wood warblers. Biotropica 13:215–223.

Haldane, J. B. S. and J. S. Huxley. 1927. Animal biology. Oxford: Oxford University Press.

Herre, E. A. 1993. Population structure and the evolution of virulence in nematode parasites of fig wasps. Science 259:1442–1445.

Herre, E. A., K. C. Jandér, and C. A. Machado. 2008. Evolutionary ecology of figs and their associates: recent progress and outstanding puzzles. Annual Review of Ecology, Evolution and Systematics 39:439–458.

Holmes, R. T. and F. W. Sturges. 1975. Bird community dynamics and energetics in a northern hardwood ecosystem. Journal of Animal Ecology 44:175–200.

Hutchinson, G. E. 1959. Homage to Santa Rosalia, or, why are there so many kinds of animals? American Naturalist 93:145–159.

Keast, A. and E. S. Morton, eds. 1980. Migrant birds in the neotropics: Ecology, behavior, distribution and conservation. Washington, D.C.: Smithsonian Institution.

Lack, D. 1947. Darwin's finches. Cambridge: Cambridge University Press.

Leigh, E. G., Jr. 1999. Tropical forest ecology: A view from Barro Colorado Island. New York: Oxford University Press.

Leigh, E. G., Jr. 2010. The evolution of mutualism. Journal of Evolutionary Biology 23: 2507–2528.

Leigh, E. G., Jr., and G. J. Vermeij. 2002. Does natural selection organize ecosystems for the maintenance of high productivity and diversity? Philosophical Transactions of the Royal Society of London B 357:709–718.

Losos, J. B. 2009. Lizards in an evolutionary tree: Ecology and adaptive radiation of anoles. Berkeley: University of California Press.

Lotka, A. J. 1925. Elements of physical biology. Baltimore: Williams & Wilkins.

Lovelock, J. 1988. The ages of Gaia. Oxford: Oxford University Press.

MacArthur, R. H. 1958. Population ecology of some warblers of northeastern coniferous forests. Ecology 39:599–619.

Mandel, J. T., K. L. Bildstein, G. Bohrer, and D. W. Winkler. 2008. Movement ecology of migration in turkey vultures. Proceedings of the National Academy of Sciences of the United States of America 105:19102–19107

Mangan, S. A., S. A. Schnitzer, E. A. Herre, K.M.L. Mack, M. C. Valencia, E. I. Sanchez, and J. D. Bever. 2010. Negative plant-soil feedback predicts tree species relative abundance in a tropical forest. Nature 466:752–755.

May, R. M. 1976. Models for two interacting populations. In: R. M. May, ed. Theoretical ecology: Principles and applications. Philadelphia: Saunders, 49–70.

McNaughton, S. J. 1985. Ecology of a grazing ecosystem: The Serengeti. Ecological Monographs 55:259–294.

Moore, R. P., W. D. Robinson, I. J. Lovette, and T. R. Robinson. 2008. Experimental evidence for extreme dispersal limitation in birds. Ecology Letters 11:960–968.

Morton, E. S. 1977. Intertropical migration in the Yellow-Green Vireo and Piratic Flycatcher. Auk 94:97–106.

Muller-Landau, H. C. 2010. The tolerance-fecundity trade-off and the maintenance of diversity in seed size. Proceedings of the National Academy of Sciences of the United States of America 107:4242–4247.

Nason, J. D., E. A. Herre, J. L. Hamrick. 2008. The breeding structure of a tropical keystone plant resource. Nature 391:685–687.

Nathan, R., W. M. Getz, E. Revilla, M. Holyoak, R. Kadmon, D. Saltz, and P. E. Smouse. 2008. A movement ecology paradigm for unifying organismal movement research. Proceedings of the National Academy of Sciences of the United States of America 105:19052–19058.

Odum, E. P. 1959. Fundamentals of ecology, 2nd ed. Philadelphia: W. B. Saunders.

Odum, H. T. 1971. Environment, power and society. New York: Wiley-Interscience.

Paine, R. T. 1977. Controlled manipulations in the marine intertidal zone, and their contributions to ecological theory. In: C. E. Goulden, ed. The Changing Scenes in Natural Sciences, 1776–1976. Philadelphia: Academy of Natural Sciences, 245–270.

Paine, R. T., and R. L. Vadas. 1969. The effects of grazing by sea urchins, *Strongylocentrotus* spp., on benthic algal populations. Limnology and Oceanography 14:710–719.

Paine, R. T., T. Wootton, and C. A. Pfister. 2010. A sense of place: Tatoosh. In: I. Billick, and M. V. Price, eds. The ecology of place: Contributions of place-based research to ecological understanding. Chicago: University of Chicago Press, 229–250.

Pastor, J. 2008. Mathematical ecology of populations and ecosystems. Chichester, U.K.: Wiley-Blackwell.

Polis, G. A., W. B. Anderson, and R. D. Holt. 1997. Toward an integration of landscape and food web ecology: the dynamics of spatially subsidized food webs. Annual Review of Ecology and Systematics 28:289–316

Robertson, D. R. 1996. Interspecific competition controls abundance and habitat use of territorial Caribbean damselfishes. Ecology 77:885–899.

Russo, S. E., P. Brown, S. Tan, and S. J. Davies. 2008. Interspecific demographic trade-offs and soil related habitat associations of tree species along resource gradients. Journal of Ecology 96(1):192–203.

Seeley, T. D. 1995. The wisdom of the hive. Cambridge, Mass.: Harvard University Press.

Sinclair, A.R.E., and M. Norton-Griffiths, eds. 1979. Serengeti: Dynamics of an ecosystem. Chicago: University of Chicago Press.

Skellam, J. G. 1951. Random dispersal in theoretical populations. Biometrika 38:196–218.

Smith, A. 1759. The theory of moral sentiments. London: A. Millar.

Smith, A., 1776. An inquiry into the nature and causes of the wealth of nations. London: Strahan and Cadell.

Smith, D. C., and A. E. Douglas. 1987. The biology of symbiosis. London: Edward Arnold.

Vermeij, G. J. 2009. Comparative economics: Evolution and the modern economy. Journal of Bioeconomics 11:105–134.

Vermeij, G. J. 2011. A historical conspiracy: Competition, opportunity, and the emergence of direction in history. Cliodynamics 2:187–207.

Vermeij, G. J. 2013. On escalation. Annual Review of Earth and Planetary Sciences 41:1–19.

Vermeij, G. J., and E. G. Leigh Jr. 2011. Natural and human economies compared. Ecosphere 2(4):Article 39.

Vogt, K. A., C. C. Grier, C. S. Meier, and R. L. Edmonds. 1982. Mycorrhizal role in net primary production and nutrient cycling in *Abies amabilis* ecosystems in western Washington. Ecology 63:370–380.

Volterra, V. 1931. Leçons sur la théorie mathématique de la lutte por la vie. Paris: Gauthier-Villars.

Von Neumann, J., and O. Morgenstern. 1944. Theory of games and economic behavior. Princeton, N.J.: Princeton University Press.

Williamson, M. 1972. The analysis of biological populations. London: Edward Arnold.

Wilson, D. S. 1980. The natural selection of populations and communities. Menlo Park, Calif.: Benjamin Cummings.

Wright, S. J., H. C. Muller-Landau, R. Condit, and S. P. Hubbell. 2003. Gap-dependent recruitment, realized vital rates, and size distribution of tropical trees. Ecology 84:3174–3185.

Wulff, J. L. 1997. Mutualism among species of coral reef sponges. Ecology 78:146–157.

Wulff, J. L. 2005. Trade-offs in resistance to competitors and predators, and their effects on the density of tropical marine sponges. Journal of Animal Ecology 74:313–321.

Ziegler, C., and E. G. Leigh Jr. 2002. A magic web. New York: Oxford University Press.

Ecology "through the Looking Glass"

What Might Be the Ecological Consequences of Stopping Mutation?

Robert D. Holt

> No practical biologist interested in sexual reproduction would
> be led to work out the detailed consequences experienced by
> organisms having three or more sexes; yet what else should
> he do if he wishes to understand why the sexes are,
> in fact, always two?
> —R. A. Fisher, *The Genetical Theory of
> Natural Selection*, 1930, p. ix[1]

What Are Scientists Doing?

Although some philosophers of science (e.g., Kuhn 1970) might disagree, most workaday scientists largely assume that a principal goal of their discipline is to craft increasingly truthful statements about the world. But as Fisher noted in the quote at the beginning of this chapter, sometimes the way to get to the truth might be through the deliberate consideration of blatant untruths, thinking through the logical consequences of altering, possibly radically, some known and pervasive feature of the world. All models of course involve a degree of simplification about the world, a kind of deliberate lie taken to explore some avenue of inquiry. Richard Levins (1966) once famously quipped that truth is the intersection of independent lies. But usually these are little white lies, and involve domain-specific assumptions. Physics has at its conceptual core (at least in introductory classes) what May (1973) called "perfect crystals," such as frictionless pendula. No such things actually exist, but acting as if they do helps us understand physical processes and patterns in the messier world. The two-body problem was thoroughly worked through in classical mechanics, even though in reality there are almost always three, four, . . . , n bodies at play in the swirling motions of celestial objects. So in like manner, in teaching community ecology, I of course work through the Lotka–Volterra model of competition. But no one in their right mind thinks that this model literally describes competition between any two real species, as there are no time-lags, age structure, spatial location, individual variability, and so forth, included in the model; resources and other limiting factors are not explicit; per capita growth rates are perfectly linear functions of densities; and, density itself is treated as a continuum, ignoring the fact that individuals are discrete entities. Yet by ignoring these other factors, one gains a conceptual hook on the complexity of the world, for instance, arriving at a clearer sense of the need to consider within- as well as between-species density dependence in determining coexistence of competing species, an insight that is embedded in modern coexistence theory (Meszéna et al. 2006, Adler et al. 2018, Chesson 2018). In like manner, Steve Hubbell does not really believe that all species of tropical trees on Barro Colorado Island have absolutely identical niches (personal communication), but one can explain surprisingly many (but not all) features of rain forest community structure with this counterfactual assumption of neutral dynamics (see, e.g., May et al. 2015).

I would like to suggest that sometimes it is valuable to assume counterfactual claims about the world that are real whoppers, not restricted to some narrow domain. So for example, if there were no parasites and pathogens, tropical tree diversity might start to collapse (if Janzen–Connell effects are all that important) over millennia (Levi et al. 2019), and sex might never have evolved in the first place (and so the world would indeed be a dull place) if John Jaenike and Bill Hamilton's hypothesis for the evolution of sex holds (see Holt 2010, Wood and Johnson 2015 for elaborations of this thought experiment). As another example, if life played out in a spatial 1-dimensional Lineland (created in Abbott 1884), rather than in our 3-dimensional world, the basic model of population growth arguably might not be exponential growth at all, but rather additive, given conservation of mass (Holt 2009).

One of the great intellectual challenges of our time is fusing ecology and evolutionary biology into a seamless discipline. This has so many dimensions one hardly knows where to start. There is a great deal of exciting work at present on how evolution at times occurs across time scales much shorter than traditionally believed, and on how such evolution drives ecological change, with feedbacks in both directions (Schoener 2011, Hendry 2017). Richard Levins (1966) presciently noted back in 1966 that "demographic time and evolutionary time are commensurate," and the current ferment of activity focused on eco-evolutionary dynamics is a fulfillment of that vision from half a century ago. Considering processes over much grander time scales, phylogenetic perspectives have enriched community ecology (Cavender-Bares et al. 2009, Cadotte and Davies 2016), and a consideration of speciation has permitted deeper insights into the factors governing the structure of ecological communities (Ricklefs 2004) and, more broadly, global patterns in the spatial distribution of biodiversity (Schluter and Pennell 2017). Speciation is at times tied to the origination of adaptation to environments, so ecological processes can help drive speciation (Nosil 2012).

What I want to do in the next several paragraphs is to carry out a massive thought experiment: Let's turn evolution off by stopping the font of heritable variation—mutation—and see what happens. By "turn evolution off," one could mean several things, at different scales, but I will principally focus on a few issues in microevolution, at ecological time scales.

All evolution ultimately depends upon heritable variation (Lewontin 1970), a principal source of which is mutation. Sex and recombination reshuffle existing variation that arises in various ways in DNA and RNA, but ultimately heritable variation depends upon mutations, which arise because of mistakes in replication (Bernstein and Bernstein 1991). In addition to mutations in the production of gametes, each cell replication event by mitosis within a multicellular organism allows the opportunity for mutation to arise, so all such organisms are genetic mosaics. So the way we will slowly turn evolution "off" is to eliminate different sources of variation via mutation.

Turning Off Somatic Mutation: Would This Matter?

The first place we might start our massive thought experiment is within individual organisms: Let's eliminate somatic mutations, in organisms where there is a sharp distinction between germ line and soma (e.g., many taxa in *Animalia*). (Somatic mutations have a more complex evolutionary role in clonal organisms; see Otto and Orive 1995, Orive 2001.) A somatic mutation is "a mutation occurring in the general body cells (as opposed to the germ cells and hence not transmitted to progeny)" (Farlex 2012). Because replication is imperfect, potentially an evolutionary process plays out within the lifespan of each individual organism. What if multicellular organisms like ourselves were able to have mitotic cell divisions with completely faithful replication of nuclear, mitochondrial, and chloroplast DNA—and at no cost? The second law of thermodynamics (at least) of course would prevent this from happening in reality. But let's pretend for a second that this is not an issue. We would want to prevent DNA damage of all sorts (which can accumulate even in quiescent stem cells; see Moehrle et al. 2015) and heritable epigenetic damage would be not happening, as well. Would there be any systematic ecological consequences of shutting off all sources of (short-term, within-organism) heritable variation?

Part of the answer to this question depends on whether or not senescence is quantitatively important in natural populations. Most organisms (but not all—Cohen 2017) seem to senesce. One of the principal theories for aging is Thomas Kirkwood's "disposable soma" theory, which pertains most cleanly to organisms with clearly differentiated germ lines and somatic tissues (Kirkwood and Austad 2000, Kirkwood 2017). The basic idea is that as an organism ages, its cells accumulate damage which can be self-reinforcing, including deleterious mutations to DNA (Bernstein and Bernstein 1991), and because repairing such damage is costly, the cost of repair increases with age. As from an evolutionary perspective the soma exists to boost the number of descendants, this cost eventually exceeds the benefit of further repair, making senescence well-nigh inescapable. Part of this cumulative damage (which we call aging) may be due to somatic DNA mutations and other heritable forms of cellular degradation (e.g., protein misfolding and epigenetic change), which can lead to cellular lineages that compete selfishly with other cells in multicellular organisms (Nelson and Masel 2017). From the perspective of the cell carrying them, such mutations (broadly defined to include all heritable cellular phenotypes) could be neutral, with no effect on function, or be deleterious, or be beneficial. By the latter I mean that the cell boosts its number of descendants, which is what we know as a neoplasm or tumor, which, if it becomes malignant, is cancer. It is now widely accepted that the accumulation of somatic mutations through an individual's life are basically what leads to cancer (Frank 2007, Martincorena and Campbell 2015, Rozhok and DeGregori 2015), given appropriate within-host environments (Maley et al. 2017). Because of the cumulative nature of mutational input, and the multistage development of cancer, the risk of cancer and the number of somatic mutations in the genome of cancerous cells increases with age (Milholland et al. 2015), particularly in postreproductive stages of life. This evolutionary process plays out in the ecology defined by the traits of an individual organism. There is a rapidly growing ferment of interest in this, the "eco-evolutionary" dynamics of cancer, potentially giving insights into fresh approaches to slow the growth or even prevent the occurrence of malignancies (e.g., Ibrahim-Hashim et al. 2017, Maley et al. 2017, Gatenby and Brown 2017, Hochberg and Noble 2017, Scott and Marusyk 2017; see essays by Michael Hochberg and Andrew Read, this volume).

If somatic mutations were completely suppressed, then cancer would disappear, and senescence would likely be moderated (albeit maybe not entirely eliminated). Though of great importance for human health, one might well wonder what impact this reduction in within-individual genetic variation would have more broadly on natural communities. Hochberg and Noble (2017) note for instance that species observed in their "normal" environments have a baseline rate of all cancers of less than 5%, and that cancer seems very rare in some species. There are examples of relatively high cancer rates in some natural populations (e.g., coral trout in the Great Barrier Reef: Sweet et al. 2012), but in most such cases, there seems to be some kind of environmental perturbation. For instance, in beluga whales in the St. Lawrence estuary, nearly 20% of dead whales were found to have cancer of some sort, implying a cancer rate comparable to humans (Martineau et al. 2002). The authors suggest that this is because of environmental contamination, namely polycyclic aromatic hydrocarbons generated by local aluminum smelters. Indeed, humans living nearby also had elevated cancer rates.

In like manner, animal demographers traditionally viewed senescence to be largely absent from natural populations, so that when senescence was observed, it was because environmental changes permitted individuals to live much longer than found in the evolutionary environment of their ancestors (Comfort 1954: e.g., as in the coddled lives of many pets and zoo animals, compared to their wild relatives). In stable environments, many researchers believe that extrinsic sources of mortality determine average lifespan, and patterns of aging simply reflect how the force of selection declines with age (Joel Parker, personal communication).

So, maybe demographic costs arising from cancers and senescence are in general so small in most natural populations that the immediate impacts of the disappearance of somatic mutations would be negligible to modest. However, there are reasons to think that somatic mutation imposes a large enough demographic load that it might be marked, at least in some cases. Gaillard et al. (2017), for instance, report that detailed long-term studies of natural mammalian populations reveal clear patterns of senescent deterioration in organismal function across many species and habitats (see also Nussey et al. 2013). Vittecoq et al. (2015) note that early neoplastic developments, which might not be recognized as clinical cancers, might nonetheless have significant ecological consequences. Such precancerous states are increasingly being observed in natural populations (Vittecoq et al. 2015). Even small decreases in flight speed, vigilance, or ability to hide could magnify predation risk, and reductions in foraging efficiency could make individuals more vulnerable to climatic stress, food limitation, or interference competition.

Hochberg and Noble (2017) may be correct that species might be expected to have low rates of cancer in the normal environment of their ancestors, but even in the absence of anthropogenic perturbations, many individuals within a species can be found in environments that differ from the ancestral evolutionary environment of their lineage. This is likely, for instance, at geographical range limits or in marginal habitats within a species' range, where immigrants in effect experience a sudden change in the environment and deterioration in fitness, relative to their ancestral conditions, and a fitness load due to somatic mutation might tip the balance towards local extinction rather than persistence. In this case, a disappearance of somatic mutation might facilitate population increase or range expansion.

Maybe the most dramatic effect of the suppression of somatic mutations would occur only over longer evolutionary time scales. Considerable molecular machinery in the cell is devoted to minimization of errors during DNA replication, DNA repair, and configuration of appropriate epigenetic patterns, as well as whole-organism responses to tumors, such as immune reactions (Nunney 2013, Beerman et al. 2014, Seluanov et al. 2018). Leo Buss (1987) argued that the evolution of stable developmental systems in multicellular organisms involved the suppression of the opportunity for competition among cell lineages, and it is likely that multicellularity itself depends upon the continual effectiveness of such suppression (Szathmáry 2015). How costly are these defense mechanisms against somatic mutations, the machinery that prevents self from dissolving into non-self? It is likely that there are real trade-offs with other aspects of survival and reproduction (Jacqueline et al. 2017), and one can find claims in the literature for costs of repair to mutations (e.g., Breivik and Gaudernack 2004), but I do not think anyone has a handle on what the magnitude of these costs might be. These costs might be larger in larger-bodied and longer-lived organisms (which have more chances for somatic mutations to occur), and in certain habitats (e.g., high mountains with greater ultraviolet light exposure). DNA repair capacity has been shown to correlate with mammalian lifespan in numerous comparative studies, as has the level of poly(ADP-ribose) polymerase 41, an enzyme that is important in the maintenance of genomic integrity (Priya et al. 2017). Eliminating these costs would provide competitive advantages to a wide range of taxa, I suspect. In any case, the issue of gauging the demographic load of somatic mutation in natural populations, and the knock-on effects on populations and communities, seems to me to be an interesting and significant open question at the interface of ecology and evolution.

Ecological Impacts of Deleterious Mutations

Now let's put somatic mutations aside, and consider mutations to the germline itself. What if one could just put a stop to novel mutations? Most mutations that affect fitness are largely assumed to be deleterious (Keightley and Lynch 2003; but see Shaw and Shaw 2014), and I will focus on them for a second, and assume that the environment is constant through time. Even though selection weeds such mutations out of a population, new mutations continuously arise, leading to a degree of persistent maladaptation in populations (Crespi 2000, Kondrashov 2017). The depression in fitness caused by the standing crop of such mutations is known as genetic load (Muller 1950). It is likely (but not yet certain) that deleterious mutations impose substantial costs in individual fitness. Kondrashov (2017, p. 260), in concluding his wide-ranging review of deleterious mutations in humans, states, "It is plausible, but not certain, that getting rid of individually rare deleterious alleles would lead to a large increase in fitness." Burt (1995) notes that lab populations of microbes and fruit flies, though grown in favorable, constant conditions with *ad libitum* food resources, nonetheless experience a degradation in mean fitness of 0.1% to 3% per generation due to mutation accumulation.[2] Even if there is no directional evolution, there must be a constant action of natural selection to purge the burgeoning load of deleterious mutations—a cryptic evolutionary process underlying seemingly static ecological systems (the general issue of cryptic evolution was raised first, I believe, by Kinnison et al. 2015). Agrawal and Whitlock (2012) have provided an excellent recent review of the issue of mutational load in natural populations, including potential ecological impacts, so I can be relatively brief in my thoughts in this section of the essay. The magnitude of mutational load in a population depends on a wide range of genetic, life history, and ecological factors. As with somatic mutation, there are surely substantial costs of maintaining repair mechanisms to reduce mutations in the germline. Mirzaghaderi and Horandl (2016) argue that the restoration of DNA via the repair of oxidative DNA damage is a primary adaptive function of meiotic sex. Sex in turn has many costs associated with it (Bell 1982), which maybe should be viewed as massive indirect ecological costs of DNA repair.

Beyond that basic cost of attempting to suppress degradation of the genome, one ecological consequence of genetic load is that deleterious mutations can increase in frequency and even become fixed in small populations, due to drift vitiating the action of natural selection. This combination of mutational and drift load can cause the extinction of these populations (Lande 1994, Gabriel et al. 1991, Whitlock 2000), for both asexual (Lynch et al. 1993, Soll et al. 2007) and sexual species (Lynch et al. 1995). In general, the ability of a population to purge deleterious mutations increases with its effective population size (Lynch 2007, p. 76). One implication of this buildup of load can be a vicious positive feedback where a population is initially rare, and becomes even rarer as it becomes progressively more maladapted, leading to a yet larger load.[3] Genetic processes (e.g., compensatory mutations, Poon and Otto 2000) as well as life history details (e.g., the timing of density dependence relative to selection, Agrawal and Whitlock 2012) can mitigate this effect to some extent but are unlikely to make it go away. Does this kind of load have ecological consequences? At local scales, communities often show substantial species turnover (Williamson 1981). In some instances, this is not obviously related to average abundance (e.g., Thuiller et al. 2007), making it unlikely that the load of deleterious mutations is a major driver of extinction (relative to, say, stochastic environmental variation). But in other cases (e.g., Eastern Wood; see Williamson 1981, pp. 93–104) turnover is largely among species with quite low populations, making the impact of deleterious mutation plausibly part of the reason for their local extinctions. Moreover, many species may be locally common, but rare over the majority of their geographical range (Brown et al. 1995); many populations could therefore have heightened extinction risk because of drift load. How important is this potential effect, quantitatively? If one could turn off new deleterious mutations, then over time selection should weed out the initial standing crop of deleterious mutations. Would extinction rates then decline, so that species would be found, e.g., over a wider range of low-quality habitats? Some mutations may have deleterious effects in some habitats, but be neutral in others. Mutational load in this case could lead to persistent habitat specialization (Kawecki et al. 1997, Holt 1996). Eliminating mutation would lead to a gradual diminution in this habitat-specific load, possibly permitting greater generalization in habitat use.

The efficient elimination of deleterious mutations is believed to be one of the advantages of sexual reproduction (Kondrashov 1988, Orive, in press). Given the ubiquity of sex across many kinds of taxa and environments, one needs an equally pervasive causal driver to explain why sex is so predominant in living forms. Few factors are as universal as deleterious mutations. Eliminate the genetic load from such mutations, and then over evolutionary time, one might see the gradual replacement of sexual taxa by similar, but asexual, forms—and the eventual loss of many ecologically important features of taxa, such as aggressive competition among males leading to costly sexual conflict, and the costs, benefits, and specific adaptations having to do with reproductive isolation—not to mention losing the demographic drag of devoting about half of all reproductive investment into males in the first place.

Moreover, as Agrawal and Whitlock (2012) observe, the load of deleterious mutations can have knock-on effects on interspecific interactions. They present a model of two species competing for shared resources, and show that even a small amount of load present in one species can tilt it towards competitive exclusion. Small populations are at risk of extinction anyway due to mutational meltdown, and negative interactions of any sort just heighten this risk (Coron et al. 2013). There could also be consequences of the load of deleterious mutations for food-web interactions. Predators are typically less abundant than their prey (e.g., Jonsson et al. 2005). Do they also thereby carry a larger load of deleterious mutations, which impairs to a degree their ability in prey capture, or boosts the metabolic costs that must be met by foraging? In other words, do asymmetries in effective population size as one ascends in a food web tend to vitiate top-down effects?

Now, On to Beneficial Mutations

Up until now, I have focused on mutations that harm fitness. Adaptations by a species to particular environments of course reflect how its lineage has captured, via natural selection, advantageous mutations that although rare compared to deleterious mutations, nonetheless have occurred sufficiently often in that lineage's history. "Adaptation is caused by selection continually winnowing the genetic variation created by mutation" (Bell 2008). Moreover, the effects of these mutations on fitness must be sufficiently great that a species can persist. Not all lineages are so lucky. As John Maynard Keynes once memorably said, "In the long run, we are all dead." In like manner, most species that have ever existed have gone extinct, in part because the physical environment of the earth is always changing. And those lineages that *have* persisted (possibly via descendent sublineages) have surely done so either because they moved to track environments to which they were already adapted or managed to adapt sufficiently via natural selection—the theme of evolutionary rescue (Gomulkiewicz and Holt 1995) . Bell (2017) has recently reviewed work on evolutionary rescue, so again I will be brief on this important theme. Most species harbor a substantial standing pool of genetic variation, and following abrupt environmental change, rescue may be sustained entirely by variation that is already present in the initial population (e.g., as in the experiments of Killeen et al. 2017). But in the absence of new mutations, the combination of selection and drift will eventually deplete heritable variation in fitness. Moreover, large populations are more likely to harbor (or generate) adaptive genetic variation that can allow them to persist in radically changed environments (Hoffmann et al. 2017). Experimental evolution studies show that the capacity of populations to adapt to new environments can be limited by having scant initial genetic variation, and that an infusion of genetic variation (via gene flow or mutation) may be essential for pronounced and sustained evolutionary responses to environmental change (Stewart et al. 2017). To what extent, and over what time scale, does the persistence of a species in a changing environment depend upon novel variation via mutation, rather than its drawing on a preexisting, standing stock of variation? Bell (2017) in his review suggests that in answering this question "there is a distinction between large populations of short-lived asexual microbes, in which rescue depends primarily on novel mutation, and small populations of long-lived sexual animals and plants, in which standing genetic variation is likely to be the sole basis of adaptation." This generalization surely breaks down in the long run, but how long is "long"? In the Red

Queen dynamics of interactions between antagonists, how long does it take for the coevolutionary dance to come to a standstill, and is there a systematic advantage for one side of the interaction that would emerge as evolution stalled? Hiltunen et al. (2015) report that in predator–prey interactions in microcosms, fluctuating population dynamics and periods of low resource availability constrain the evolution of prey defense, enhancing the top-down effect of the predator on prey numbers. They explain this as due to a lower effective population size in the prey, hampering input of novel mutations. Extrapolating such dynamics to the limit of a negligible input of novel mutations, another nod to Lewis Carroll (as filtered through the Disney animated movie "Alice in Wonderland") seems apt as a description of what we would expect to eventually observe: The White Rabbit said: "Don't just do something—stand there."

We do not know the answers to these (and many other) important questions (Hendry 2017) at the interface of ecology and evolution. How much of the stability and emergent structure of natural ecosystems rests on ongoing evolutionary dynamics of the constituent species, and to what extent do such dynamics require the constant infusion of novel mutational variation, rather than the churning and reshuffling of preexisting variation? In answering this question, I suggest that thinking through counterfactual worlds, where mutation is imagined to be cut off, could illuminate our understanding.

Acknowledgments

I thank the University of Florida Foundation for support, and Mike Barfield and Vitrell Sherif for assistance. I would like to acknowledge and thank from the bottom of my heart the late Henry Horn, who provided much inspiration and support (not to mention witty repartee) over the years, and the late John Bonner, who deftly and graciously took on the oversight of my Princeton senior thesis in 1973 on "The Evolution of Multicellularity," after the untimely death in the Fall of 1972 of my previous advisor, Robert MacArthur. I also thank Steve Austad, Charles Baer, Michael Hochberg, Tom Kirkwood, Alexey Kondrashov, and Joel Parker for useful insights, Lewis Carroll for perennial inspiration, and Andy Dobson for hinting at the title.

Note

1. There is an ambiguity in the quote from Fisher. Did he mean the number of distinct mating types in a species, or the number of parents each individual has? If the latter were what Fisher had in mind, then the number almost universally is either one (for asexual or clonal organisms), or two (for the rest). Parker (2004) has argued that if one thinks of social insect colonies as "superorganisms," then in several species of

Pogonomyrmex, more than two distinct mating types are required to create a viable colony. Each individual ant, nonetheless, still has just one to two parents. But if the former were what Fisher meant, we now know some taxa have multiple distinct mating types (indeed, thousands for some fungal species; Kothe 1996). These are almost entirely isogamous. There are plausible evolutionary arguments for why with anisogamy, species with three or more sexes are expected to evolve towards a two-sex state (e.g., Feigel et al. 2009), and even with isogamy, recent evolutionary theory (Constable and Kokko 2018) suggests that mating type diversity will stay low (even just at two types), if sex is facultative and rare.

2. A recent review suggests that the per generation degradation of fitness via mutation accumulation ranges from 0.1% to 2% (Hendry et al. 2018).

3. In sexual species, selection should drive mutation rates to the lowest physically feasible rate, and the same holds for asexual species in constant environments (Leigh 1970). However, the efficacy of selection is hampered by drift, and more so in small populations. Both theory and evidence indeed suggest that mutation rates are higher in a small population (Lynch 2011), which should thus tend to experience greater genetic loads.

References

Abbott, E. A. 1884. Flatland: A romance of many dimensions. Reprinted 1991, with a new introduction by T. Banchoff. Princeton, N.J.: Princeton University.

Adler, P. B., D. Smull, K. H. Bear, R. T. Choi, T. Fumiss, A. Kulmatiski, et al. 2018. Competition and coexistence in plant communities: intraspecific competition is stronger than interspecific competition. Ecology Letters 21:1319–1329.

Agrawal, A. F., and M. C. Whitlock. 2012. Mutation load: The fitness of individuals in populations where deleterious alleles are abundant. Annual Review of Ecology, Evolution, and Systematics 43:115–135.

Beerman, I., J. Seita, M. A. Inlay, I. L. Weissman, and D. J. Rossi. 2014. Quiescent hematopoietic stem cells accumulate DNA damage during aging that is repaired upon entry into cell cycle. Cell Stem Cell 15:37–50.

Bell, G. 1982. The masterpiece of nature: The evolution and genetics of sexuality. Berkeley: University of California Press.

Bell, G. 2008. Selection: The mechanism of evolution. 2nd ed. Oxford: Oxford University Press.

Bell, G. 2017. Evolutionary rescue. Annual Review of Ecology, Evolution, and Systematics 48:605–627.

Bernstein, C., and H. Bernstein. 1991. Aging, sex, and DNA repair. San Diego: Academic Press.

Breivik, J., and G. Gaudernack. 2004. Resolving the evolutionary paradox of genetic instability: A cost-benefit analysis of DNA repair in changing environments. FEBS Letters 563:7–12.

Brown, J. H., D. W. Mehlman, and G. C. Stevens. 1995. Spatial variation in abundance. Ecology 76:2028–2043.

Burt, A. 1995. Perspective: The evolution of fitness. Evolution 49:1–8.

Buss, L. W. 1987. The evolution of individuality. Princeton, N.J.: Princeton University Press.

Cadotte, M. W., and T. J. Davies. 2016. Phylogenies in ecology: A guide to concepts and methods. Princeton, N.J.: Princeton University Press.

Cavender-Bares, J., K. H. Kozak, P.V.A. Fine, and S. W. Kembel. 2009. The merging of community ecology and phylogenetic biology. Ecology Letters 12:693–715.

Chesson, P. 2018. Updates on mechanisms of maintenance of species diversity. Journal of Ecology 106:1773–1794.

Cohen, A. A. 2017. Taxonomic diversity, complexity, and the evolution of selfing. In: R.P. Shefferson, O. R. Jones, and R. Salguero, eds. The evolution of senescence in the tree of life. 2017. Cambridge: Cambridge University Press, 83–101.

Comfort, A. 1954. Biological aspects of senescence. Biological Reviews 29:284–329.

Constable, G.W.A., and H. Kokko. 2018. The rate of facultative sex governs the number of expected mating types in isogamous species. Nature Ecology & Evolution 2:1168–1175. doi: https://doi.org/10.1038/s41559-018-0580-9.

Coron, C., S. Meleard, E. Porcher, and A. Robert. 2013. Quantifying the mutational meltdown in diploid populations. American Naturalist 181:623–636.

Crespi, B. L. 2000. The evolution of maladaptation. Heredity 84:623–629.

Farlex Partner Medical Dictionary. 2012. https://medical-dictionary.thefreedictionary.com /somatic+mutation.

Fisher, R. A. 1930. The genetical theory of natural selection. Reprint. 1958. New York: Dover Publications.

Frank, S. A. 2007. Dynamics of cancer: incidence, inheritance, and evolution. Princeton, N.J.: Princeton University Press.

Gabriel, W., R. Burger, and M. Lynch. 1991. Population extinction by mutational load and demographic stochasticity. In: A. Seitz, and V. Loeschcke, eds. Species conservation: A population-biological approach. Basel: Birkhauser Verlag, 49–59.

Gaillard, J.-M., M. Garratt, and J.-F. Lemaitre. 2017. Senescence in mammalian life history traits. In: R. P. Shefferson, O. R. Jones, and R. Salguero-Gomez, eds. The evolution of senescence in the tree of life. Cambridge: Cambridge University Press, 126–155.

Gatenby, R. A., and J. Brown. 2017. Mutations, evolution and the central role of a self-defined fitness function in the initiation and progression of cancer. BBA: Reviews on Cancer 1867:162–166.

Gomulkiewicz, R., and R. D. Holt. 1995. When does evolution by natural selection prevent extinction? Evolution 49:201–207.

Heimpel, G. E., and N. J. Mills. 2017. Biological control: Ecology and applications. Cambridge: Cambridge University Press.

Hendry, A. P. 2017. Eco-evolutionary dynamics. Princeton N.J.: Princeton University Press.

Hendry, A. P., D. J. Schoen, M. E. Wolak, and J. M. Reid. 2018. The contemporary evolution of fitness. Annual Review of Ecology, Evolution, and Systematics. 49:457–476.

Hiltunen, T., G. B. Ayan, and L. Becks. 2015. Environmental fluctuations restrict eco-evolutionary dynamics in predator-prey system. Proceedings of the Royal Society B 282:20150013.

Hochberg, M. E., and R. J. Noble. 2017. A framework for how environment contributes to cancer risk. Ecology Letters 20:117–134.

Hoffmann, A. A., C. M. Sgro, and T. N. Kristensen. 2017. Revisiting adaptive potential, population size, and conservation. Trends in Ecology and Evolution 32:506–517.

Holt, R. D. 1996. Demographic constraints in evolution: Towards unifying the evolutionary theories of senescence and niche conservatism. Evolutionary Ecology 10:1–11.

Holt, R. D. 2009. Darwin, Malthus, and movement: A hidden assumption in the demographic foundations of evolution. Israel Journal of Ecology and Evolution 55:189–198.

Holt, R. D. 2010. A world free of parasites and vectors: would it be heaven, or would it be hell? Israel Journal of Ecology and Evolution 56:239–250.

Ibrahim-Hashim, A., M. Robertson-Tessi, P. M. Enriquez-Navas, M. Damaghi, Y. Balagurunathan, J. W. Wojtkowiak, et al. 2017. Defining cancer subpopulations by adaptive strategies rather than molecular properties provides novel insights into intratumoral evolution. Cancer Research 77:2242–2254.

Jacqueline, C., P. A. Biro, C. Beckmann, A. P. Moller, F. Renaud, G. Sorci, et al. 2017. Cancer: A disease at the crossroads of trade-offs. Evolutionary Applications 10:215–225.

Jonsson, T., J. E. Cohen, and S. R. Carpenter. 2005. Food webs, body size, and species abundance in ecological community description. Advances in Ecological Research 36:1–84.

Kawecki, T. J., N. H. Barton, and J. D. Fry. 1997. Mutational collapse of fitness in marginal habitats and the evolution of ecological specialization. Journal of Evolutionary Biology 10:407–429.

Keightley, P. D., and M. Lynch. 2003. Toward a realistic model of mutations affecting fitness. Evolution 57:681–685.

Killeen, J. A., C. Gougat-Barbera, S. Krenek, and O. Kaltz. 2017. Evolutionary rescue and local adaptation under different rates of temperature increase: A combined analysis of changes in phenotype expression and genotype frequency in *Paramecium* microcosms. Molecular Ecology 26:1734–1746.

Kinnison, M. T., N. G. Hairston Jr., and A. P. Hendry. 2015. Cryptic eco-evolutionary dynamics. Annals of the New York Academy of Science. 1360:120–125.

Kirkwood, T.B.L., and S. N. Austad. 2000. Why do we age? Nature 408:233–238.

Kirkwood, T. 2017. The disposable soma theory. In: R. Shefferson, O. Jones, and R. Salguero-Gómez, eds. The evolution of senescence in the tree of life. Cambridge: Cambridge University Press, 23–29.

Kondrashov, A. S. 1988. Deleterious mutations and the evolution of sexual reproduction. Nature 336:435–440.

Kondrashov, A. S. 2017. Crumbling genome: The impact of deleterious mutations on humans. Hoboken: Wiley-Blackwell.

Kothe, E. 1996. Tetrapolar fungal mating types: Sexes by the thousands. FEMS Microbiology Review 18:65–87.

Kuhn, T. S. 1970. The structure of scientific revolutions. Chicago: University of Chicago Press.

Lande, R. 1994. Risk of population extinction from fixation of new deleterious mutations. Evolution 48:1460–1469.

Leigh, E. G., Jr. 1970. Natural selection and mutability. American Naturalist. 104:301–305.

Levi, T., M. Barfield, S. Barrantes, C. Sullivan, R. D. Holt, and J. Terborgh. 2019. Tropical forests can maintain hyperdiversity because of enemies. Proceedings of the National Academy of Sciences of the United States of America 116:581–586.

Levins, R. 1966. The strategy of model building in population biology. American Scientist 54:421–431.

Lewontin, R. C. 1970. The units of selection. Annual Review of Ecology and Systematics. 1(1):1–18.

Lynch, M. 2007. The origins of genome architecture. Sunderland, Mass.: Sinauer Associates.

Lynch, M. 2011. The lower bound to the evolution of mutation rates. Genome Biology and Evolution 3:1107–1118.

Lynch, M., R. Burger, D. Butcher, and W. Gabriel. 1993. The mutational meltdown in asexual populations. Journal of Heredity 84:339–344.

Lynch, M., J. Conery, and R. Burger. 1995. Mutation meltdowns in sexual populations. Evolution 49:1067–1080.

Maley, C., A. Aktipis, T. A. Graham, A. Sottoriva, A. M. Boddy, M. Janiszewska, et al. 2017. Classifying the evolutionary and ecological features of neoplasms. Nature Reviews: Cancer 17:605–619.

Martincorena, I., and P. J. Campbell. 2015. Somatic mutation in cancer and normal cells. Science 349:1483–1489.

Martineau, D., K. Lemberger, A. Dalaire, P. Labelle, T. P. Lipscomb, P. Michel, and I. Mikaelian. 2002. Cancer in wildlife, a case study: Beluga from the St. Lawrence estuary, Quebec, Canada. Environmental Health Perspectives 110:285–292.

May, F., A. Huth, and T. Wiegand. 2015. Moving beyond abundance distributions: Neutral theory and spatial patterns in a tropical forest. Proceedings of the Royal Society B 282(1802):20141657. doi: https://doi.org/10.1098/rspb.2014.1657.

May, R. M. 1973. Stability and complexity in model ecosystems. Princeton, N.J.: Princeton University Press.

Medawar, P. B. 1952. An unsolved problem of biology. London: Lewis Press.

Meszéna, G., M. Gyllenberg, L. Pasztor, and J.A.J. Metz. 2006. Competitive exclusion and limiting similarity: a unified theory. Theoretical Population Biology 69:68–87.

Milholland, B., A. Auton, Y. Suh, and J. Vijg. 2015. Age-related somatic mutations in the cancer genome. Oncotarget 6:24627–24635.

Mirzaghaderi, G., and E. Hörandl. 2016. The evolution of meiotic sex and its alternatives. Proceedings of the Royal Society B 283:20161221.

Moehrle, B. M., K. Nattami, A. Brown, M. C. Florian, M. Ryan, M. Vogel, et al. 2015. Stem cell-specific mechanisms ensure genomic fidelity within HSCs and upon aging of HSCs. Cell Reproduction 13:2412–2424.

Muller, H. J. 1950. Our load of mutations. American Journal of Human Genetics 2:111–176.

Nelson, P., and J. Masel. 2017. Intercellular competition and the inevitability of multicellular aging. Proceedings of the National Academy of Sciences of the United States of America 114:12982–12987.

Nosil, P. 2012. Ecological speciation. Oxford: Oxford University Press.

Nunney, L. 2013. The real war on cancer: The evolutionary dynamics of cancer suppression. Evolutionary Applications 6:11–19.

Nussey, D. H., H. Froy, J.-F. Lemaitre, J. M. Gaillard, and S. N. Austrad. 2013. Senescence in natural populations of animals: Widespread evidence and its implications for biogerontology. Ageing Research Review 12:214–225.

Orive, M. E. 2001. Somatic mutations in organisms with complex life histories. Theoretical Population Biology 59:235–249.

Orive, M. E. In press. The evolution of sex. In: S. Scheiner, ed. The theory of evolution. Chicago: University of Chicago Press.

Otto, S. P., and M. E. Orive. 1995. Evolutionary consequences of mutation and selection within an individual. Genetics 141:1173–1187.

Parker, J. D. 2004. A major evolutionary transition to more than two sexes? Trends in Ecology and Evolution 19:83–86.

Poon, A., and S. P. Otto. 2000. Compensating for our load of mutations: Freezing the meltdown of small populations. Evolution 54:1467–1479.

Priya S., L. A. Sepúlveda, J. A. Halliday, J. Liu, M. A. Bravo Núñez, I. Golding, et al. 2017. The transcription fidelity factor GreA impedes DNA break repair. Nature 550:214–218

Ricklefs, R. E. 2004. A comprehensive framework for global patterns in biodiversity. Ecology Letters 7:1–15.

Rozhok, A. I., and J. DeGregori. 2015. Toward an evolutionary model of cancer: Considering the mechanisms that govern the fate of somatic mutations. Proceedings of the National Academy of Sciences of the United States of America 112:8914–8921.

Schluter, D., and M. W. Pennell. 2017. Speciation gradients and the distribution of biodiversity. Nature 546:48–55.

Schoener, T. W. 2011. The newest synthesis: Understanding the interplay of evolutionary and ecological dynamics. Science 331:426–429.

Scott, J., and A. Marusyk. 2017. Somatic clonal evolution: A selection-centric perspective. Biochimica et Biophysica Acta 1867:139–150.

Seluanov, A., V. N. Gladyshev, J. Vijg, and V. Gorbunova. 2018. Mechanisms of cancer resistance in long-lived mammals. Nature Reviews: Cancer 18:433–441.

Shaw R. G., and F. W. Shaw. 2014. Quantitative genetic study of the adaptive process. Heredity 112:13–20.

Soll, S. J., C. D. Arenas, and N. Lehman. 2007. Accumulation of deleterious mutations in small abiotic populations of RNA. Genetics 175:267–275.

Stewart, G. S., M. R. Morris, A. B. Genis, M. Szucs, B. A. Melbourne, S. J. Tavener, and R. A. Hufbauer. 2017. The power of evolutionary rescue is constrained by genetic load. Evolutionary Applications 10:731–741.

Sweet, M., N. Kirkham, M. Bendall, L. Currey, J. Bythell, and M. Heupel. 2012. Evidence of melanoma in wild marine fish populations. PLOS One 7(8):e41989, doi: https://doi .org/10.1371/journal.pone.0041989.

Szathmáry, E. 2015. Toward major evolutionary transitions theory 2.0. Proceedings of the National Academy of Sciences of the United States of America 112:10104–10111.

Thuiller, W., J. A. Slingsby, S.D.J. Privett, and R. M. Cowling. 2007. Stochastic species turnover and stable coexistence in a species-rich, fire-prone plant community. PLOS One 2(9):e938. doi: https://doi.org/10.1371/journal.pone.0000938.

Vittecoq, M., B. Roche, S. P. Daoust, H. Ducasse, D. Missé, J. Abadie, et al. 2013. Cancer: A missing link in ecosystem functioning. Trends in Ecology and Evolution 28:628–635.

Whitlock, M. C. 2000. Fixation of new alleles and the extinction of small populations: Drift load, beneficial alleles, and sexual selection. Evolution 54: 1855–1861.

Williamson, M. 1981. Island Populations. Oxford: Oxford University Press.

Wood, C. L., and P.T.J. Johnson. 2015. A world without parasites: Exploring the hidden ecology of infection. Frontiers in Ecology and the Environment 13:425–434.

Ecology and Evolution Is Hindered by the Lack of Individual-Based Data

Tim Coulson

One of the biggest challenges in ecology is the collection of large amounts of high-quality, individual-based data from a range of species living in a spectrum of communities (Clutton-Brock and Sheldon 2010). Technology has led to a revolution in the data that can be collected on individuals in the laboratory and the wild, but there is still much to do before we have databases that will allow us to address some of the key questions in ecology and evolution.

Demography—the study of factors influencing birth, death, and dispersal—is at the heart of all of biology (Caswell 2001). Individuals are born, develop throughout life, reproduce when the appropriate opportunities arise producing offspring in a range of states, and then die. Populations change in size when the numbers added via birth and immigration differ from the numbers lost via death and emigration. Evolution occurs when demography varies across individuals with different genotypes. Community structure changes when change in the size or structure of one population influences the demography of other, interacting species. Energy flows through ecosystems change with community structure. In recent decades, the majority of biologists have accepted that demography is fundamental to the dynamics of the natural world, and this has been reflected in a surge of demographic models describing population dynamics, evolution, and species interactions.

These models map properties of individuals, be it their genotype or phenotype, to the dynamics of the population and community via their propensities to survive, reproduce, and disperse. These models are consequently most easily parameterized with individual-based data (Coulson 2012). If you can collect data on individuals—be they plants, animals, fungi, or microbes—it is generally a very sensible thing to do. For example, given the option of collecting data on just the total number of animals in a population, or collecting information on properties of each individual living within the population, it is preferable, whenever possible, to collect data from individuals. However, that is often hard: Tracking individual microbes is, currently, for all intents and purposes, impossible. Nonetheless, we are currently experiencing something of a revolution in how individual-based data are collected. In this opinionated rant I explain why individual-based data are so important, explain how they can be used once collected, describe some of the key approaches used to collect them, and call for engineers to engage in developing biodiversity technology to aid the collection of data at all levels of biological organization from individuals in both the lab and the wild.

Birth, Death, and Movement

It is universally accepted, in the world beyond accountancy firms, that accountants are some of the dullest people on the planet. I once received a damning referee's report on a grant (that was funded) that my contribution to ecology and evolution was nothing more than accountancy. The referee, who clearly relished their task of pouring scorn on every syllable that I had written, felt that describing my work as accountancy was surely the blow that would sink the application. Fortunately, the committee that decided to fund it, were substantially better acquainted with the importance of accountancy than the hapless referee. They understood that all ecological and evolutionary change comes about through the birth, death, and movement of individuals. If you can construct a ledger of individual-level attributes of your population, community, or ecosystem, you can start to understand why population size and structure changes with time, why allele frequencies may trend following an environmental perturbation, or what the causes and consequences of an increased average body mass index may mean for humanity's future.

The reason that the individual is key is that it is the fundamental biological unit. A genome cannot exist except within an individual; a phenotype without an individual is nonsensical; and behaviors have to be expressed by individuals. A genome cannot hunt prey until it is expressed as an individual. Individuals can, of course, combine into pairs or groups, but the dynamics that emerge depend upon the individuals and the way they interact. Many people before me have championed the need to focus on individuals, and the value of individual-based datasets (for example, Clutton-Brock and Sheldon 2010), but all too often, as a new technology comes along, the focus can shift away from the individual. The development of molecular methods led to a revolution in our understanding of evolution, but it was not perhaps the huge step forward it was initially expected to be, partly because molecular genetic data is most useful when coupled with additional information about individual phenotypes, behaviors, and life histories (Houle et al. 2010). We currently hear overblown claims about the advances that genomics, proteomics, metabolomics, and epigenomics will deliver. These are fantastic tools that provide important insight, but their great power comes, and will come, when these tools are coupled with data across the performance of individuals within a population. The individual is central to all biology, so we should be studying individuals whenever we can. So why is it so hard to study individuals? Why do grant funding bodies so often refuse to fund studies that are individual-focused? Perhaps it is just not cool to go out to collect data from individuals via observation with binoculars and scopes. It is, however, enjoyable—even if I am not perhaps the right person to advocate field biology as a discipline.

I have some of the worst eyesight in ecology, and almost certainly the worst in field ecology. Before the advent of flat computer screens, you could hear the static on the end of my nose when I coded in GLIM on a mainframe or typed up a manuscript. I am a peculiar advocate for more focus on collecting more individual-based data as no one in their right mind would send me into the field. The St Kilda Soay sheep and Rum red deer projects tried that once. On St Kilda I ended up as general skivvy, mainly carrying equipment up and down hills, whereas on Rum I simply held up the field team. So, I am not championing individual-based data collection because I think I should be in the field collecting it, although it is crucial that theoreticians get to know the details of the system they are modelling. I am championing it because I believe it is central to all of ecology and evolution, and I believe that such data are necessary to parameterize models that hold the key to ecology and evolution becoming a predictive science. Well, a better predictive science than it is now.

There are two dominant modelling paradigms to link from individual-based data to population, community and ecosystem level dynamics: individual-based modelings (IBM) and structured population modeling. The logic behind individual-based models is that individuals consist of a set of attributes, and these attributes underpin rules that describe how individuals survive, reproduce, disperse, and interact. These rules result in emergent dynamics at the level of the population or community (DeAngelis et al. 1992). Statistics summarizing the structure or size of the population or community may feed back to influence the performance of individuals with particular attributes. Depending upon how the attributes of individuals are passed from parents to offspring, models can be evolutionarily explicit, or not (Coulson et al. 2017). Individual-based models simulate what goes on within a population or community. A good model will capture key features of the dynamics of the system, both in terms of the distribution of individual attributes and higher levels of biological organization.

Structured models map individuals to populations and communities by classifying individuals into classes and modeling the dynamics of these classes. Individuals can move between classes, or they can stay in the same class throughout life. The class might be a life history stage, an age, a size, or any other attribute you can measure on individuals. Leslie matrices, Lefkovich matrices, and integral projection models are all types of structured model (Caswell 2001, Easterling et al. 2000).

Density-dependence, environmental stochasticity, demographic stochasticity, and species interactions can all be incorporated into IBMs and structured models (DeAngelis et al. 1992, Ellner et al. 2017). Both single-sex and two-sex models can be constructed (Schindler et al. 2015). Models can be developed to simultaneously explore the dynamics of life history, populations, and genetic or phenotypic characters (Coulson et al. 2011). Both types of model have been used to address questions in life history theory, adaptive dynamics, population ecology, quantitative genetics, and population genetics. They are remarkably flexible, and they allow the investigation of how key quantities used in ecology and evolution are linked. They have been championed as tools to link ecological and evolution in theory and practice.

Both types of model can be parameterized from data collected in the field and the lab from individuals. In order to construct these models, you need to know how traits develop—the domain of developmental biology. You also need to know how traits are inherited—the motivation behind evolutionary genetics. And you need to know how the traits influence survival and reproduction. Ideal datasets for building these datasets will be multigenerational studies that have recorded the complete life histories of large numbers of individuals, while simultaneously recording salient features of the environment. Such datasets exist for species ranging from guppies to wolves, and from thistles to albatross. They are challenging datasets to collect, but once they have been collected, they allow a vast number of questions to be answered through the construction, validation, and analysis of IBMs and structured models.

The parameterization of IBMs and structured models is typically phenomenological. In other words, functions associating individual attributes to their survival, reproduction, and development are data-driven (Ellner, Childs and Rees, 2016). However, if a mechanistic understanding of development or inheritance is known, this can be included in models in the place of phenomenological associations identified from the statistical analysis of data (Coulson et al. 2011, 2017). As genomics, epigenomics, metabolomics, and transcriptomics provide more mechanistic insight into the role of genes and the environment on the development and inheritance of phenotypes, phenomenological models will be replaced with more mechanistic models. Whether this will improve their predictive ability is an open question.

The best individual-based dataset I am aware is that masterminded by David Reznick, Ron Bassar, Andres Lopez-Sepulcre, and Joe Travis. In

2007 and 2008 they seeded four natural streams, each with 100 guppies of known genetic background. The guppies were collected from streams where they coexisted with predators, and they were moved to streams where predators were absent. Every month since, a team of interns catches the fish, returns them to the lab for processing, marks any unmarked individuals, before returning each fish to a point in the focal stream where it was captured. Processing involves weighing each fish, measuring them, photographing them for landmark analysis, and taking a scale for genotyping. David Reznick and his team began the project to investigate a fascinating pattern that had been observed multiple times in Trinidad. When guppies are released from predation, they begin an evolutionary journey from a high-predation phenotype to a low-predation one. High-predation guppies live fast and die young. They reach sexual maturity at a young age and small size and produce large litters of small young. The fish feed almost exclusively on a high-quality diet of invertebrates, and the fish are very rapid swimmers. The low-predation life history they evolve to consists of more sluggish fish with a slower life history: They reach sexual maturity at a larger size and greater age, and they produce smaller litters of larger offspring (Travis et al. 2014). This evolutionary journey is a repeatable pattern that has been observed each time guppies have been released from predators (Reznick and Endler 1982). David Reznick and colleagues realized that the most powerful route to understanding this evolutionary pattern was to study individuals. And nearly ten years on they have data on upwards of 60,000 marked individuals. Each year they conduct a common garden experiment to characterize divergence between fish from the ancestral stream and fish in the introduction streams. In addition, mesocosm experiments are conducted to assay the behavior and demography of fish, and to control for factors that are thought to play a key role in driving the evolutionary change under nearly natural conditions (Bassar et al. 2013). Genomic and epigenetic work has revealed evidence for adaptive and nonadaptive plasticity, and shown how nonadaptive plasticity can accelerate rates of evolution (Ghalambor et al. 2015). Guppies can also be kept in the laboratory, with various teams using them to conduct experiments on their behavior. Guppies are rapidly becoming the animal equivalent of *Arabidopsis* in that they can be studied in detail in both the lab and in the field, and there are a growing number of genetic resources.

Reznick's guppy study is large-scale and costly to run. It requires field vehicles, a field lab, teams of interns, field managers, common garden experiments, a database manager, and large amounts of genotyping. It is a clear demonstration that ecology has moved on from being a cheap science to requiring considerable investment. However, the returns are enormous. Reznick and his colleagues are generating an unprecedented dataset they will make available to the scientific community. But guppies are just one species. We need individual-based studies across the board. Fortunately, there are several other species that have been studied extensively at the individual level, but they are relatively few and far between.

The world's longest running individual-based study is based at Wytham Woods. In 1947, David Lack started studying individually banded great tits. Chris Perrins took over running the project from Lack, and Ben Sheldon took over from Perrins. Recently Sheldon has taken the study high tech. Birds are passive integrated transponder (PIT)-tagged, and the wood is dotted with PIT tag readers. Data on foraging and incubation behavior can now be monitored remotely. Nest boxes can warmed or cooled, and individual birds can be allowed to, or prevented from, feeding at specific feeding stations (Aplin et al. 2015). Biodiversity technology, from PIT tags to satellite collars that monitor vital signs and metabolism, are revolutionizing the way we can collect data on individuals. Ecology is now a big data science: We can collect vast amounts of data from huge numbers of individuals. But just as we arrive at the point of being able to collect big data on large numbers of individuals from numerous species across the world and construct powerful, realistic models that link from the individual to the population, community, and ecosystem, our field seems to have started to abandon individual-based studies. Growing numbers of large-scale individual-based studies are ending, and precious few new ones are being set up. The focus instead has moved to biodiversity research, some of which is rather poorly defined. Much of this research relies on making inferences on some underlying process from large-scale data on some surrogate proxy of what individuals are up to. The questions being asked are often interesting, but they could often be answered more convincingly with detailed datasets of individuals.

Not all species are easily studied at the individual level. Catching, marking, and frequently monitoring microbes is impossible, and working with eukaryotes in the bathypelagic environment would require multiple deep-sea submersibles being used nearly continuously. That would be cool, but expensive. The challenge of tracking individuals and repeatedly monitoring them has led to a bias in the focus of individual-based studies towards species living in closed populations that are easy to capture, and which are found at relatively high density. Historically most individual-based studies have consequently been conducted on plants, such as the famous Barro Colorado study in Panama (Conti et al. 1999), or on vertebrates like the guppies and great tits discussed previously. However, the focus of individual-based studies is starting to change: Large insects can now be successfully marked and followed in the field, and migratory vertebrates can be studied in their winter and breeding ranges and tracked as they move between them. It seems that a bias towards multicellular eukaryotes is probably always inevitable in individual-based studies, but as engineers start to shrink technology to monitor multiple attributes of ever-smaller individuals, the portion of the 9 million or so species available to study should start to increase.

So the question I want to answer is: Why is ecology still thought of as a cheap science? Good field ecology requires investment and, often, recourse to cutting-edge technology. The data that can be collected from a system, particularly when multiple interacting species are studied at the individual level, are invaluable in answering questions at the population, community, and ecosystem level. I encourage those ecologists and evolutionary biologists who have never worked with individual-based data to explore those datasets that are now freely available, and to think big about individual-level data might help them answer questions that cannot be satisfactorily answered with broad-scale data. Second, why are ecologists prepared to watch long-term individual-based data studies peter out? Long-term, individual-based datasets provide some of the richest data that ecology and evolutionary biology has. There are a few dinosaurs who are not prepared to openly share their data because they view them as too precious for general consumption (Mills et al. 2015), but data from a growing number of studies are now freely available (Jones et al. 2008). Long-term, detailed, individual-based datasets are rapidly becoming the public property of science. And as they do so we will be able to answer many questions that have proven intractable when challenged with courser-scale data. But we need more of these studies. We need funding to set up new studies, we need resources to ensure that our existing studies continue, and that data collected are made freely available (Whitlock et al. 2015). And when this is done, we can start to construct mechanistic models linking from systems biology to the ecosystem. When that happens, the fields of ecology and evolution will have come of age.

References

Aplin, L. M., et al. 2015. Experimentally induced innovations lead to persistent culture via conformity in wild birds. Nature 518:538–541.

Bassar, R. D., et al. 2013. Experimental evidence for density-dependent regulation and selection on Trinidadian guppy life histories. American Naturalist 181:25–38.

Caswell, H. 2001. Matrix population models. Hoboken: Wiley.

Clutton-Brock, T., and B. C. Sheldon. 2010. Individuals and populations: The role of long-term, individual-based studies of animals in ecology and evolutionary biology. Trends in Ecology & Evolution 25:562–573.

Condit, R., et al. 1999. Dynamics of the forest communities at Pasoh and Barro Colorado: Comparing two 50–ha plots. Philosophical Transactions of the Royal Society of London B: 354:1739–1748.

Coulson, T. 2012. Integral projection models, their construction and use in posing hypotheses in ecology. Oikos 121:1337–1350.

Coulson, T., et al. 2011. Modeling effects of environmental change on wolf population dynamics, trait evolution, and life history. Science 334:1275–1278.

Coulson, T., B. E. Kendall, J. Barthold, F. Plard, S. Schindler, A. Ozgul, and J.-M. Gaillard. 2017. Modeling adaptive and nonadaptive responses of populations to environmental change. American Naturalist 190:313–336.

DeAngelis, D. L., and L. J. Gross. 1992. Individual-based models and approaches in ecology: Populations, communities and ecosystems. London: Chapman & Hall.

Easterling, M. R., S. P. Ellner, and P. M. Dixon. 2000. Size-specific sensitivity: Applying a new structured population model. Ecology 81:694–708.

Ellner, S. P., D. Z. Childs, and M. Rees. 2016. Data-driven modelling of structured populations: A practical guide to the integral projection model. Berlin: Springer Nature.

Ghalambor, C. K., et al. 2015. Non-adaptive plasticity potentiates rapid adaptive evolution of gene expression in nature. Nature 525:372–375.

Houle, D., D. R. Govindaraju, and S. Omholt. 2010. Phenomics: The next challenge. Nature Reviews Genetics 11:855–866.

Jones, O. R., et al. 2008. A web resource for the UK's long-term individual-based time-series (LITS) data. Journal of Animal Ecology 77:612–615.

Mills, J. A., et al. 2015. Archiving primary data: solutions for long-term studies. Trends in Ecology & Evolution 30:581–589.

Reznick, D., and J. A. Endler. 1982. The impact of predation on life history evolution in Trinidadian guppies (*Poecilia reticulata*). Evolution 36:160–177.

Schindler, S., P. Neuhaus, J.-M. Gaillard, and T. Coulson. 2013. The influence of nonrandom mating on population growth. American Naturalist 182:28–41.

Travis, J., D. N. Reznick, R. D. Bassar, A. López-Sepulcre, R. Ferriere, and T. Coulson. 2014. Do eco-evo feedbacks help us understand nature? Answers from studies of the Trinidadian guppy. Eco-Evolutionary Dynamics 50:1–40.

Whitlock, M. C., et al. 2015. A balanced data archiving policy for long-term studies. Trends in Ecology & Evolution 31:84–85.

Do Temperate and Tropical Birds Have Different Mating Systems?

Christina Riehl

Socially monogamous songbirds in the temperate zone have notoriously high rates of extra-pair mating, but recent studies suggest that this is not always true of their Neotropical counterparts. Extra-pair mating in tropical passerines is variable: Although some species are as promiscuous as temperate-zone passerines, others are genetically monogamous. What ecological or phylogenetic factors explain this variation and why are some tropical passerines faithful to their mates? One prominent hypothesis posits that extra-pair mating is constrained by breeding synchrony, as asynchronous breeding in the tropics might limit the number of receptive females in an area and, by extension, the opportunities for extra-pair fertilizations. However, comparative tests of the effects of synchrony have generally been inconclusive, and the influence of synchrony on extra-pair mating is probably minimal. In this essay I review the recent literature on Neotropical mating systems and find that low rates of extra-pair mating are better explained by year-round territory defense by both sexes. Both year-round territoriality and genetic monogamy are more common in understory insectivores (antbirds, antshrikes, and wrens) than in canopy or open-country species (swallows, tanagers, and emberizids). It is associated with egalitarian parental care, coordinated male and female singing (duetting), and long-term pair bonds. Whereas most studies of extra-pair mating emphasize the role of *male* behaviors in pursuing copulations outside the pair bond, I argue that *female* behaviors—including female song and territory defense—may be equally important. Our understanding of genetic mating patterns in tropical birds is still in its infancy, and rigorous tests of these hypotheses await more data.

Extra-Pair Mating in New World Temperate and Tropical Songbirds

Songbirds in the temperate zone are famously unfaithful to their mates. Although most species form pair-bonds and share parental care of the nestlings, between 10% and 30% of those nestlings are the result of extra-pair matings (Griffith et al. 2002). In some socially monogamous species, extra-pair young are as common as within-pair young: depending on the population, between 35% and 70% of tree swallow nestlings (*Tachycineta bicolor*) are sired by extra-pair males (Barber et al. 1996). In fact, promiscuity is so widespread in temperate birds that Griffith et al. (2002) noted that "levels of extra-pair paternity below 5% of offspring are now considered worthy of explanation."

It came as something of a surprise, therefore, when the first molecular analyses of parentage in tropical birds found little evidence of infidelity, suggesting that extra-pair mating might be generally rare (Telecky 1989, Robertson and Kikkawa 1994, Fleischer et al. 1997). Monogamy seemed to be consistent with other Neotropical life-history traits, such as small clutch sizes, small testes, and low levels of circulating testosterone (Stutchbury and Morton 1995, 2008; Wikelski et al. 2003). As more studies were published, though, genetic mating patterns proved frustratingly inconsistent. Unlike clutch size, which predictably increases with latitude across taxa and habitats (Skutch 1985, Jetz et al. 2008), mating patterns in the tropics are highly variable. Some species do indeed have extremely low rates of extra-pair mating, but others are as just as promiscuous as temperate songbirds (Macedo et al. 2008).

Why are some—but not all—tropical species genetically monogamous? Or, from a different perspective, why are temperate species so promiscuous? Despite twenty-five years of research, genetic mating patterns are known from fewer than twenty Neotropical species. This dearth of information seriously hampers our understanding of the evolutionary ecology of tropical birds and impedes our ability to compare life-history strategies across latitudes. Not only do genetic mating patterns determine individual reproductive fitness, they also influence every question related to sexual selection, including the evolution of male and female song, territoriality, plumage dimorphism, display traits, and even parental care and the division of labor between males and females (Møller and Birkhead 1994). Without a better understanding of the mating strategies of tropical birds, we are unlikely to fully understand the role that extra-pair mating plays in sexual selection and the evolution of avian mating systems.

Is Extra-Pair Mating Constrained by Available Mates?

The most intuitive hypotheses to explain variation in extra-pair mating assume that it is adaptive and should be common unless it is limited by a local scarcity of extra-pair mates (Stutchbury and Morton 1995). If so, infidelity is predicted to be most frequent in populations with high breeding synchrony and/or high breeding density (Møller and Birkhead 1994), since males in either situation should be in close proximity to receptive females on neighboring territories. Might tropical species, with their relatively aseasonal habitats, long breeding seasons, and low population densities, simply lack opportunities to mate outside the pair bond? Early analyses across species showed a promising correlation between breeding synchrony and extra-pair paternity (Stutchbury and Morton 1995, Stutchbury et al. 2007, Stutchbury et al. 1998), but subsequent comparative studies have failed to support the link (reviewed in Griffith et al. 2002 and Macedo et al. 2008). In the Neotropics, several studies have failed to find a relationship between extra-pair paternity and breeding synchrony either within a population (Krueger et al. 2008, Cramer et al. 2011, Douglas et al. 2012, Tarwater et al. 2013) or across temperate and tropical populations of the same species (LaBarbera et al. 2010, Eikenaar et al. 2013). This may be partly because, contrary to *a priori* expectations, tropical populations are not necessarily less synchronous than their temperate counterparts. As Macedo et al. (2008) pointed out, many tropical habits are highly seasonal, including deciduous and montane forests. Furthermore, long tropical breeding seasons do not necessarily translate into low synchrony (Wikelski and Wingfield 2003): humid-forest insectivores such as spotted antbirds (*Hylophylax naeivoides*) may be reproductively active for most of the year, with individual pairs attempting as many as 8 clutches in a season (Willis 1972).

If reproductive opportunities don't limit extra-pair mating, what does? The available (albeit scanty) evidence suggests that Neotropical passerines fall into two categories (Table 1). Low (<5%) rates of extra-pair mating are observed in just a few species, all of which defend stable all-purpose territories throughout the year (often the same territory for many years; Tarwater et al. 2013). Interestingly, with the exception of the yellow-bellied elaenia, all of these are understory insectivores. By contrast, non-territorial or seasonally territorial species tend to have much higher rates of extra-pair copulations, comparable to or higher than those observed in the temperate zone. Females are largely silent (with the exception of house wrens) and do not duet with their partners or participate in territory defense. The available data do not permit many generalizations, but these species are primarily found in edge, second-growth, and grassland habitats. None are insectivores of the forest understory.

Neotropical understory insectivores—including members of ant-following guilds like antshrikes, antbirds, antwrens, antpittas, ant-tanagers, wrens, and woodcreepers—typify the "slow" end of the tropical life-history spectrum, characterized by a suite of interrelated traits including low adult mortality, delayed reproduction, small clutches, and metabolic and immunological adaptations to long life (Ricklefs and Wikelski 2002). Although far less is known about behavioral traits than life-history traits of tropical birds, these species also tend to share long-term pair bonds, coordinated male and female duetting, female participation in territory defense, and—apparently—low rates of extra-pair mating. At least part of the

Table 1. Extrapair Paternity Rate and Life-History Characteristics of 16 Socially Monogamous Neotropical Passerines

SPECIES NAME	% EP YOUNG (N)	% EP BROODS (N)	TERRITORIALITY	DUETTING?	HABITAT	SOURCE
Year-round territory defense						
Yellow-bellied Elaenia (*Elaenia flavogaster*)	4% (24)	7.6% (13)	Year-round, MF	yes	scrub/grass	Stutchbury et al. 2007
Dusky Antbird (*Cercomacra tyrannina*)	0% (15)	0% (9)	Year-round, MF	yes	humid forest	Fleischer et al. 1997
Black-crowned Antshrike (*Thamnophilus atrinucha*)	3% (89)	2% (50)	Year-round, MF	yes	humid forest	Tarwater et al. 2013
Banded Wren (*Thyrophilus pleurostictus*)	4% (156)	10% (50)	Year-round, MF	no	dry forest	Cramer et al. 2011
Rufous-and-white Wren (*Thyrophilus rufalbus*)	2% (158)	6% (51)	Year-round, MF	yes	humid forest	Douglas et al. 2012
Bicolored Wren (*Campylorhynchus griseus*)	4.6% (222)	? (99)	Year-round, MF	yes	dry forest	Haydock et al. 1996
Buff-breasted Wren (*Thyrothorus leucotis*)	4% (53)	3% (31)	Year-round, MF	yes	humid forest	Gill et al. 2005

Nonterritorial or seasonally territorial

Species						
Mangrove Swallow (*Tachycineta albilinea*)	15% (97)	26% (31)	Breeding, M(F?)	no	Coastal/water	Moore et al. 1999
White-rumped Swallow (*Tachycineta leucorrhoa*)	56% (342)	77% (78)	Nonterritorial	no	Grassland	Ferretti et al. 2011
Clay-colored Thrush (*Turdus grayi*)	38% (37)	53% (19)	Breeding, M	no	2° forest edge	Stutchbury et al. 1998
House Wren (Southern) (*Troglodytes aedon*)	15.7% (166)	32.5% (40)	Year-round?, M	no	2° forest edge	LaBarbera et al. 2010
Cherrie's Tanager (*Ramphocelus costaricensis*)	49% (59)	55% (32)	Nonterritorial	no	Humid forest	Krueger et al. 2008
White-banded Tanager (*Neothraupis fasciata*)	27.9% (59)	39.4% (66)	Breeding, MF	no	Scrub/grass	Moreira 2014
Lesser Elaenia (*Elaenia chiriquensis*)	37% (38)	67% (15)	Breeding, M	no	scrub/grass	Stutchbury et al. 2007
Rufous-collared Sparrow (*Zonotrichia capensis*)	47% (34)	62% (21)	Breeding, M	no	arid scrub	Eikenaaret al. 2013
Blue-black Grassquit (*Voltinia jacarina*)	50% (20)	63% (11)	Breeding, M	no	scrub/grass	Carvalho et al. 2006

Note: "% EP young" refers to the percentage of nestlings sampled that were sired by extra-pair males, whereas "% EP Broods" refers to the percentage of broods sampled that contained at least one nestling sired by an extrapair male. Total sample sizes are given in parentheses. Species endemic to neotropical islands are excluded.

confusion about whether extra-pair copulation rates really do differ across latitudes probably stems from the tendency to lump all "tropical" species into one category, disregarding what appear to be important differences in territoriality, song, and pair-bonding behavior.

Female Song, Territory Defense, and Genetic Monogamy

Assuming that the limited data in Table 1 are representative of a real pattern, genetically monogamous and promiscuous tropical species differ strikingly in the extent to which females sing and participate in territory defense. The two behaviors are linked: duetting species typically sing in response to territorial intrusion, and experiments in two species (the bay wren, *Thryothorus nigricapillus*, and the dusky antbird) showed that females were able to defend their territories by singing even when their mates were removed (Levin 1996). Because duets are easily elicited by playbacks, an extensive literature already exists on the structure and function of duetting in tropical birds and this theory may yield insights into extra-pair mating behavior. One possibility is that duetting reflects cooperative territory defense, since joint defense might be more effective against intruders than a solo song, and both members of a pair should benefit from deterring rivals that could oust them. An alternative hypothesis, though, is that the primary function of duetting is to safeguard one's *mate* against same-sex rivals seeking extra-pair copulations—an essentially selfish function that has been called "acoustic mate-guarding" (Langmore 1998). Members of a pair might answer each other's songs to prevent them from singing alone, effectively advertising their mated status and repelling same-sex rivals.

Whether duetting is a product of sexual conflict or cooperation is still an open question, but a few studies have provided fascinating evidence in support of the mate-guarding hypothesis. Levin (1996) found that, contrary to prior expectations, duets were always initiated by female bay wrens, suggesting that males might sing in response to their mates in order to advertise their mated status and prevent them from solo singing. Further support for the mate-guarding hypothesis comes from playback experiments on Peruvian warbling antbirds (*Hypocnemis peruviana*; Seddon and Tobias 2006, Tobias and Seddon 2009), barred antshrikes (*Thamnophilus doliatus*; Koloff and Mennill 2011), and black-bellied wrens (*Thryothorus fasciatoventris*; Logue and Gammon 2004). In all three of these species, individuals respond more strongly to solo songs of same-sex rivals than to opposite-sex solos or duetting pairs, indicating that lone same-sex rivals are perceived to be a greater threat. Interestingly, in all cases females respond as—or more—strongly to same-sex songs as males do, suggesting that, if duets do indeed guard against extra-pair copulations, both sexes are equally vigilant.

Even if duetting does limit extra-pair copulations, however, it is likely that female song initially arose in the context of territory defense rather than (or in addition to) mate-guarding. Odom et al. (2015) recently used phylogenetic analysis to reconstruct the evolution of duetting in the New World blackbird clade (Icteridae) and found strong evidence that it has arisen only in lineages in which female song was ancestral. Female song, in turn, was correlated with year-round territoriality—a result supported by analyses of other tropical bird lineages (Slater and Mann 2004). Female song appears to have been lost not only in temperate blackbirds which have evolved long-distance migration (and hence lost year-round territorial defense), but also in non-migratory tropical blackbirds in which females do not defend territories, such as the polygynous oropendolas and caciques, and the brood-parasitic cowbirds (Price et al. 2009). These analyses provide support to the hypothesis that temperate-tropical differences in female song and territoriality are not driven by latitude per se, but by an interrelated set of natural history traits that are favored in some Neotropical habitats and lineages but largely lost in the temperate zone.

Future Directions

Virtually every review of mating behaviors of tropical birds ends with a plea for more data (Stutchbury and Morton 2008, Macedo et al. 2008, Neodorf 2004), and this essay is no exception. Here I have argued that genetic monogamy is correlated with a specific suite of life-history traits shared by many Neotropical understory insectivores. It seems likely that when extra-pair mating is rare, it is because it has not been favored by natural selection and not because it is constrained by a local scarcity of receptive females. Without more data on extra-pair parentage rates across species, though, it is impossible to know how general this conclusion is. Equally importantly, for the vast majority of species we lack the natural history information that would allow inferences about the fitness advantages of fidelity or the potential costs of extra-pair mating. We still lack systematic analyses of how genetic mating patterns affect parental investment in tropical birds, or how behavioral traits such as territoriality and pair-bonding interact with ecological variables such as density, food abundance, and nest predation. Is it true, as often claimed, that sexual selection is really weaker in tropical birds than in temperate ones (Badyaev and Hill 2003)? Or are display traits such as duetting and female song indicative of sexual selection by both sexes? How does longevity affect extra-pair paternity rate, and are these patterns consistent with classical ecological models of r and K selection? Many questions remain open; their answers will shed light not just on tropical birds but on the evolution of global avian life-histories.

Acknowledgments

I thank Ioana Chiver, Valentina Ferretti, Janeene Touchton, and Stefan Woltmann for sharing ideas, enthusiasm, and unpublished data that contributed to this review.

References

Badyaev, A. V., and G. E. Hill. 2003. Avian sexual dimorphism in relation to phylogeny and ecology. Annual Review of Ecology and Systematics 34:27–49.

Barber C. A., R. J. Robertson, and P. T. Boag. 1996. The high frequency of extra-pair paternity in tree swallows is not an artifact of nest boxes. Behavioral Ecology and Sociobiology 38:425–430.

Carvalho C.B.V., R. H. Macedo, and J. A. Graves. 2006. Breeding strategies of a socially monogamous Neotropical passerine: extra-pair fertilizations, behavior, and morphology. Condor 108: 579–590.

Cramer, E.R.A., M. L. Hall, S. R. de Kort, I. J. Lovette, and S. L. Vehrencamp. 2011. Infrequent extra-pair paternity in the banded wren, a synchronously breeding tropical passerine. Condor 113:637–645.

Douglas, S. B., D. D. Heath, and D. J. Mennill. 2012. Low levels of extra-pair paternity in a Neotropical duetting songbird, the rufous-and-white wren (*Thyrothorus rufalbus*). Condor 114:393–400.

Eikenaar, C., F. Bonier, P. R. Martin, and I. T. Moore. 2013. High rates of extra-pair paternity in two equatorial populations of rufous-collared sparrow, *Zonotrichia capensis*. Journal of Avian Biology 44:600–602.

Ferretti, V., V. Massoni, F. Bulit, D. W. Winkler, and I. J. Lovette. 2011. Heterozygosity and fitness benefits of extrapair mate choice in white-rumped swallows (*Tachycineta leucorrhoa*). Behavioral Ecology 22:1178–1186.

Fleischer, R. C., C. L. Tarr, E. S. Morton, A. Sangmeister, and K. C. Derrickson. 1997. Mating system of the dusky antbird, a tropical passerine, as assessed by DNA fingerprinting. Condor 99:512–514.

Gill, S. A., M. J. Vonhof, J. M. Stutchbury, E. S. Morton, and J. S. Quinn. 2005. No evidence for acoustic mate-guarding in duetting buff-breasted wrens (*Thryothorus leucotis*). Behavioral Ecology and Sociobiology 57:557–565.

Griffith, S. C., I.P.F. Owens, and K. Thuman. 2002. Extra pair paternity in birds: A review of interspecific variation and adaptive function. Molecular Ecology 11:2195–2212.

Haydock, J., P. G. Parker, and K. N. Rabenold. 1996. Extra-pair paternity uncommon in the cooperatively breeding bicolored wren. Behavioral Ecology and Sociobiology 38:1–16.

Jetz, W., C. H. Sekercioglu, and K. Böhning-Gaese. 2008. The worldwide variation in avian clutch size across species and space. PLOS Biology 6:e303.

Koloff, J., and D. Mennill. 2011. Aggressive responses to playback of solos and duets in a Neotropical antbird. Animal Behaviour 82:587–593.

Krueger, T. R., D. A. Williams, and W. Searcy. 2008. The genetic mating system of a tropical tanager. Condor 110:559–562.

LaBarbera, K., P. E. Llambías, E.R.A. Cramer, T. D. Schaming, and I. J. Lovette. 2010. Synchrony does not explain extrapair paternity rate variation in northern or southern house wrens. Behavioral Ecology 21:773–780.

Langmore, N. E. 1998. Functions of duet and solo songs of female birds. Trends in Ecology and Evolution 13:136–140.

Levin, R. 1996. Song behaviour and reproductive strategies in a duetting wren, *Thyrothorus nigricapillus*: I. Removal experiments. Animal Behaviour 52:1093–1106.

Logue, D. M., and D. E. Gammon. 2004. Duet song and sex roles during territory defence in a tropical bird, the black-bellied wren, *Thryothorus fasciatoventris*. Animal Behaviour 68:721–731.

Macedo, R.H.F., J. Karubian, and M. S. Webster. 2008. Extrapair paternity and sexual selection in socially monogamous birds: are tropical birds different? Auk 125:769–777.

Møller, A. P., and T. R. Birkhead. 1994. The evolution of plumage brightness in birds is related to extrapair paternity. Evolution 48:1089–1100.

Moore, O. R., B.J.M. Stutchbury, and J. S. Quinn. 1999. Extrapair mating system of an asynchronously breeding tropical songbird: The mangrove swallow. Auk 116:1039–1046.

Moreira, P. M. 2014. Reprodução cooperativa e paternidade extra-par em *Neothraupis fasciata*. PhD dissertation, Universidade de Brasília, Brazil.

Neodorf, D.L.H. 2004. Extrapair paternity in birds: Understanding variation among species. Auk 121:302–307.

Odom, K. J., K. E. Omland, and J. J. Price. 2015. Differentiating the evolution of female song and male-female duets in the New World blackbirds: Can tropical natural history traits explain duet evolution? Evolution 69:839–847.

Price, J. J., S. E. Lanyon, and K. E. Omland. 2009. Losses of female song with changes from tropical to temperate breeding in the New World blackbirds. Proceedings of the Royal Society of London B 276:1971–1980

Ricklefs, R. E., and M. Wikelski. 2002. The physiology-life history nexus. Trends in Ecology and Evolution 17:462–468.

Robertson, B. C., and J. Kikkawa. 1994. How do they do it? Monogamy in *Zosterops lateralis chlorocephala*. Journal of Ornithology 135:459.

Seddon, N., and J. A. Tobias. 2006. Duets defend mates in a suboscine passerine, the warbling antbird (*Hypocnemis cantator*). Behavioral Ecology 17:73–83.

Skutch, A. F. 1985. Clutch size, nesting success, and predation on nests of Neotropical birds, reviewed. Ecological Monographs 36:575–594.

Slater, P.J.B., and N. I. Mann. 2004. Why do the females of so many bird species sing in the tropics? Journal of Avian Biology 35:289–294.

Stutchbury, B.J.M., and E. S. Morton. 1995. The effect of breeding synchrony on extra-pair mating systems in songbirds. Behaviour 132:675–690.

Stutchbury, B.J.M., E. S. Morton, and W. H. Piper. 1998. Extra-pair mating system of a synchronously breeding tropical songbird. Journal of Avian Biology 29:72–78.

Stutchbury, B.J.M., E. S. Morton, and B. Woolfenden. 2007. Comparison of the mating systems and breeding behavior of a resident and a migratory tropical flycatcher. Journal of Field Ornithology 78:40–49.

Stutchbury, B.J.M., and E. S. Morton. 2008. Recent advances in the behavioral ecology of tropical birds. Wilson Journal of Ornithology 120:26–37.

Tarwater, C. E., J. D. Brawn, and J. D. Maddox. 2013. Low extrapair paternity observed in a tropical bird despite ample opportunities for extrapair mating. Auk 130:733–741.

Telecky, T. 1989. The breeding biology and mating system of the common myna (*Acridotheres tristis*). Ph.D. diss., University of Hawaii, Honolulu.

Tobias, J. A., and N. Seddon. 2009. Signal jamming mediates sexual conflict in a duetting bird. Current Biology 19:577–582.

Wikelski, M., M. Hau, W. D. Robinson, and J. C. Wingfield. 2003. Reproductive seasonality of seven Neotropical passerine species. Condor 105:683–695.

Willis, E. O. 1972. The behavior of spotted antbirds. Ornithological Monographs 10:1–162.

Leaf Structure and Function

Peter J. Grubb

Leaf form and function are thought to be critical determinants of the ecology of plants. Before considering the many unknowns about leaf structure and function it will be helpful to set out some of the facts that are well established. The mean size of leaf blades in a plant community decreases along gradients of increasing soil dryness and increasing cold (Richards 1996). In the tropics and subtropics deciduousness is characteristic of plant species that experience seasonally dry soils (Richards 1996), whereas in the temperate regions deciduous species are associated with soils from which it is seasonally difficult to withdraw water because at low temperatures the viscosity of water increases and the permeability of root cell membranes declines (Kramer 1969). In the Northern Hemisphere, in regions where the degree of winter cold is moderate, it is common to find evergreens on the most infertile soils, and deciduous species on the richer soils (Monk 1966, Walter 1968). In any forest composed of both evergreen and deciduous species the leaves of evergreens are usually thicker and tougher than those of the deciduous species (Chabot and Hicks 1982, Grubb 1986); the greater toughness provides greater generalized defence against physical and biological damage (Grubb 1986). More generally, species with longer-lived leaves have a higher dry mass per unit area of leaf blade, while per unit dry mass they have lower maximum rates of net assimilation (photosynthesis less respiration) (Grubb 1984, Field and Mooney 1986).

The "design" of the leaf of an average land plant as a machine that captures light and absorbs CO_2 while minimizing loss of H_2O is impressive (Taiz and Zeiger 2010). The internal structure of the leaf resembles that of a lung in that there is a high ratio of internal surface to external surface. The skin (epidermis) of the leaf is typically colorless. The green cells in the tissue between the upper and lower epidermis (the *mesophyll*) are arranged around air spaces (Fig. 1) and large fractions of the cell surfaces are in contact with air spaces rather than abutting other cells. Usually there is a more compact arrangement of cells in the upper part of the mesophyll (the *palisade*, Fig.1). In the currently dominant group of plants on land (the flowering plants) water is supplied to the photosynthesizing cells

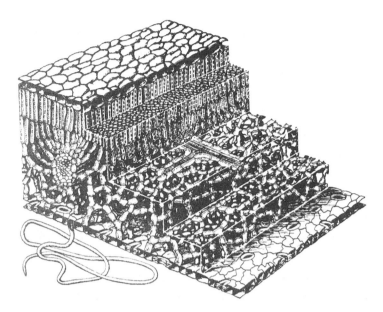

FIGURE 1. A three-dimensional view of the internal structure of an apple leaf. Note particularly the contrast between the tightly packed palisade cells (in the upper half of the leaf) and widely separated cells of the spongy mesophyll (in the lower half), also the pores (stomata) confined to the lower epidermis. Reprinted from Eames and MacDaniels, *Introduction to Plant Anatomy* (McGraw-Hill, 1951) by permission of McGraw-Hill Education.

by a dense network of canals made up of tiny empty pipes, the xylem cells, which lose their protoplasmic contents at the end of their development. These canals in the "veins" are accompanied by arrays of living cells (the phloem) that transport away the products of photosynthesis from mature leaves and many chemical components of leaves that are senescing. Loss of water from the leaf occurs through minute pores and is regulated by the cells surrounding the pores (the stomata); these close at night and when the leaf begins to dry. Commonly the stomata of species of moist sites shut after a slight-to-modest water loss, and those of dry sites close only after greater water loss.

Despite so much being found out about the structure and function of leaves, there is still a huge amount to be discovered. In my chapter on leaf structure and function in *The Encyclopaedia of Ignorance* (Grubb 1977a) I set out 20 questions, very various in kind. In Table 1 I have summarized the findings in the last 40 years regarding the ten questions on which most progress has been made, and in Table 2 the more modest advances regarding the ten questions on which least progress has been made.

Table 1. Answers to the Ten Questions Set by Grubb (1977) on Which There Has Been Greatest Progress in Understanding

1. Why is the form of modern-day leaves so different from what was seen in the distant past?

   The general rarity of leaves of the dissected fern-frond type among modern seed plants (which contrasts with its dominance in pteridosperms in the Carboniferous) reflects greater conductance of modern xylem elements, closer spacing of veins and networking together making possible wide areas of lamina (Brodribb et al. 2005, Zwieniecki and Boyce 2014). Confinement of very long, wide strap-shaped leaves to wet tropics and wet soils in drier tropics (*Dracaena* and more widely *Pandanus* (Grubb 2003)) paralleling confinement of lepidodendrids to peat forests in the Carboniferous period possibly reflects persistent problem in conductivity and/or susceptibility to cavitation.

2. What are the proportions of the flow of liquid water across the lamina through (a) the lumina of the xylem conduits of the fine veins, (b) cell walls and protoplasts of the mesophyll, (c) cell walls and protoplasts of the epidermides?

   Still the subject of debate and modeling (Buckley 2015, Buckley et al. 2015); it seems that more water moves through walls than protoplasts, and much water moves as vapor; the proportion of liquid water movement that is through the epidermides is still unclear. Much of the water flows through the xylem to the photosynthesizing cells of the mesophyll; the conductance of this pathway is generally related to the maximum net assimilation rate per unit dry mass of leaf (Brodribb et al. 2005, Sack and Holbrook 2006); it depends on at least three variables: the radius of the broadest conducting elements (Aasamaa et al. 2001), the vein length per unit area (Sack et al. 2013) and the length of the path between the end of a vein and the point where the water evaporates (Brodribb et al. 2007).

3. Why do most leaves have spongy mesophyll as well as palisade?

   Having spongy mesophyll increases the lamina area per unit dry mass and nitrogen N, and that is of value in intercepting more light and casting more shade on competitors. Currently, ideas are being tested using single-gene mutants of *Arabidopsis* with different amounts of air space in the leaf (Lohmeier et al. 2017). Having more air space in leaves leads to greater proportion of water movement being in the vapor phase and greater sensitivity to gradients of temperature within the leaf (Buckley 2015, Buckley et al. 2015).

4. Bundle sheath extensions (BSEs); why do some plant types have them and others not? Some are fibrous and structural, others thin-walled.

   Those not primarily structural provide low-resistance pathways to the upper and lower epidermides; a mutant of tomato with no BSEs had reduced leaf hydraulic conductance, stomatal conductance and net assimilation rates (Zsögön et al. 2015). This finding is consistent with the results of modelling and experiments comparing herbaceous and woody species (Buckley et al. 2015, Buckley and Gilbert 2011).

Table 1. (*Continued*)

5. Why is the epidermis colorless, apart from the stomatal guard cells?

For 25 species of varying life form and habitat the epidermis was found to transmit less than 10% of incoming ultraviolet radiation, which has the potential to damage the process of photosynthesis (Robberecht and Caldwell 1978); flavonoid and related pigments were responsible for much of the absorptance. This function of the epidermis could work alongside its role (suggested long ago (Haberlandt 1914)) as a short-term store of water which can move to the mesophyll and make possible continued photosynthesis when there is a risk of over-heating and/or desiccation as in a shade leaf in a sunfleck or a sun leaf when the air becomes very still.

6. Why do C4 plants have the C4 and C3 pathways in separate cells rather than in a single cell type?

Genetic manipulation has proven that the resistance to backward diffusion from the bundle sheath is vital to maintaining a ratio of CO_2 to O_2 within the sheath that makes the overall process more effective than the C_3 system alone (Ludwig et al. 1998). A very few seed plants (Chenopodiaceae in southwest Asia) carry out C_4 photosynthesis with intracellular compartmentation (Edwards et al. 2004) but this is less effective than compartmentalization among cells.

7. Why are some species amphistomatous, i.e., have stomata in both lower and upper epidermis? Why have many small stomata or few large? The negative correlation between size and density was already known.

Amphistomatous species are typically fast-growing and herbaceous (Muir 2015); the benefit of reduction in resistance to inward diffusion of CO_2 outweighs the hazard of occasional waterlogging of the intercellular spaces by infiltration during heavy or prolonged rainfall. The strong negative correlation between stomatal size and number per unit area means that different species may have the same epidermal conductance to CO_2 with large or small stomata (Sack et al. 2003). There is a strong correlation between stomatal size and 2c DNA content (Beaulieu et al. 2008), but in a given clade there can also be large variation in stomatal size with little difference in 2c DNA (Jordan et al. 2015).

8. Why do the more-or-less pendent soft young leaves of many species of tropical and subtropical lowland rain forests have delayed chloroplast development and red coloration based on anthocyanins?

A critical review in 2002 (Dominy) concluded that the most likely explanation of the red color is that it makes the leaves invisible to herbivores, which cannot see red. One hypothesis for the origin of trichromacy in primates is that it overcame that problem (Dominy and Lucas 2001), but this hypothesis has been contested and trichromacy related primarily to detection of ripe fruit against a background of green leaves (Sumner and Mollon 2003). Very recently new evidence for a fungistatic role of the anthocyanins has been published (Tellez et al. 2016). When in two forests species that have red young leaves were compared with species that have green young leaves, the former had lower rates of mortality as seedlings but lower relative growth rates of their trunks as small trees, implying a benefit and a cost (Queenborough et al. 2013).

(Continued)

Table 1. (*Continued*)

9. Are the physical and chemical properties of leaves and visiting ants that are supposed to deter herbivores, effective?

There is much evidence now for the effectiveness of physical (Lucas et al. 2000) and chemical (Mithoefer and Boland 2013) defenses and visiting ants lured by extrafloral nectaries and sometimes nest sites (Rosumek et al. 2009).

10. Why do some plants change the inclination of a lamina under particular conditions; e.g., leaflets of Oxalis *species bend down in bright light, whereas pairs of leaflets of many tropical legumes bend down and become firmly appressed at night?*

"Solar tracking" (heliotropism) is widespread in light-demanding species and can increase yield either by maximizing the radiation absorbed or by minimizing it, depending on species and situation (Ehleringer and Forseth 1989); by hypothesis the highly shade-tolerant *Oxalis* species of northern temperate forests have retained this behavior when evolving from light-demanding ancestors (Grubb 1988). The behavior of tropical legumes has been hypothesized to reduce herbivory on leaves with a high nitrogen concentration by increasing the difficulty of eating a leaf when thickness is doubled, and to do so at the time of day when most herbivory by inverte-brates occurs (Grubb and Jackson 2007).

In this new review I consider five questions that have come to the fore since 2000. These questions can all be studied from the point of view of the field ecologist, interested in the coexistence of species and in the ways in which plants are suited to their habitats, and that of the laboratory ecologist, interested in the details of leaf structure and/or the physical processes and chemical mechanisms that are vital to leaf function.

Do the Length of the Leaf Stalk and the Shape of the Leaf Blade Really Matter?

This issue relates to the recent fashion for collecting data on easily mea-sured properties of leaves (traits) for large collections of species, and using analyses of such large data sets to draw conclusions about the value of each variable measured. Consider leaf form, and specifically (a) the length of the stalk (petiole) relative to the length or area of the blade (lamina), and (b) the shape of the blade.

In the wet tropical lowlands, the great majority of species that have leaves with stalks that are long relative to the length of the blade, and the blades are wide relative to length (often heart-shaped or umbrella-shaped), require gaps in the forest canopy or forest edges for establishment and

Table 2. Answers to the Ten Questions Set by Grubb (1977) on Which There Has Been the Least Progress

1. Why are all leaves the size they are, and not an order of magnitude larger or smaller? How is the upper limit on leaf size set?

Few attempts have been made to answer the question; a recent model (Jensen and Zwieniecki 2013) hypothesizes that limits are set by intrinsic properties of the carbohydrate transport network.

2. Why is the percentage of current shoot dry mass in leaves commonly 60%–70%, the rest in subtending stem?

The most recent comprehensive review of allocation to leaves, stems, and roots in different kinds of plants and plants of different size (Usoltsev et al. 2015) does not answer this question.

3. What are relative leaf sizes adapted to? Usual growing conditions or the most unfavorable seasons or years?

This is of greatest interest and unresolved for evergreens that survive only moderately less-favorable conditions for a period each year; e.g., in warm temperate rain forest (Grubb et al. 2013). Where the unfavorable period is more severe (e.g., in predominantly deciduous forest in the dry tropics), the smaller leaf sizes of evergreens suggest that adaptation to the unfavorable season is more important.

4. Is the ratio of CO_2 taken in to H_2O lost likely to drive evolution of leaf sizes?

Likely to be important only where water is commonly in short supply; i.e., not in rain forests, especially in tropical montane and subalpine rain forests, which have successively smaller leaves and are commonly in cloud; there is a need for more models on maximizing CO_2 uptake alone (Givnish 1984).

5. Why are certain leaf shapes associated with particular life forms, notably cordate (heart-shaped) blades on long stalks (petioles) with climbers?

A model based on balancing photosynthetic gains and metabolic costs of replacing water lost, incorporating mechanical efficiency and appropriate lamina inclination, predicted large cordate leaves with long petioles in well-lit situations, and narrow-based small leaves with short petioles at more shaded sites (Givnish and Vermeij 1976).

6. Why is lobing seen mostly in thinner (deciduous or herbaceous) leaves? Is the answer related to resistance to water flow through the lamina and especially resistance in thin walls of epidermis in thin leaves?

This suggestion was not supported by an experimental study (Sisó et al. 2001), which found that the hydraulic conductance of the lamina in *Quercus* species is positively related to the degree of lobing, suggesting that the mesophyll of highly lobed leaves benefits from high rates of water supply that help maintain water status and photosynthetic activity despite the fact that the thinner still air layer on lobed leaves increases loss of water as well as loss of heat.

(Continued)

Table 2. (*Continued*)

7. What is the function of the hypodermis (the layer immediately inside the epidermis, most often found below the upper epidermis)?

    Function is presumably correlated with cell wall thickness. If thick-walled, it gives support and toughness that may inhibit herbivores (Dominy et al. 2008); if thin-walled it may act as a short-term water reservoir as hypothesized previously for the epidermis. In either case it may attenuate ultraviolet B radiation (Flenley 1992).

8. What is the function of the wax deposited in the stomatal pores of some species?

    Plugs of wax or cutin have been best studied in Winteraceae, but are also found in other dicots and in monocots; in some conifers there are plugs of resinous material (Wilkinson 1979); in at least one species of Winteraceae plugs are associated with stomata remaining open at night (Feild and Holbrook 2000); in one drought-tolerant *Quercus* species a greater degree of plug development was seen as adaptive (Roth-Nebelsick et al. 2013).

9. In flapping leaves how large is the bellows effect on the movements of CO_2 into the leaf and H_2O out?

    This question seems to have been ignored completely even though the issue must be a real one for large soft leaves easily deformed by wind.

10. What is the function of the drip tips, i.e., elongated narrowing apices, found on the leaves of many tropical rain forest plants?

    This question is unresolved. They are characteristic of tropical and subtropical lowland and lower montane rain forests, rare or absent in upper montane forests, and absent in subalpine rain forests (Grubb 1977, Grubb and Stevens 1985, Goldsmith et al. 2017). They are also rare in tropical dry forests. Within the Amazon basin they are most abundant in the areas with the highest mean values for rainfall in the wettest trimester (Malhado et al. 2012). They are commonest in shade-inhabiting species and characteristic of juveniles in the case of tall trees that regenerate in shade (Panditharathna et al. 2008). Leaf wettability is not, on average, different in species with and without drip tips (Goldsmith et al. 2017). Experiments have shown that drip tips do facilitate run-off but do not inhibit colonization by epiphyllous bryophytes, and have yielded contradictory results with respect to colonization by fungi that might damage the leaf (Burd 2007).

onward growth. These features increase competitive ability where light is abundant; the long stalks increase the chance of placing a leaf above that of a neighbor, and an individual leaf with a wider blade is more effective than one with a narrower blade in shading competitors below. In contrast the many species that can establish in deep shade and persist as juveniles in the understorey do so because of their tolerance of shade rather than their ability to cast shade; their leaves typically have short stalks and the blades are egg-shaped to oblong, narrow relative to their

Figure 2. Photographs of cleared dried leaves of four gap-demanding tree species found in tropical lowland rain forest in northern Queensland, two with the "right" kind of leaf (above, *Mallotus mollissimus*, Euphorbiaceae, left, petiole bent, and *Acronychia acidula*, Rutaceae), and two with the "wrong" kind of leaf (below *Glochidion hylandii*, Euphorbiaceae, left, and *Alphitonia incana*, formerly *A. philippinensis*, Rhamnaceae). The bar is 5-cm long. Reproduced from Christophel and Hyland (1993), with permission from CSIRO Publishing.

length. But there are exceptions among both gap-demanders and shade-tolerators. In northern Queensland, to take one example, there are three genera (*Alphitonia, Glochidion,* and *Trema*) in different families that are abundant and widespread at the edges of primary forest and in secondary forest but have the "wrong" leaf form; i.e., in silhouette the leaves appear to be like the leaves of shade tolerators (Fig. 2). Likewise, there are at least twelve genera in eight families containing strongly shade-tolerant species with the "wrong" leaf type having notably long stalks; e.g., *Elaeocarpus* and *Franciscodendron*.

How are we to explain the species with the "wrong" leaf form? Is it the case that the competitive advantages of long stalks and wide blades for gap-demanders are slight relative to other properties that take more time and more resources to measure, such as the allocation of dry mass in the whole plant, maximum net assimilation rate and dark respiration rate at relevant temperatures, and rate of height increase? Or are there certain circumstances in gaps where the possession of an array of leaves of the shade-tolerator type has an advantage over the majority type? Similarly is the expenditure of resources on long petioles in deep shade a trivial matter relative to the properties that are time-demanding to measure (the gas exchange properties just mentioned, plus measurement of the flexibility of these properties in relation to shade and the degree of persistence of physical and chemical defences against diseases and herbivores in shade) or are they of advantage under certain circumstances; e.g., when a gap opens in the canopy above?

One approach would be to grow light-demanding species exemplifying "right" and "wrong" leaf types at different densities and different irradiances, and record both growth rates and the patterns of shading among species, thus revealing how important petiole length and blade size can be relative to absolute rate of height growth by whole plants. A similar experiment could be run with shade-tolerant species but conducted over a longer period (say, 5–10 years) with species having the "right" and "wrong" leaves grown in deep shade at different densities, and then given a canopy gap or simply more light if the experiment is done under replicated shade screens.

A particularly interesting problem with respect to shade-plants with the "wrong" type of leaf arises in the northern temperate deciduous forests (NTDF) of Eurasia and North America where among the most shade-tolerant species there are various species of maple (*Acer* spp.) that all have long petioles and wide blades with several points—a strong contrast with the various species of beech (*Fagus* spp.) that have the "right" kind of leaf judged by the standards of tropical lowland rain forest (TLRF). Species of lime or basswood (*Tilia* spp.) present a similar problem but they tend not to be as tolerant as the maples, and their heart-shaped blades are not matched by such very long stalks as are seen in the maples.

NTDF differs importantly from TLRF in that each spring there is a large increase in the irradiance intercepted by seedlings and saplings that are heavily shaded for the summer because they leaf out before the taller trees (Gill et al. 1998). Measurements and experiments with beech (*Fagus grandifolia*), buckeye (*Aesculus glabra*), sugar maple (*Acer saccharum*) and other species have established that the benefit that the summer-shaded saplings have as a result of the spring window is substantial (Gill et al. 1998, Augspurger et al. 2005, Augspurger 2008). Insofar as the saplings are crowded the possession of long petioles may confer a significant advantage in competition between individuals. It is also possible that the same benefit is significant when a canopy gap is opened up by treefall, as suggested previously for TLRF species, but it is known that at least in the well-studied case of sugar maple versus beech the rate of growth in overall plant height is the most important factor in determining the different outcomes of competition in gaps of different size (Canham 1988).

Finally, we may reflect on the contrast between the great majority of species of tree in TLRFs being strongly shade tolerant at the seedling and sapling stages, whereas only a small proportion of tree species in NTDF are so tolerant. It seems that the evolution of strong shade tolerance plus tolerance of winter cold has not been easy. Why?

Does Variation in Foliar Nitrogen Concentration Contribute to Maintenance of Species Richness?

The background to this question is the very general issue as to whether or not interspecific variation in the properties of the leaves in one functional type of plant, e.g., TLRF shade-tolerant tree or late-successional semidesert shrub, contributes to the maintenance of species richness in a community. The foliar concentration of nitrogen often varies by a factor of from three to five within one functional type of plant in a reasonably species-rich community, whether that be a forest, a woodland, or a semidesert (Wright et al. 2001, Grubb 2002), whereas the difference between the mean for one kind of forest and that for a very different one may involve a factor only two or less (Vitousek and Sanford 1986). It is important to understand that the wide variation cannot be explained away as a result of high concentrations in gap demanders, and low concentrations in shade tolerators. For example, in TLRF on basalt in northeast Australia (Grubb et al. 2008) the mean concentration of nitrogen in mature leaves on saplings of shade-tolerant trees was found to vary with species from 9.2 mg g^{-1} in *Cardwellia sublimis* to 42 in *Melicope vitiflora*, and among the accompanying species requiring canopy gaps for establishment the range was from 18 mg g^{-1} in *Elaeocarpus grandis* to 49 mg g^{-1} in *Polyscias murrayi*.

One hypothesis to explain the variety would be that the nitrogen-rich species grow more quickly, but suffer more herbivory, and so a balance is achieved. But various studies show that the species that accumulate dry

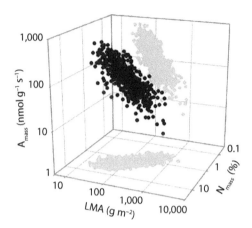

FIGURE 3. The three-way relationships among three of the six variables used by Wright et al. (2004) to define the "worldwide leaf economics spectrum," based on data for 706 species; the direction of the data cloud in three-dimensional space is shown in the shadows projected on the wall and floor of the space. Reprinted by permission from Nature Publishing Group.

mass most quickly are not necessarily the ones richest in nitrogen, and there is often no correlation between foliar nitrogen concentration and losses to herbivores. More work is needed not only on the community ecology side but also on the physiology. Leaving aside the special case of plants with nitrogen-fixing bacteria in their roots or leaves, we do not know how some species accumulate more nitrogen per unit leaf mass than others. Is there a correlation with the maximum or achieved rate of uptake of nitrogen per unit area of root surface, or do nitrogen-rich species have higher ratios of root surface to root mass?

Understanding the Worldwide Leaf Economics Spectrum

Much attention has been paid to the now-established concept of the "worldwide leaf economics spectrum" (Wright et al. 2004). As illustrated in Figure 3, there are statistically significant linear relationships among various leaf attributes plotted on log–log scales. I have pointed out elsewhere (Grubb 2016) that the existence of variation by an order of magnitude in attribute y at a given value of x shows that we cannot understand the variety in nature by thinking in terms of "simple trade-offs," as advocated by some authors. For example, the maximum rate of assimilation per unit dry mass (A_{mass}) of leaf can vary by an order of magnitude at a leaf dry mass per unit fresh area (LMA) of 100 g m^{-2}, and the LMA can vary by an order of magnitude at a concentration 1% nitrogen in the leaf dry mass (N_{mass}). For the evolutionist there is the question of whether any sense can be made of the variation in terms of what is apparently advantageous for different kinds of plants defined by their phylogeny or their ecological distribution. Ideally, investigation of this question should be linked with work by mechanists who explain how the variation in y can possibly arise at a given value of x.

In my recent treatment (Grubb 2016) I have suggested various possible answers to the mechanistic question. For example, N_{mass} may be reduced at a given LMA by a leaf having a substantial covering of nonphotosynthetic hairs, the epidermis impregnated with a large amount of silica, a layer of nonphotosynthetic cells (sometimes thick-walled) under the epidermis, a layer of large water-storing colorless cells within the mesophyll, or substantial numbers of fibres or thick-walled nonphotosynthetic idioblasts of various shapes within the mesophyll (most often columnar or star-shaped). A_{mass} at a given N_{mass} can be lowered by a lower specific activity of the enzyme, ribulose bisphosphate carboxylase-oxygenase, catalyzing incorporation of CO_2, a lesser proportion of nitrogen in the chloroplasts being in that enzyme and more in the chlorophyll-protein complex (as in species growing habitually in deep shade), and the use of significant amounts of nitrogen in chemicals that are defensive against herbivores (e.g. alkaloids, cyanogenic glycosides, and glucosinolates) or in compatible osmotica used in maintaining the water content of the protoplasm in the leaves of plants on soils that are dry and/or salt-rich (e.g., glycine betaine and proline). What is crying out to be done is the study of habitat, leaf structure, and leaf chemistry for species that occur along a series of transects through the cloud of points in Figure 3, taken at right angles to the trend line and at different points along it.

Explaining Lower Maximum Instantaneous Rates of Photosynthesis in Species with Longer-Lived Leaves

The question of why species with longer-lived leaves have lower maximum instantaneous rates of net assimilation per unit dry mass did not attract attention until the 1980s, when it was first suggested that the reason is the need to invest a higher proportion of leaf nitrogen in defensive compounds (Grubb 1984, Field and Mooney 1986). There was then a delay of about 20 years until evidence of the kind of defence involved was published (Takashima et al. 2004)—not compounds in the protoplast but increased amounts of nitrogen in the cell walls of the mesophyll, which are significantly thicker in longer-lived leaves. Various studies have provided the following ranges: 0.1–0.2 μm for herbs, 0.15–0.3 μm for deciduous broad-leaved trees, and 0.25–0.5 μm for broad-leaved evergreens. I have reviewed the recent literature in this field briefly elsewhere (Grubb 2016).

Unknowns that remain are as follows. Precisely how do the thicker walls of the mesophyll provide increased defence? Is the nitrogen concentration in the thicker walls the same as in the thinner? What proportion of the reduction in maximum rate of net assimilation is simply a reflection of the longer diffusion path in water in the thicker cell walls? A very recent study has shown that the effect of greater wall thickness on diffusion can be about as large as that of the use of nitrogen in the walls (Onoda et al. 2017). There are several other unknowns about mesophyll resistance (a measure of the inability of the cells to use CO_2): length of diffusion path in the cytosol and chloroplast, abundance of aquaporins, involvement of carbonic anhydrase and specific activity of the CO_2-fixing enzyme ribulose bisphosphate carboxylase. These are being actively studied at the present time, and might throw up differences between longer- and shorter-lived leaves.

Testing the hypothesis that thicker cell walls in some sense protect the mesophyll seems to me difficult. Are leaf-mining insects really so widespread that defence against them is generally important, and would thicker walls really be protective? Do a large proportion of leaves suffer infections by fungi spreading through the mesophyll, and again would thicker walls help? Protection of the functioning of the mesophyll when the leaf is partially dried is a possibility. One possible scenario is as follows. The chloroplasts just inside the cell walls of fully expanded mesophyll cells, especially in the palisade, are closely spaced, and during cell shrinkage presumably become crowded with a proportion being forced inward into the cytoplasm; for these chloroplasts there could be a big increase in the resistance to diffusion of CO_2 toward them, diffusion in water being 10,000-fold slower in water than in air. A recent study (Scoffoni et al. 2014) comparing woody species with shorter-lived leaves (expected to have thinner mesophyll cell walls) and with longer-lived leaves (expected to have thicker mesophyll cell walls) found no marked differences in the rate of loss of cell volume per unit decrease in water potential above turgor loss point (i.e., when the walls cease to be stretched by water being absorbed into the protoplast inside). It remains possible that thicker-walled mesophyll cells are more resistant to shrinkage in response to decline in water potential below turgor loss point; more experiments are needed.

Development of Toughness While Leaves Are Still Expanding

A different area of ignorance about the cell walls of leaves has become apparent in the context of leaf toughness and herbivory. It had generally been thought that leaves cannot become markedly toughened until

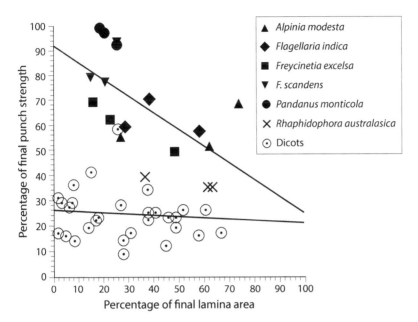

FIGURE 4. Percentage of final punch strength as a function of percentage final lamina area for individual leaves of ten species of dicots (open circles) and six species of monocots (different symbol for each species) in Australian tropical lowland rain forest. The regressions are $y = 90.6 - 0.664x$ ($r = 0.629$), and $y = 26.0 - 0.052x$ ($r = 0.105$). Reproduced from Figure 1, p. 1366, *Annals of Botany*, vol. 101, by Dominy, Grubb, Jackson, et al. (2008), by permission of Oxford University Press, http://global.oup.com.

they stop expanding, or—put another way—markedly toughened leaves cannot go on expanding (Coley and Kursar 1996). However, a surprising result was obtained during part of a study on herbivory and defence in tropical lowland rain forests comparing monocots (e.g., palms, pandans, aroids, gingers, heliconias, and broad-leaved shade-tolerant grasses) and the most speciose group of flowering plants, the eudicots. A study on monocot and dicot leaves at different stages of development found that when the leaves of monocots had reached only 10%–50% of final lamina area their punch strength (in Newtons) was already 40%–100% of the value for fully mature leaves, whereas with the same percentage of final area the dicots had only 10%–30% of the punch strength of fully mature leaves (Fig. 4). It seems that the very high toughness already present in young monocot leaves must arise from the cellulose in the bands of fibres, but new work in biochemistry is needed to explain how this toughness is compatible with expansion distributed across the whole lamina.

Looping back to the field situation, we find that in TLRF at a number of sites around the globe monocots suffer smaller losses of leaf area than neighboring dicots, and a much greater proportion of monocots have notably high toughness and/or rolled or folded young leaves, which are thus protected against herbivores—especially those that attack through the leaf margin (Grubb et al. 2008). However, it has not yet been shown critically that the greater toughness and/or leaf rolling and leaf folding are responsible for a lower rate of loss of leaf area rather than any differences in leaf chemistry. Perhaps experiments using dicot leaves perfused with potentially defensive chemicals from various monocots would help. Any experiment that involves unrolling young leaves of monocots to see whether herbivory is increased should be done under persistently shaded conditions so as to avoid complications caused by photodamage; unrolling adult leaves of the ginger plant *Amomum villosum* that had curled up in response to intense insolation led to increased photo-damage (Feng et al. 2002).

Conclusions

There is a need for evolutionists and mechanists to explain not just the overall broad trends in leaf size, shape, composition, structure, and functioning but also the contradictory findings in some interspecific comparisons and the great variation around trend lines found in others. How often is the "wrong" design actually adaptive under certain circumstances? How far does the wide scatter around some trendlines reflect benefits to plants that live under different conditions? Or do very many plant species evolve and persist with features that are in the minds of ecologists "suboptimal"? Not every species is widespread or locally dominant.

There are other unknowns different in kind, illustrated above: the question of whether or not the wide range in leaf nitrogen concentration in the leaves of one functional type of plant in one community type in one place plays any part in the maintenance of species richness, exactly how can thicker walls protect mesophyll cells in longer-lived leaves, and how can leaves with toughened walls still expand?

One point I made in 1977 certainly remains true. Researchers with many different interests and talents can make valuable contributions to dispersing the remaining clouds of ignorance around the structure and function of leaves.

Acknowledgments

I am indebted to Jamie Males for help in compiling the answers in Tables 1 and 2, Abbas Hashemi for making the figures, Charles Canham for help with respect to saplings in the "spring window," Lawren Sack on water relations, and the editors and an anonymous reviewer for very helpful suggestions.

References

Aasamaa, K., A. Sōber, and M. Rahi. 2001. Leaf anatomical characteristics associated with shoot hydraulic conductance, stomatal conductance and stomatal sensitivity to changes in leaf water status in temperate deciduous trees. Australian Journal of Plant Physiology 28:765–774.

Augspurger, C. K. 2008. Early spring leaf out enhances growth and survival of saplings in temperate deciduous forest. Oecologia 156:281–286.

Augspurger, C. K., J. M. Cheesman, and C. F. Salk. 2005. Light gains and physiological capacity of understorey woody plants during phenological avoidance of canopy shade. Functional Ecology 19:537–546.

Beaulieu, J. M., I. J. Leitch, S. Patel, A. Pendharkar, and C. A. Knight. 2008. Genome size is a strong predictor of cell size and stomatal density in angiosperms. New Phytologist 179:975–986.

Brodribb, T. J., T. S. Field, and G. J. Jordan. 2007. Leaf maximum photosynthetic rate and venation are linked by hydraulics. Plant Physiology 144:1890–1898.

Brodribb, T. J., N. M. Holbrook, M. A. Zwieniecki, and B. Palma. 2005. Leaf hydraulic capacity in ferns, conifers and angiosperms: impacts on photosynthetic maxima. New Phytologist 165:839–846.

Buckley, T. N. 2015. The contributions of apoplastic, symplastic and gas phase pathways for water transport outside the bundle sheath in leaves. Plant, Cell & Environment 38:7–22.

Buckley, T. N., G. P. John, C. Scoffoni, and L. Sack. 2015. How does leaf anatomy influence water transport outside the xylem? Plant Physiology 168:1616–1635.

Buckley, T. N., L. Sack, and M. E. Gilbert. 2011. The role of bundle sheath extensions and life form in stomatal responses to leaf water status. Plant Physiology 156:962–973.

Burd, M. 2007. Adaptive function of drip tips: a test of the epiphyll hypothesis in *Psychotria marginata* and *Faramea occidentalis* (Rubiaceae). Journal of Tropical Ecology 23:449–455.

Canham, C. D. 1988. Growth and canopy architecture of shade-tolerant trees: response to canopy gaps. Ecology 69:786–795.

Chabot, B. F., and D. J. Hicks. 1982. The ecology of leaf life spans. Annual Review of Ecology, Evolution, and Systematics 13:229–259.

Christophel, D. C., and B.P.M. Hyland. 1993. Leaf atlas of Australian tropical rain forest leaves. Melbourne: CSIRO Publications.

Coley, P. D., and T. A. Kursar. 1996. Anti-herbivore defenses of young tropical leaves: physiological constraints and ecological trade-offs. In: S. S. Mulkey, R. L. Chazdon, and A. P. Smith, eds. Tropical forest plant ecophysiology. New York: Chapman & Hall, 305–336.

Dominy, N. J., P. J. Grubb, R. V. Jackson, P. W. Lucas, D. J. Metcalfe, J.-C. Svenning, and I. M. Turner. 2008. In tropical lowland rain forests monocots have tougher leaves than dicots, and include a new kind of tough leaf. Annals of Botany 101:1363–1377.

Dominy, N. J., and P. W. Lucas. 2001. Ecological importance of trichromatic vision to primates. Nature 410:363–366.

Dominy, N. J., P. W. Lucas, L. W. Ramsden, P. Riba-Hernandez, K. E. Stoner, and I. M. Turner. 2002. Why are young leaves red? Oikos 98:163–167.

Eames, A. J., and L. H. MacDaniels. 1951. Introduction to plant anatomy. 2nd ed. New York: McGraw-Hill.

Edwards, G. E., V. R. Franceschi, and E. V. Voznesenskaya. 2004. Single cell C_4 photosynthesis versus the dual cell (Kranz) paradigm. Annual Review of Plant Biology 55:173–196.

Ehleringer, J. R., and I. N. Forseth. 1989. Diurnal leaf movements and productivity in canopies. In: G. Russell, B. Marshall, and P. G. Jarvis, eds. Plant canopies: Their growth, form and function. Cambridge: Cambridge University Press, 129–142.

Feild, T. S., and N. M. Holbrook. 2000. Xylem sap flow and stem hydraulics of the vesselless angiosperm *Drimys granadensis* (Winteraceae) in a Costa Rican elfin forest. Plant, Cell & Environment 23:1067–1077.

Feng, Y.-L., K.-F. Cao, and Z.-L. Feng. 2002. Thermal dissipation, leaf rolling and inactivation of PSII reaction centres in *Amomum villosum*. Journal of Tropical Ecology 18:865–876.

Field, C., and H. Mooney. 1986. The photosynthesis-nitrogen relationship in wild plants. In: T. J. Givnish, ed., On the economy of form and function. Cambridge: Cambridge University, 25–55.

Flenley, J. R. 1992. Ultraviolet-B insolation and the altitudinal forest limit. In: P. A. Furley, J. Proctor, and J. A. Ratter, eds. Nature and dynamics of forest-savanna boundaries. London: Chapman & Hall, 273–282.

Gill, D. S., J. S. Amthor, and F. H. Bormann. 1998. Leaf phenology, photosynthesis, and the persistence of saplings and shrubs in a mature northern hardwood forest. Tree Physiology 18:281–289.

Givnish, T. J. 1984. Leaf and canopy adaptations in tropical forests. In: E. Medina, H. A. Mooney, and C. Vázquez-Yánes, eds. Tasks for vegetation science. Vol. 12. Physiological ecology of plants of the wet tropics. Dordrecht: Dr W. Junk. 51–84.

Givnish, T. J., and G. J. Vermeij. 1976. Sizes and shapes of liane leaves. The American Naturalist 110:743–778.

Goldsmith, G. R., L. B. Bentley, A. Shenkin, N. Salinas, B. Blonder, R. E. Martin, R. Castro-Cosco, P. Chambi-Porroa, S. Diaz, B. J. Enquist, G. P. Asner, and Y. Malhi. 2017. Variation in leaf wettability traits along a tropical montane elevation gradient. New Phytologist 214:989–1001. doi: https://doi.org/10.1111/nph.14121.

Grubb, P. J. 1977a. Leaf structure and function. In: R. Duncan, and M. Weston-Smith, eds. The encyclopaedia of ignorance, Vol. 2. Life sciences and earth sciences. Oxford: Pergamon, 317–330.

Grubb, P. J. 1977b. The control of forest growth and distribution on wet tropical mountains: with special reference to mineral nutrition. Annual Review of Ecology, Evolution, and Systematics 8:83–107.

Grubb, P. J. 1984. Some growth points in investigative plant ecology. In: J. H. Cooley, and F. B. Golley, eds. Trends in ecological research for the 1980s. New York: Plenum, 51–74.

Grubb, P. J. 1986. Sclerophylls, pachyphylls and pycnophylls: The nature and significance of hard leaf surfaces. In: B. Juniper and R. Southwood, eds. Insects and the plant surface. London: Edward Arnold, 137–150.

Grubb, P. J. 1988. A reassessment of the strategies of plants which cope with shortages of resources. Perspectives in Plant Ecology, Evolution and Systematics 1:3–31.

Grubb, P. J. 2002. Leaf form and function—towards a radical new approach [commentary]. New Phytologist 155:317–320.

Grubb, P. J. 2003. Interpreting outstanding features of the flora and vegetation of Madagascar. Perspectives in Plant Ecology, Evolution and Systematics 6:125–146.

Grubb, P. J. 2016. Trade-offs in interspecific comparisons in plant ecology and how plants overcome proposed constraints. Plant Ecology & Diversity 9:3–33.

Grubb, P. J., P. J. Bellingham, T. S. Kohyama, F. I. Piper, and A Valido. 2013. Disturbance regimes, gap-demanding trees and seed mass related to tree height in warm temperate rain forests worldwide. Biological Reviews 88:701–744.

Grubb, P. J., and R. V. Jackson. 2007. The adaptive value of young leaves being tightly folded or rolled on monocotyledons in tropical lowland rain forest: an hypothesis in two parts. Plant Ecology 192:317–327.

Grubb, P. J., R. V. Jackson, I. M. Barberis, J. N. Bee, D. A. Coomes, N. J. Dominy, M.A.S. De la Fuente, P. W. Lucas, D. J. Metcalfe, J.-C. Svenning, I. M. Turner, and O. Vargas. 2008. Monocot leaves are eaten less than dicot leaves in tropical lowland rain forests: correlations with toughness and leaf presentation. Annals of Botany 101:1379–1389.

Grubb, P. J., and P. F. Stevens. 1985. The forests of the Fatima Basin and Mt Kerigomna, Papua New Guinea with a review of montane and subalpine rainforests in Papuasia. Department of Biogeography and Geomorphology Publication BG/5, Australian National University, Canberra.

Haberlandt, G. 1914. Physiological plant anatomy. London: Macmillan.

Jensen, K. H., and M. A. Zwieniecki. 2013. Physical limits to leaf size in tall trees. Physical Review Letters 110:018104

Jordan, G. J., J. Gregory, R. J. Carpenter, A. Koutoulis, A. Price, and T. J. Brodribb. 2015. Environmental adaptation in stomatal size independent of the effects of genome size. New Phytologist 205:608–617.

Kramer, P. J. 1969. Plant and soil water relationships: A modern synthesis. New York: McGraw-Hill.

Lohmeier, C., R. Pajor, M. R. Lundgren, A. Mathers, J. Sloan, M. Bauch, et al. 2017. Cell density and air space patterning in the leaf can be manipulated to increase leaf photosynthetic capacity. The Plant Journal 92:981–994.

Lucas, P. W., I. M. Turner, N. J. Dominy, and N. Yamashita. 2000. Mechanical defences to herbivory. Annals of Botany 86:913–920.

Ludwig, M., S. von Cammerer, G. D. Price, M. R. Badger, and R. T. Furbank. 1998. Expression of tobacco carbonic anhydrase in the C_4 dicot *Flaveria bidentis* leads to increased leakiness of the bundle sheath and a defective CO_2 concentrating mechanism. Plant Physiology 117:1071–1081.

Malhado, A.C.M., Y. Mahli, R. J. Whittaker, R. J. Ladle, H. ter Steege, N. N. Fabre, O. Phillips, W. F. Laurance, L.E.O.C. Aragao, et al. 2012. Drip-tips are associated with intensity of precipitation in the Amazon rain forest. Biotropica 44:728–737.

Mithoefer, A., and W. Boland. 2012. Plant defense against herbivores: chemical aspects. Annual Review of Plant Biology 63:431–450.

Monk, C. D. 1966. An ecological significance of evergreenness. Ecology 47:504–505.

Muir, C. D. 2015. Making pore choices: repeated regime shifts in stomatal ratio. Proceedings of the Royal Society B: Biological Sciences 282:20151498.

Onoda, Y., I. J. Wright, J. R. Evans, K. Hikosaka, K. Kitajima, Ü. Niinemets, H. Poorter, T. Towns, and M. Westoby. 2017. Physiological and structural tradeoffs underlying the leaf economics spectrum. New Phytologist 214:1447–1463.

Panditharathna, P.A.K.A.K., B.M.P. Singhakumara, H. P. Griscom, and M. S. Ashton. 2008. Change in leaf structure in relation to crown position and size class for tree species within a Sri Lankan tropical rain forest. Botany-Botanique 86:633–640.

Poorter, H., A. M. Jagodinsky, R. Ruiz-Peinado, S. Kuyah, Y. Luo, J. Oleksyn, V. A. Usoltsev, T. N. Buckley, P. B. Reich, and L. Sack. 2015. How does biomass distribution change with size and differ among species? An analysis for 1200 plant species from five continents. New Phytologist 208:736–749.

Queenborough, S. A., M. R. Metz, R. Valencia, and S. J. Wright. 2013. Demographic consequences of chromatic leaf defence in tropical tree communities: Do red young leaves increase growth and survival? Annals of Botany 112:677–684.

Richards, P. W. 1996. The tropical rain forest. 2nd ed. Cambridge: Cambridge University Press.

Robberecht, R., and M. M. Caldwell. 1978. Leaf epidermal transmittance of ultraviolet-radiation and its implications for plant sensitivity to ultraviolet-radiation induced injury. Oecologia 32:277–287.

Rosumek, F. B., F.A.O. Silveira, F. de S. Neves, N. P. de U. Barbosa, L. Diniz, Y. Oki, F. Pezzini, G. Wildon Fernandes, and T. Cornelissen. 2009. Ants on plants: A meta-analysis of the role of ants as plant biotic defences. Oecologia 160:537–549.

Roth-Nebelsick, A., V. Fernández, J. J. Peguero-Pina, D. Sancho-Knapik, and E. Gil-Pelegrín. 2013. Stomatal encryption by epicuticular waxes as a plastic trait modifying gas exchange in a Mediterranean evergreen species (Quercus coccifera L.). Plant, Cell & Environment 36:579–589.

Sack, L., and N. M. Holbrook. 2006. Leaf hydraulics. Annual Review of Plant Biology 57:361–381.

Sack, L., P. D. Cowan, N. J. Jaikumar, and N. M. Holbrook. 2003. The 'hydrology' of leaves: coordination of structure and function in temperate woody species. Plant, Cell & Environment 26:1343–1356.

Sack, L., C. Scoffoni, G. P. John, H. Poorter, C. M. Mason, R. Mendez-Alonzo, and L. A. Donovan. 2013. How do leaf veins influence the worldwide leaf economic spectrum? Review and synthesis. Journal of Experimental Botany 64:4053–4080.

Scoffoni, C., C. Vuong, S. Diep, H. Cochard, and L. Sack. 2014. Leaf shrinkage with dehydration: coordination with hydraulic vulnerability and drought tolerance. Plant Physiology 164:1772–1778.

Sisó, S., J. J. Camarero, and E. Gil-Pelegrin. 2001. Relationship between hydraulic resistance and leaf morphology in broadleaf Quercus species: A new interpretation of leaf lobation. Trees: Structure and Function 15:341–345.

Sumner, P., and J. D. Mollon. 2003. Did primate trichromacy evolve for frugivory or folivory? In: J. D. Mollon, J. Pokorny, K. Knoblauch, eds. Normal and defective colour vision. Oxford: Oxford University, 21–30.

Taiz, L., and E. Zeiger, eds. 2010. Plant physiology. 5th ed. Sunderland, Mass.: Sinauer Associates.

Takashima, T., K. Hikosaka, and T. Hirose. 2004. Photosynthesis or persistence: nitrogen allocation in leaves of evergreen and deciduous Quercus species. Plant, Cell & Environment 27:1047–1054.

Tellez, P., E. Rojas, and S. Van Bael. 2016. Red coloration in young tropical leaves associated with reduced fungal pathogen damage. Biotropica 48:150–153.

Usoltsev, V. A., T. N. Buckley, P. B. Reich, and L. Sack. 2015. How does biomass distribution change with size and differ among species? An analysis for 1200 plant species from five continents. New Phytologist 208:736–749.

Vitousek, P. M., and R. L. Sanford. 1986. Nutrient cycling in moist tropical forest. Annual Review of Ecology, Evolution, and Systematics 17:137–167.

Walter, H. 1968. Die Vegetation der Erde in ökophysiologischer Betrachtung, Vol. 2. Die gemässigten und arktischen Zonen. Jena, Germany: Gustav Fischer.

Wilkinson, H. P. 1979. The plant surface (mainly leaf): Part I, Stomata. In: Anatomy of the dicotyledons. 2nd ed. Vol. 1. Systematic anatomy of leaf and stem, with a brief history of the subject. Oxford: Clarendon Press, 97–165.

Wright, I. J., P. B. Reich, and M. Westoby. 2001. Strategy shifts in leaf physiology, structure and nutrient content between species of high- and low-rainfall and high- and low nutrient habitats. Functional Ecology 15:423–434.

Wright, I. J, P. B. Reich, M. Westoby, D. D. Ackerly, Z. Baruch, F. Bongers, J. Cavender-Bares, T. Chapin, J. H. C. Cornelissen, M. Diemer, et al. 2004. The worldwide leaf economics spectrum. Nature 428:821–827.

Zsögön, A., A. C. Negrini, L. E. Peres, H. T. Nguyen, and M. C. Ball. 2015. A mutation that eliminates bundle sheath extensions reduces leaf hydraulic conductance, stomatal conductance and assimilation rates in tomato (Solanum lycopersicum). New Phytologist 205:618–626.

Zwieniecki, M. A., and C. K. Boyce. 2014. Evolution of a unique anatomical precision in angiosperm leaf venation lifts constraints on vascular plant ecology. Proceedings of the Royal Society B: Biological Sciences 281:20132829.

PART III

COEXISTENCE

The Dimensions of Species Coexistence

Jonathan M. Levine and Simon P. Hart

Over the last several decades, ecologists' understanding of species coexistence has crystalized. The maturation of the field did not occur because of the discovery of important new theoretical coexistence mechanisms or a major empirical endeavor. Instead, the near-simultaneous publication of Chesson's Annual Review (2000a), and Hubbell's (2001) unified neutral theory forced ecologists to reconsider how we organize our thinking on the maintenance of species diversity. The framework for understanding coexistence that emerged from this period has proven quite powerful, but also exposes fundamental uncertainty about the spatial and temporal scales of coexistence in nature, and how coexistence depends on the complexity of the component interactions. Here, we introduce a series of relationships between coexistence and these three major dimensions—space, time, and interaction complexity—that form the basis of a theoretically grounded empirical research agenda for the coming decades.

Interest in species coexistence, or the maintenance of species diversity, is as old as ecology itself (Kingsland 1991). The basic problem—how to explain the coexistence of species despite the fact that some competitors are better than others—emerges in Darwin (1859) but is most cleanly framed by Hutchinson (1961) a century later. The answer to this problem, at least since the time of Darwin, has been that species coexist because they interact with the environment in different ways, causing each to be limited by different factors. Over the twentieth century, this principle was formalized and elaborated on in a large body of mathematical models and experiments. This work included quantitative theory and empirical validation of resource competition dynamics (MacArthur 1970, Tilman 1982), elegant models of fluctuation-dependent (Chesson and Warner 1981, Armstrong and McGehee 1980) and predator-mediated (Holt et al. 1994) coexistence and clever empirical tests of these and other processes (Paine 1966, Sousa 1979). Nonetheless, by the end of the century, there were, by some counts, more than 100 theories for the maintenance of species diversity (Palmer 1994), and the field was ripe for synthesis.

The rise of neutral theory (Hubbell 2001), which posited that species coexistence could be reasonably explained by the equivalence of competitors, forced ecologists to think more precisely about how species differences influence the outcome of competition. For many, this thinking was shaped by the publication of Chesson's (2000a) Annual Review, which concisely explained how the outcome of competition depends on the relative magnitude of niche differences that stabilize coexistence, and average fitness differences that drive competitive exclusion (Adler et al. 2007, Levine and Hille Ris Lambers 2009). Importantly, Chesson (2000a) showed how stabilizing niche differences, those between-species trade-offs that cause intraspecific limitation to exceed interspecific limitation, can be categorized into six classes of mechanism (Table 1). These include variation-independent mechanisms that operate in systems with no extrinsic spatial or temporal heterogeneity, and variation-dependent mechanisms that require species-specific interactions with environmental heterogeneity in space or time (Table 1). This framework is synthetic because nearly all previously described theoretical mechanisms for stable coexistence can be organized into the six categories, clarifying both their common foundations and interrelationships.

We believe that ecology is now at a point where our theoretical understanding of how coexistence works has matured. Further theoretical work resting on the traditional assumptions of pairwise interactions and stationary conditions will certainly add depth to our knowledge but will more than likely prove to be specific cases of the mechanisms laid out by Chesson (2000a). By contrast, ecologists' modern framing of the coexistence problem has exposed fundamental empirical questions that remain unanswered, questions relevant to coexistence in any community, and with answers potentially general across habitats: (1) To what extent is the coexistence we observe in nature derived from spatial environmental variation, temporal environmental variation, or variation-independent mechanisms? (2) What are the spatial and temporal scales of species coexistence? (3) How does coexistence depend on complex species interactions that only emerge in diverse communities?

Answering these questions requires empirically measuring coexistence along three major dimensions: space, time, and interaction complexity. Here, we introduce three curves that allow ecologists to do so. We first describe the theoretical foundations of these relationships, and then use the curves to frame an empirical research agenda. Our approach, which exploits the broad classes of mechanism outlined by Chesson (2000a) (Table 1), represents a departure from more traditional investigations of individual coexistence mechanisms such as resource partitioning or specialist-enemy control. The reason is one of practicality: Questions about the dimensions of coexistence involve the aggregate effect of individual mechanisms. Very few ecologists are in a position to quantify the separate contributions of multiple coexistence mechanisms in a field system, nor is this level of resolution necessarily required for addressing the specific questions outlined here.

Table 1. Mechanisms Stabilizing Species Coexistence Following Chesson (2000a)

Variation-independent mechanisms: All mechanisms that stabilize coexistence that do not require spatial or temporal environmental heterogeneity, such as resource partitioning.

Variation dependent mechanisms: Temporal

Temporal storage effect: Species-specific responses to environmental fluctuations combined with a life stage that is invulnerable to competition allow species that drop to low density to avoid competition under their favored environmental conditions, a benefit not experienced when species are common.

Temporal relative nonlinearity: Species differences in the nonlinearity of their growth response to a fluctuating environmental factor cause them to differentially respond to and affect those fluctuations. Coexistence is favored when each species promotes a regime of environmental fluctuations that is more favorable to their competitor than to themselves.

Variation dependent mechanisms: Spatial

Spatial storage effect: Species-specific responses to spatial environmental variation allow species that drop to low density to avoid competition in their most favored locations, a benefit not experienced when species are common.

Spatial relative nonlinearity: Species differences in the nonlinearity of their growth response to a spatially varying environmental factor cause them to differentially respond to, and affect, spatial environmental variation. Coexistence is favored when each species promotes a pattern of environmental variation more favorable to their competitor than themselves.

Fitness-density covariance: Species-specific responses to spatial environmental variation coupled with local dispersal allow species to accumulate in locations favorable to them, concentrating intraspecific relative to interspecific interactions in space.

The Spatial Scales of Species Coexistence

We would argue that a large fraction of the coexistence we observe in nature depends on spatial environmental heterogeneity at some scale. Countless observational studies have shown that different species are favored in different locations in spatially heterogeneous environments (Whittaker

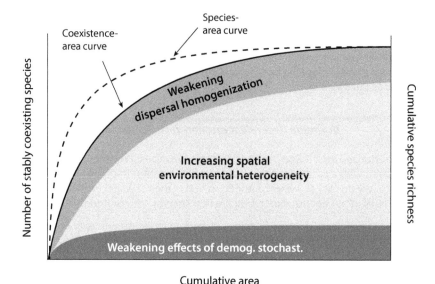

Cumulative area

FIGURE 1. A coexistence–area curve shows how the number of coexisting species increases with increasing area, resulting from the combined effects of increases in environmental heterogeneity, weakening of the homogenizing effects of dispersal, and weakening effects of demographic stochasticity. A species–area curve, where more species are found at each scale than can stably coexist at that scale, is shown for comparison.

1967, Chase and Leibold 2003). Moreover, we take as trivial the fact that species from different bioclimatic zones can coexist at continental scales (Holdridge 1947) yet spatial coexistence mechanisms are likely to be similarly powerful forces at much smaller scales as well (Silvertown et al. 1999). If spatial environmental heterogeneity is responsible for much of the coexistence we observe in nature, at what spatial scale does most of this coexistence arise? What fraction of the species in a meadow, for example, could coexist in 10 m² of meadow, or 50 m²? What fraction of the species in the meadow are only in that meadow due to coexistence mechanisms operating at even larger spatial scales than the meadow itself? Understanding how the number of *coexisting* species increases with area motivates our proposition for a *coexistence–area curve* (Hart et al. 2017). The coexistence–area curve quantifies the number of species that can coexist via mechanisms operating within a given area, and its curvature depicts the rate at which the number of coexisting species increases with area (Fig. 1).

Because of its clear theoretical underpinnings as detailed in the next section, the coexistence–area curve connects the mechanisms of species diversity maintenance to the species–area curve, one of the classic patterns in ecology (Chave et al. 2002). The main shortcoming of the species–area curve for the questions posed here is that the number of species found in a given area does not readily translate into the number of species that can stably coexist or even co-occur for long periods of time at that spatial scale (Laurance et al. 2011). Heterogeneity that allows for stable coexistence at larger spatial scales subsidizes more local diversity when dispersal allows for transient or sink populations at the local scale (Shmida and Ellner 1984). Therefore, particularly informative is the quantitative difference between the species–area curve and coexistence–area curve for a particular habitat (Fig. 1), which reveals the fraction of species found at a given spatial scale that is dependent on coexistence mechanisms operating at larger scales.

What Makes the Coexistence–Area Curve Rise?

Existing theory identifies three reasons that increasing spatial scale leads to greater coexistence (Hart et al. 2017): weakening effects of demographic stochasticity, increasing environmental variation, and weakening of the homogenizing effects of dispersal (Fig. 1). Demographic stochasticity is well known to act more powerfully on small populations, depleting species diversity when populations drift to extinction (Hubbell 2001, Vellend 2010). Thus, even if many species can coexist in a large, spatially homogeneous landscape, only a fraction of those species would persist in small areas, due to the overwhelming effects of demographic stochasticity (Tilman 2004). As area increases and more individuals per species are supported, this depressive effect of demographic stochasticity on coexistence declines. Though the contribution of weakening drift is likely to be greatest for the initial rise of the coexistence–area curve, the spatial scale at which these effects become negligible is unknown (Gilbert and Levine 2017).

The rise in the coexistence–area curve due to greater environmental variation present in larger areas (Fig. 1) almost certainly exceeds that due to weakening drift. Differences in how species respond to environmental variation can stabilize coexistence when these differences have the effect of concentrating intraspecific relative to interspecific interactions. In Chesson's (2000a) framework, this concentration can arise from the spatial storage effect, fitness-density covariance, or spatial relative nonlinearity (see Table 1) and is modulated by dispersal (Chesson 2000b, Snyder and Chesson 2003). Although environmental heterogeneity accumulating at different scales can promote diversity maintenance even with global dispersal (e.g., via spatial storage effects), dispersal generally erodes coexistence. It does so by forcing different species to interact, homogenizing intra- and interspecific interactions (weakening fitness-density covariance).

Heterogeneity at spatial scales small enough to be averaged over by dispersing organisms will therefore have weaker benefits for coexistence than the same heterogeneity at larger scales (Snyder and Chesson 2003). This interaction between dispersal and heterogeneity partly contributes to the rise of the coexistence–area curve because the larger the area, the greater the opportunity for heterogeneity on a scale exceeding the homogenizing capacity of dispersal.

The Temporal Scales of Species Coexistence

Just as we can ask how increasing environmental heterogeneity with area promotes species coexistence, we can ask the same with respect to time. Greater numbers of coexisting species with the greater temporal environmental variation expected over longer time series results from two mechanisms: the temporal storage effect and temporal relative nonlinearity (defined in Table 1). Temporal storage effects, which concentrate intraspecific relative to interspecific interactions in time, have been shown to contribute to coexistence in systems ranging from desert annuals to perennial grasslands to tropical forest tree communities (Angert et al. 2009, Usinowicz 2012, Chu and Adler 2015). Both temporal coexistence mechanisms (explained in Table 1) are potentially amplified by environmental variation that accumulates over time, generating a rising coexistence–time curve.

More formally, the coexistence–time curve (Fig. 2) describes how the number of coexisting species increases with the variability associated with increasingly long intervals of time. Of course, this curve needs to be defined at a given spatial scale and so the y-intercept indicates the number of species that coexist on variation-independent or purely spatial mechanisms of coexistence at that spatial scale. The y-intercept thus allows one to quantify how many species can, on average, coexist if the environmental conditions of any one year were to continue indefinitely into the future (no temporal variation). More interesting from a temporal perspective, is how many more species could coexist if the variable conditions experienced over two, three, or ten years, for example, were to repeat themselves. The curve approaches an asymptote at the number of species that coexist given the temporal variability experienced in a very long time series. The curve also therefore quantifies the fraction of total coexistence dependent on temporal environmental variation: It is simply the difference between the asymptotic species number and the y-intercept, divided by the asymptotic species number.

Though the coexistence–time relationship should always monotonically increase, its curvature quantifies the dependence of coexistence on infrequent events, a central unknown in the study of species diversity maintenance (Angert et al. 2009). For example, steeply rising and then saturat-

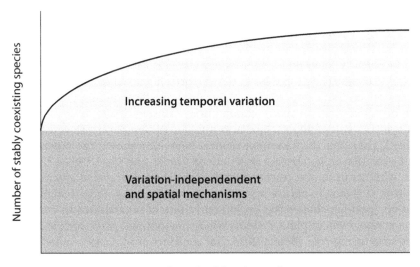

FIGURE 2. A coexistence–time curve shows how the number of coexisting species increases with the variability associated with increasingly long time intervals. The *y*-intercept shows the average number of species that coexist if the conditions experienced in any one year were to occur indefinitely; this coexistence results from variation-independent and spatial mechanisms of coexistence. The rise of the curve reflects the greater possibilities for coexistence with increasing environmental variation that occurs in increasingly long intervals of time.

ing curves indicate that most of the temporally mediated coexistence arises from the variability predictably occurring in just a few years of time. By contrast, more gradually rising curves would indicate that relatively rare climate events play important roles for long-term species coexistence.

What Are the Timescales of Extinction and Evolutionary Change in Competing Populations?

Fully considering the temporal scales of species coexistence requires quantifying how fast drift depletes the number of species we observe co-occurring today, and how evolution operating on ecological timescales shapes competitive outcomes, two processes ignored in our coexistence–time curve. With regards to the first point, mathematical theory on coexistence almost always considers populations that can drop to, and recover from, vanishingly small densities. But given the finite nature of individuals in real populations, ecologists know that in principle, even species with positive invasion growth rates eventually drift to extinction due to demographic stochasticity (Turelli 1980). What is not known, however, is over what timescales drift depletes the number of species we observe co-occurring in ecological communities today (Gilbert and Levine 2017).

If the timescales of competitive exclusion and ultimately local extinction are sufficiently long, then evolutionary changes that occur over the course of the ecological dynamic have the opportunity to alter competitive outcomes. Given that competition is the major force determining the ecological fate of populations in many communities (Harper 1977), it is difficult to believe that competition is not also a major contemporary selective force. Indeed, character displacement models have long shown the potential for trait evolution in competing populations (Taper and Case 1992). Nonetheless, whether evolution rescues populations otherwise headed towards competitive exclusion in nature remains poorly understood. Empirical evidence that competition shapes the evolutionary fate of populations in ecological time is rare, even in plant systems where competition plays such an important structuring role (Beans 2014; but see Hart et al. 2019). Whether this sparse evidence reflects the true irrelevance of evolution in ecological time, or simply the fact that few studies are designed to detect the role of contemporary evolution in shaping competitive dynamics, is unknown.

The Interaction Complexity Required for Species Coexistence

Although the study of coexistence is motivated by the large number of species around us, nearly all our understanding comes from studying competition between pairs of species in isolation (Levine et al. 2017). But how effectively can ecologists understand the coexistence of, say, 50 species based simply on pairwise niche differences between them? Are pairwise interactions even the driver of competitive outcomes in diverse systems where the interaction between species may depend on the presence of other species in the system (Billick and Case 1994)? Theory suggests two pathways by which the coexistence of two species is dependent on the presence of a third, fourth, or any number of additional competitors: interaction chains and higher-order interactions (Wootton 1993, Billick and Case 1994). Both of these pathways contribute to the complexity of interactions in species-rich systems.

Do Chains of Pairwise Competitive Interactions Stabilize Coexistence?

The outcome of competition between species depends not only on their direct interactions, but also their indirect interactions mediated by changes in the abundance of other species in the system (Levine 1976, Wootton 1993, Billick and Case 1994). Interaction complexity can thereby arise when pairwise interactions are embedded into networks of other (still pairwise) competitive interactions (Levine et al. 2017). Rock/paper/scissors-type dynamics between three competitors illustrate the potential effects of this interaction complexity: the coexistence of any two species depends on the interaction chain involving the third competitor (Kerr et al. 2002, Allesina and Levine 2011). But does adding species to communities necessarily benefit coexistence via the creation of more complex and less transitive interaction networks? It may, but any positive effect is likely to be bounded given that decades of niche theory (Case 1991) and a large empirical literature (Levine et al. 2004) show that higher species diversity typically makes it more difficult for additional species to coexist with residents. Nevertheless, the prevalence of indirect interaction chains, their dependence on species diversity, and their contribution to observed coexistence in nature remain poorly understood (Allesina and Levine 2011, Saavedra et al. 2017).

How Many Species Must Be Included in an Interaction to Explain Coexistence?

The second pathway by the which the coexistence of two species (or any subset of the community) depends on the presence of other competitors involves higher-order interactions (Wootton 1993, Billick and Case 1994, Levine et al. 2017). Here, each competitor modifies the per capita effects that other competitors have on one another. Put another way, the effects of multiple species on the growth of a focal species cannot be understood as the sum of independent effects of each competitor. The fundamental unit of competition is no longer the two-way "pairwise" interaction, but is instead the three-way, four-way, or even higher-order interaction that involves nonadditive per capita interaction terms. As with intransitive competition, higher-order interactions have been shown in theory to cause more diverse systems to be more stable (Bairey et al. 2016), illustrating their potential impact upon community dynamics. Abrams (1983) suggests that higher-order interactions frequently emerge under reasonable assumptions in mathematical models of competition, suggesting that these interactions are likely to be the rule rather than the exception in ecological systems. The question remaining is how these higher-order interactions contribute to species coexistence in nature.

We formalize this problem with a coexistence–interaction complexity curve (Fig. 3). This curve quantifies how the inclusion of interactions of increasingly higher order contribute to species coexistence. Envision a community containing n coexisting species. One can in principle ask how many species can coexist under the assumption that interactions are fundamentally pairwise, and then do the same under the assumption of three-way, four-way, and up to n-way interactions. The resulting coexistence–interaction complexity curve (Fig. 3) should generally

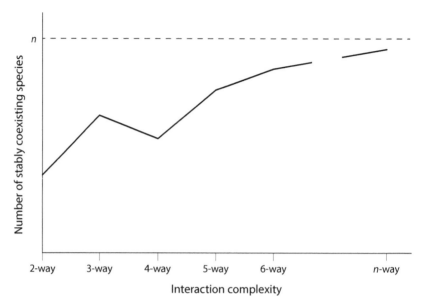

FIGURE 3. A coexistence–interaction–complexity curve shows the dependence of coexistence in a community on interactions of increasingly higher order. Given that the number of coexisting species in this curve is always a prediction from a field-parameterized model, even the highest-order interactions may not explain all the coexistence observed in a system (*n*) due to errors in model structure and fitted parameters.

approach its maximum value with *n*-way interactions, but the rate at which this value is approached is informative; the more gradual the rise, the more interaction complexity is necessary for coexistence. For an extreme example, a curve predicting the coexistence of all *n* species with just two-way (pairwise) interactions indicates that interaction complexity via higher-order interactions is simply unimportant for the observed level of coexisting species. We note that higher-order interactions may sometimes predict less coexistence than lower-order interactions (e.g., as illustrated from three-way to four-way interactions in Fig. 3), though the theoretical conditions under which this occurs requires further investigation.

Empirical Progress

Part of our motivation for introducing curves relating coexistence to area, time, and interaction complexity is to precisely frame questions about how each of these factors influences the maintenance of species diversity. In doing so, these curves also motivate a theoretically justified empirical research agenda to identify the spatial and temporal scales of species coexistence, and the role of interaction complexity in shaping diversity maintenance. Empirical progress towards achieving these aims can come from field-parameterizing mathematical competition models and then manipulating these models to evaluate how coexistence varies along the three dimensions identified here. The general approach we outline in the remainder of this section involves phenomenological models of competition, but these approaches are also compatible with more mechanistic approaches (Tilman 1982). As noted in the introduction, however, our sense is that the effort required to empirically parameterize multiple mechanistic competition models is beyond the scope of most individual research programs, and the inferences made would only reflect the consequences of the particular mechanisms studied.

The first step in developing a field-parameterized competition model for a given system involves fitting functions that quantitatively describe how individual demographic performance (e.g., growth, survival, fecundity) depends on the density and composition of one's neighbors (Levine et al. 2017, Hart et al. 2018). The requisite data can come from observational datasets quantifying the demographic performance of individuals, as long as these individuals experience significant variation in the identity and abundance of competitors. Alternatively, the same models can be fit by quantifying the demographic performance of individuals subjected to an experimentally imposed gradient in competitor density. With the fitted model one then simulates system dynamics with individual-based or integral projection (for size-structured populations) approaches. Further details on simulating competition models fit to empirical data are available in Adler et al. (2010). In what follows we describe how this general approach can be tailored to quantify coexistence–area, –time, and –interaction complexity curves for a given habitat.

Although the processes governing the rise of the coexistence–area curve are well-understood theoretically, the empirical challenge of generating such a curve is great (Hart et al. 2017). In principle, the approach of classic habitat fragmentation studies (Laurance et al. 2011), where areas of differing size are isolated from their surroundings, could be used to evaluate how many species can coexist as a function of area. However, such an approach is unlikely to be feasible as one needs to somehow account for edge effects and wait the many generations necessary to see the exclusion and drift processes play out. Fortunately, this experiment can be simulated with the field-parameterized competition models outlined in the prior paragraph. By fitting these models to spatially mapped data and allowing for x- and y-coordinate variation in demographic rates (that reflects underlying environmental heterogeneity), one can produce a spatial map of fitted vital rates and/or interaction coefficients for multiple species in a study area. Species dynamics and coexistence can then be simulated based on the spatial map of these rates observed at the 1-m^2 scale, 10-m^2 scale, and so on. Obtaining the required data would of course represent a tremendous empirical effort, and would further require estimates of species' dispersal kernels, but the payoff is an understanding of the scales at which diversity is maintained.

The same general approach can be used to produce a coexistence–time curve, but here, rather than subsetting the area of a habitat for the simulations, one would subset the time series of year-specific demographic and competitive rates. Specifically, one would simulate the number of coexisting species if the environmental conditions of any one year were to continue indefinitely into the future (as in Adler et al. 2006) anchoring the y-intercept of the curve. One would then quantify the number of species that can coexist in simulations of community dynamics, where yearly sets of parameters are drawn from increasingly long intervals of time, and then sampled indefinitely. Such approaches also lend themselves to understanding the role of demographic stochasticity if one simulates populations that are composed of discrete numbers of individuals (Adler et al. 2010). Moreover, the area-dependence of this effect, driven by the greater number of individuals in larger areas, can be used to quantify the contribution of demographic stochasticity to the coexistence–area curve.

Finally, the same general approach can be used to empirically describe the coexistence–interaction complexity curve (relating to higher-order interactions). Doing so is an enormous challenge due to the overwhelming number of experimental treatments necessary to quantify higher-order interactions in the diverse communities that motivate the problem. However, quantifying the improvement in explanatory power from even two-way to three-way interactions would represent a major advance, especially if we expect diminishing coexistence returns from increasingly higher-order interactions. This advance is achievable by first fitting models that allow for only two-way interactions between species, and then doing the same allowing for three-way interactions (and so on if the system allows). The appropriate data for these fits are most feasibly obtained by measuring the demographic response of individuals to orthogonal manipulations of the density of multiple competitors (Weigelt et al. 2007). With fitted pairwise and higher-order interaction terms, one can then simulate the number of coexisting species allowing interactions of increasingly higher order, which is the metric of interaction complexity along the x-axis of Figure 3.

Conclusions

For some, the past efforts of ecologists to explain species coexistence have epitomized the failure of our field to generate testable hypotheses and general principles (Lawton 1994, Palmer 1994, Hubbell 2001). This could not be further from the current truth, as ecologists now understand how the fundamental processes maintaining species diversity work and combine to generate coexistence. We know, for example, the theoretical rules required to maintain species diversity via competition colonization trade-offs (Tilman 1994, Adler and Mosquera 2000), spatial environmental heterogeneity (Chesson 2000b), and predator tolerance–competitive ability trade-offs (Holt et al. 1994). The major outstanding questions are not theoretical, but empirical. Moreover, by identifying broad classes of coexistence mechanisms that work in similar ways, as in the Chesson (2000a) framework, empirical ecologists can pose questions in a manner grounded in general theory. With this mindset, we have proposed relationships between coexistence and three major dimensions—space, time, and interaction complexity—as a foundation for progress towards addressing the major scientific unknowns in this dynamic area of research. Undoubtedly, the species diversity that ecologists aim to explain presents a daunting challenge, but empirical ecologists studying coexistence are now positioned to make advances that match those of past theory.

Acknowledgments

We thank the ETH Zurich Plant Ecology group and D. Tilman for comments on the chapter. Support was provided by ETH Zurich and U.S. NSF grant DEB 1644641.

References

Abrams, P. A. 1983. Arguments in favor of higher-order interactions. *American Naturalist* 121:889–891

Adler, F. R., and J. Mosquera. 2000. Is space necessary? Interference competition and limits to biodiversity. Ecology 81:3226–3232.

Adler, P. B., and S. P. Ellner, J. M. Levine. 2010. Coexistence of perennial plants: An embarrassment of niches. Ecology Letters 13:1019–1029.

Adler, P. B., J. Hille Ris Lambers, P. Kyriakidis, Q. Guan, and J. M. Levine. 2006. Climate variability has a stabilizing effect on the coexistence of prairie grasses. Proceedings of the National Academy of Sciences of the United States of America 103:12793–12798.

Adler, P. B., J. Hille Ris Lambers, and J. M. Levine. 2007. A niche for neutrality. Ecology Letters 10:95–104.

Allesina, S., and J. M. Levine. 2011. A competitive network theory of species diversity. Proceedings of the National Academy of Sciences of the United States of America 108:5638–5642.

Angert, A. L., T. E. Huxman, P. Chesson, and D. L. Venable. 2009. Functional trade-offs determine species coexistence via the storage effect. Proceedings of the National Academy of Sciences of the United States of America 106:11641–11645.

Armstrong, R. A., and R. McGehee. 1980. Competitive exclusion. American Naturalist 115:151–170.

Bairey, E., E. D. Kelsic, and R. Kishony. 2016. High-order species interactions shape ecosystem diversity. Nature Communications 7:12285.

Beans, C. M. 2014. The case for character displacement in plants. Ecology and Evolution 4:852–865.

Billick, I., and T. J. Case. 1994. Higher order interactions in ecological communities: What are they and how can they be detected? Ecology 75:1529–1543.

Case, T. J. 1991. Invasion resistance, species build-up and community collapse in metapopulation models with interspecies competition. Biological Journal of the Linnean Society 42:239–266.

Chase, J. M., and M. A. Leibold. 2003. Ecological Niches: Linking classical and contemporary approaches. Chicago: University of Chicago Press.

Chave, J., H. Muller-Landau, and S. Levin. 2002. Comparing classical community models: theoretical consequences for patterns of diversity. American Naturalist 159:1–23.

Chesson, P. 2000a. Mechanisms of maintenance of species diversity. Annual Review of Ecology and Systematics 31:343–366.

Chesson, P. 2000b. General theory of competitive coexistence in spatially varying environments. Theoretical Population Biology 58:211–237.

Chesson, P. L., and R. R. Warner. 1981. Environmental variability promotes coexistence in lottery competitive systems. American Naturalist 117:923–943.

Chu, C., and P. B. Adler. 2015 Large niche differences emerge at the recruitment stage to stabilize grassland coexistence. Ecological Monographs 85:373–392.

Darwin, C. 1859. The origin of species. New York: Modern Library.

Gilbert, B., and J. M. Levine. 2017. Ecological drift and the distribution of diversity. Proceedings of the Royal Society 284:20170507.

Harper, J. L. 1977. The population biology of plants. New York: Academic Press.

Hart, S. P., R. P. Freckleton, and J. M. Levine. 2018. How to quantify competitive ability. Journal of Ecology 106:1902–1909.

Hart, S. P., M. M. Turcotte, and J. M. Levine. 2019. The effects of rapid evolution on species coexistence. Proceedings of the National Academy of Sciences of the United States of America. doi: https://doi.org/doi/10.1073/pnas.1816298116.

Hart, S. P., J. Usinowicz, and J. M. Levine. 2017. The spatial scales of species coexistence. Nature Ecology and Evolution 1:1066–1073.

Holt, R. D., J. Grover, and D. Tilman. 1994. Simple rules for interspecific dominance in systems with exploitative and apparent competition. American Naturalist 144:741–771.

Holdridge, L. R. 1947. Determination of world plant formations from simple climatic data. Science 105:367–8.

Hubbell, S. P. 2001. A unified neutral theory of biodiversity and biogeography. Princeton, N.J.: Princeton University Press.

Hutchinson, G. E. 1961. The paradox of the plankton. American Naturalist 95:137–145.

Kerr, B., M. A. Riley, M. W. Feldman, and B.J.M. Bohannan. 2002. Local dispersal promotes biodiversity in a real-life game of rock-paper-scissors. Nature 418:171–174.

Kingsland, S. E. 1991. Defining Ecology as a Science. In: L. A. Real, and J. H. Brown, eds. Foundations of ecology: Classic papers with commentaries. Chicago: University of Chicago Press, 1–13.

Lawton, J. H. Are there general laws in ecology? Oikos 84:177–192.

Laurance, W. F., J.L.C. Camargo, R.C.C. Luizão, S. G. Laurance, S. L. Pimm, E. M. Bruna, et al. 2011. The fate of Amazonian forest fragments: a 32-year investigation. Biological Conservation 144:56–67.

Levine, J. M., P. B. Adler, and S. G. Yelenik. 2004. A meta-analysis of biotic resistance to exotic plant invasions. Ecology Letters 7:975–989.

Levine, J. M., J. Bascompte, P. B. Adler, and S. Allesina. 2017. Beyond pairwise mechanisms of species coexistence in complex communities. Nature 546:56–64.

Levine, J. M., and J. Hille Ris Lambers. 2009. The importance of niches for the maintenance of species diversity. Nature 461:254–257.

Levine, S. H. 1976. Competitive interactions in ecosystems. American Naturalist 110: 903–910.

MacArthur, R. 1970. Species packing and competitive equilibrium for many species. Theoretical Population Biology 1:1–11.

Paine, R. T. 1966. Food web complexity and species diversity. American Naturalist 100:65–75.

Palmer, M. W. 1994. Variation in species richness: towards a unification of hypotheses. Folia Geobotanica et Phytotaxonomica 29:511–530.

Saavedra, S., R. R. Rohr, J. Bascompte, O. Godoy, N. J. B. Kraft, and J. M. Levine. 2017. A structural approach for understanding multispecies coexistence. Ecological Monographs 87:470–486.

Silvertown, J., M. E. Dodd, D. J. G. Gowing, and J. O. Mountford. 1999. Hydrologically defined niches reveal a basis for species richness in plant communities. Nature 400: 61–63.

Shmida, A., and S. Ellner. 1984. Coexistence of plant species with similar niches. Vegetatio 58:29–55.

Snyder, R. E., and P. Chesson. 2003. Local dispersal can facilitate coexistence in the presence of permanent spatial heterogeneity. Ecology Letters 6:301–309.

Sousa, W. P. 1979. Disturbance in marine intertidal boulder fields: The nonequilibrium maintenance of species diversity. Ecology 60:1225–1239.

Taper, M. L., and T. J. Case. 1992. Models of character displacement and the theoretical robustness of taxon cycles. Evolution 46:317–333.

Tilman, D. 1982. Resource competition and community structure. Princeton, N.J.: Princeton University Press.

Tilman D. 1994. Competition and biodiversity in spatially structured habitats. Ecology 75:2–16.

Tilman D. 2004. Niche trade-offs, neutrality, and community structure: A stochastic theory of resource competition, invasion, and community assembly. Proceedings of the National Academy of Sciences of the United States of America 101:10857–10861.

Turelli, M. 1980. Niche overlap and invasion of competitors in random environments: II, The effects of demographic stochasticity. In: W. Jäger, H. Rost, and P. Tautu, eds. Biological growth and spread: mathematical theories and applications. Berlin: Springer Nature.

Usinowicz, J., S. J. Wright, and A. R. Ives. 2012. Coexistence in tropical forests through asynchronous variation in annual seed production. Ecology 93:2073–2084.

Vellend, M. 2010. Conceptual synthesis in community ecology. Quarterly Review of Biology 85:183–206.

Weigelt, A., J. Schumacher, T. Walther, M. Bartelheimer, T. Steinlein, and W. Beyschlag. 2007. Identifying mechanisms of competition in multi-species communities. Journal of Ecology 95: 53–64.

Whittaker, R. H. 1967. Gradient analysis of vegetation. Biological Reviews 42:207–264.

Wootton, J. T. 1993. Indirect effects and habitat use in an intertidal community: interaction chains and interaction modifications. American Naturalist 141:71–89.

Evolution, Speciation, and the Persistence Paradox

Andrew R. Tilman and David Tilman

Contrary to what one might imagine the effects of natural selection to be, the fossil record repeatedly shows that the evolution of new species (Benton 1995, 1996), and the arrival of novel species from other realms (Vrba 1992, Webb 2006, Patzkowski and Holland 2007), has almost never led to the extinction of resident species. Rather, such novel taxa and existing resident species often persist with each other for a million or more years. This persistence is paradoxical precisely because selection should favor any traits that made a member of a species be a superior competitor or more resistant to predation or disease. No current evolutionary theory explains why and how natural selection may be constrained so as to lead to the repeated emergence species that are ecologically incapable of being sufficiently superior in their traits to cause the extinction of one or more of the established species with which they interact. The resolution of this paradox has the potential to simultaneously advance our understanding of evolution and ecology by explaining why and how the traits of species are constrained.

Persistence is so commonly observed both now and across periods of millions of years in the fossil record (e.g., Patzkowski and Holland 2007, Webb 2006, Pennington and Dick 2004, Vrba 1992), that it at times seems to be taken for granted. Most of us accept, as given, that lions, cheetahs, dogs, hyenas, and other predators; wildebeest, giraffe, zebra, elephants, buffalo, and other herbivores; and numerous species of grasses, forbs, shrubs, and trees all persist in an African savanna.

Might such persistence be simply the consequence of a survivorship bias, where the species occurring in a system are an idiosyncratic subset of those regional species that happened to be able to coexist, perhaps because of their order of appearance or other chance factors? Or, might there be deeper, more fundamental evolutionary forces that constrain speciation such that tradeoffs among the multiple traits of a novel species somehow assure that it cannot cause the extinction of existing species, be they its competitors, predators, or prey? The latter possibility seems more likely to us because, as we discuss below, both speciation and species migrations into new realms time and again have led to persistence, not extinction (Vermeij 1991, Flannery 2001). In this paper we illustrate the potential complexity of mathematically determining the suites of traits required for multiple species to persist on each of the trophic levels of a food web. We emphasize at the start the context of our discussion. For a given species, we are always referring to habitats that encompass the range of physical and biotic conditions where it can persist, and noting that its entrance into such habitats, whether by speciation or migration, is rarely associated with the extinction of any existing species.

Those who have mechanistically modeled multispecies persistence using consumer-resource models, or used field data to test such ideas, have gained significant insights into the food web tradeoffs that allow stable persistence of several species (e.g., Leibold 1989, Schmitz 1994, Holt et al. 1994, Polis and Strong 1996, McPeek 1996, McCann et al. 1998, Vos et al. 2004, Chesson and Kuang 2008, McPeek 2012, Barabás et al. 2017). However, the suite of trait-based tradeoffs that can allow stable persistence of the large numbers of species we frequently observe in food webs are yet to be discovered. Indeed, for such models it is easy to find species traits (i.e., model parameters) that allow a new species to invade a food web of coexisting species, but increasingly difficult, as the number of interacting species increases, to find the traits such that each additional predator, herbivore, or plant species persists while not causing the extinction of species on its or a different trophic level.

Nature seems to have no such difficulty. Species that evolved in one realm, separated from all other realms, at times have migrated into a new realm and the resident and invading species have consistently coexisted. Although one might imagine that new species would be superior in some way, and that their appearance might thus cause the extinction of some other species on its or another trophic level, this is rare in the fossil record. Rather, the observed persistence is so long—a million or more years—that it is difficult to assert that any changes observed later in the fossil were the result of the new species (e.g., Benton 1995, 1996, Vermeij 1991, Flannery 2001, Patzkowski and Holland 2007).

Clearly, mechanistic food web theory reveals that persistence requires interspecific tradeoffs, but the evolutionary basis for and mechanistic nature of such tradeoffs are poorly known. Other insights have come from analyses of the effects of food webs' connectance patterns (e.g., May 1973, Martinez 1992, Dunne et al. 2002), weak trophic interactions (McCann et al. 1998) and spatial coupling (McCann et al. 2005) on food web stability and persistence. Interestingly, the probability of stability in large food webs increases with the prevalence of stronger intraspecific density dependence (Barabás et. al. 2017). Similarly, tradeoffs between intraspecific density dependence and competitive ability for a limiting resource can facilitate persistence (McPeek 2012). These insightful results align with the qualitative generalization that persistence requires intraspecific limitation to outweigh interspecific limitation. However, there is, as of yet, no general theory of traits that cause i species of plants, j species of herbivores, and k species of predators to persist together for realistic values of i, j, and k. And no theory yet explains why natural selection seems to have consistently led to the emergence of such persisting species.

We hope that the ideas and evidence we present might stimulate the empirical and theoretical pursuit of a more general theory of the origins and nature of both speciation and multispecies persistence. Such a theory might be an important step toward a deeper unification of ecology and evolution.

Speciation and Persistence

At the most fundamental level, the long-term pattern of increasing biodiversity on Earth is prima facie evidence that the emergence of new species has not, on average, caused the extinction of existing species. For instance, 400 million years ago (mya) there were ~50 different families of terrestrial plants and animals. This increased to ~200 families by 200 mya, to ~500 families 100 mya, and to ~1400 families of terrestrial plants and animals by 5 mya (Benton 1995). Marine taxa had a similar pattern and rate of diversification, except that it began approximately 200 million years earlier (Benton 1995). The majority of extinctions that have occurred during these long periods of diversification were not associated with the emergence of new taxa, but rather were associated with exogenously driven mass extinction events (Raup and Sepkoski 1986, Benton 1995).

Patterns of diversification within taxonomic groups show that new and existing species coexisted for millions of years. For instance, the tetrapods increased in diversity from a single ancestral species to more than 20,000 species of birds, mammals, reptiles, and amphibians with little evidence in the fossil record that the emergence of either new taxonomic groups or new species within them led to the displacement of established species or groups (Benton 1996). Upon inspection, such persistence seems plausible if, as has been suggested, the new species that arose most often differed in their diets and/or habitats from established species (Benton 1996). Diversification of Darwin's finches provides much finer-scale evidence of ecology-based diet-driven niche differentiation, speciation, and resulting persistence (Grant 1986, Grant and Grant 2006). The radiations of large numbers of plant species within a region by a taxonomic group, such as the Dipterocarpaceae in Southeast Asia and the Proteaceae in South Africa, are further demonstrations of persistence, rather than competitive displacement, during the diversification of life.

Species Migration and Persistence

What might evolutionary theory lead us to expect to happen when ecologically similar but phylogenetically distinct taxa come into contact after millions of years of speciation and evolution in different biogeographic realms? Wouldn't novel mutations, recombination, pre-existing initial advantages associated with phylogeny, and all of the physical and biotic differences between two realms have tipped the scales of evolution in favor of species from one realm over those from the other realm during their millions of years of separate evolution? How could there have been such precisely convergent evolution in physically separated biomes that the phylogenetically distinct taxa would necessarily persist with, rather than displace, pre-existing species after migration into their realm? Imagine, for large, slow-growing animals, that natural selection had led an invading species from a different realm to have a competitive advantage over an ecologically similar established species that gave a relative growth-rate difference between two species of just 0.001 per generation (assumed here to be 30 years). This tiny competitive advantage would cause the competitively superior invading species to displace the other species in about 400,000 years, by which time the superior competitor would have increased by a factor of 10^6 relative to the inferior competitor. Calculations like these illustrate, for species that lack tradeoffs that cause stable persistence, how phenomenally similar competing species would have to be to coexist for the millions of years revealed in the fossil record in the absence of persistence-causing interspecific tradeoffs (Tilman 2011).

The outcomes of major terrestrial and marine interchanges that are recorded in the fossil record and of recent anthropogenic species movement, however, repeatedly show persistence of resident and invading species in food webs (Tilman 2011). During the past 500 years, for instance, thousands of novel plant species have been introduced into oceanic islands. Many of these species are now naturalized on particular islands. For 13 well-studied sets of islands, plant diversity has doubled due to such invasions, but only approximately 3% of the native plant species have, as of this time, gone extinct (Sax et al. 2003). New Zealand had approximately 2000 plant species become naturalized and is reported to have lost only three species. The greatest number of plant extinctions occurred on the Hawaiian Islands, which lost 71 of their original 1,294 plant species as 1,090 new species became naturalized. However approximately 400 more native Hawaiian plant species are listed as threatened or endangered with extinction, with exotic invaders being one of multiple risk factors. Large numbers of species of plants also have been introduced into new continents as part of the horticultural trade. Some of these eventually become naturalized, and a few become highly abundant, but extinction of the native plant species by invasive competitors has been extremely rare to date (Davis 2003). Similarly, experimental seed additions to established plant communities have repeatedly been found to increase plant diversity and to rarely cause the displacement of pre-existing species (Turnbull et al. 2000, Myers and Harms 2009, Tilman 1997).

Paleontological data on trans-realm species migrations show few if any signs of displacements based on studies of the Ordovician period (~450 mya) through the Tertiary period (~3 mya). An Ordovician invasion of a shallow midcontinent North American sea by corals, mollusks, and other benthic taxa increased species diversity by 40% but, a million years later, had not led to extinctions of pre-existing species (Patzkowski and Holland 2007). Similarly during the Tertiary period, greater than 200 mollusk species moved into the North Sea from the Pacific, but this massive increase in mollusk diversity was not associated with a shift in the background extinction rate of pre-existing North Sea mollusks (Vermeij 1989), leading Vermeij to conclude that extinction was "an unlikely consequence" of marine interchanges, a view supported by other marine studies (Gould and Callaway 1980, Lindberg 1991).

Similar patterns of invasion and persistence occurred during successive waves of migration of both herbivorous and predatory mammals between North America and Asia, and Asia and Africa, during the Miocene epoch, from approximately 24 to 5 mya (e.g., Flynn et al. 1991, Webb 2006, Benton 1996, Flannery 2001) and between North and South America from approximately 9 to 3 mya (Webb 1991, Vrba 1992). Whether the invaders were browsing or grazing herbivores, or scavengers or predators, and whether they were placentals or marsupials, with rare exception the pre-existing mammals of a realm persisted with the invaders for a million or more years. Similarly, large-scale movement of bird species between the two American continents did not lead to extinctions of existing bird species. Rather, in many cases, the novel taxa themselves radiated in the new realm, creating even more species (Webb 1991, 2006; Flannery 2001). Numerous plant species also migrated between realms during these geologic periods, persisting with the resident taxa in the regions they invaded (Goldblatt 1993; Morley 2003; Pennington and Dick 2004).

Although these multiple cases of post-invasion persistence are insightful, exploration of a series of related questions would let us better understand the context of these invasions and subsequent persistence. For instance, considering all the species in a realm, what are the traits of those species that did invade another realm, and of those that did not? How do the traits of the invaders relate to the traits of resident species in the invaded realm? Do successful invaders have novel suites of traits that allow them to exploit an "empty niche," thus experiencing little competition in the new realm and therefore not displacing other species? Of those species that enter a new realm, which taxa do or do not undergo rapid radiation, and why? And, in particular for the Great American Biotic Interchange, why did many of the North American mammalian invaders into South America undergo rapid speciation, whereas the South American invaders into North America rarely speciated? Webb (2006) offers the tantalizing possibility that the broad biogeographic origins of the northern invaders (having African, Eurasian, and North American origins) made them more evolutionarily successful than the native taxa of the south that had evolved in a much smaller region. Also, invasion occurs at low abundance where stochastic effects can dominate dynamics. Perhaps the species that did not successfully invade failed to do so for stochastic reasons, or perhaps some species migrated but were unable to coexist in a new food web.

Tradeoffs and Persistence in Food Webs

The competitive persistence observed in the fossil record as species radiated and as they moved from one realm to another might be explained by all competing species, from all major realms, being somehow bound to a universal interspecific competition tradeoff surface (Tilman 2011). We know that interspecific tradeoffs are required for competing species to coexist, whether it be among species of plants, or herbivores, or predators (MacArthur 1972, Tilman 1982, Tilman and Pacala 1993, Holt et al. 1994, Liebold 1996, Chesson and Huntly 1997). A recent review enumerated more than a dozen interspecific tradeoffs, each of which seems to explain particular cases of persistence of competing species (Kneitel and Chase 2003). These cases include tradeoffs between competition and colonization abilities; rapid growth rates and efficiency; the ability to forage for smaller versus larger seeds; the ability to compete for light versus soil resources, or for one kind of resource versus a second kind; competitive ability versus resistance to predation or disease; between having optimal growth at cooler versus warmer temperatures; and between the ability to live at consistently low resource levels or to exploit resource pulses.

However, speciation and persistence occur within food webs, and species encounter new food webs when migrating into a new realm. As noted by Chesson and Kuang (2008), competition and predation can be equally important in determining persistence. In a food web, species are impacted by the traits of the species they consume and the traits of the species that consume them, not just by their competitors. Plants, for instance, experience competitors as well as seed predators, herbivores, pathogens, and mutualists (or lack thereof) in a new realm.

When the isthmus of Panama formed approximately 3.5 mya, northern placental mammals and the southern marsupial mammals came into contact after more than 50 million years with minimal interchange. North American placental predators such as various species of dogs, skunks, weasels, cats, and bears entered South America and coexisted with its preexisting marsupial prey and predators (e.g., Marshall et al. 1982). Similarly, South American herbivores, such as porcupines, giant ground sloths, arboreal sloths, capybaras, and various ungulates and edentate mammals, encountered and persisted in the presence of North American herbivores and predators (Webb 1991).

These examples, and numerous similar cases, suggest that food webs must share remarkable functional similarity despite the differing phylogenetic origins and evolutionary histories of their species. These observations raise questions. How are the traits of each species evolutionarily constrained such that novel species that evolve in a food web or invade into a food web do not displace other species in that food web? Wouldn't a novel predator have top-down impacts on the full food web, and a novel plant have bottom-up effects on it?

If tradeoffs are, as we suggest, the underlying factor constraining speciation and causing multispecies persistence, the long-term persistence of resident species with species invading from other realms would seem to imply that food web tradeoffs should be universal, multidimensional, and simultaneously balance costs and benefits to competitive abilities, to predation and disease susceptibility, and to other environmental constraints. However, as of now we know little about food web tradeoffs.

Simple Models, Complex Outcomes

To explore what we knew and perhaps find what we had not expected, we modeled food webs with three trophic levels—plants, herbivores and predators—and a limiting plant resource (Box 1). Our model was a system of differential equations describing the dynamics of each variable—a limiting plant resource, two species of plants, an herbivore species, and a predator. Even this simple food web had surprising features. As expected, the two plant species were able to coexist when competing for a single limiting resource only with a tradeoff: the plant species that was a better resource competitor had to be more susceptible to herbivory (Chase et al. 2001). In particular, throughout our models, plant species 2, P_2, is totally resistant to herbivory whereas plant species 1, P_1, is susceptible to herbivory. For both plant species to persist when all plants are limited by the same single resource, the herbivore must be present and P_1 must be the superior competitor for the resource in the absence of the herbivore (Chase et al. 2001).

Box 1: Model Structure

Consider a simple model with three trophic levels and two plant species. There is a single resource, R that grows according to $f(R)$ and is depleted by the plants that require it. The first plant, P_1, grows as a function of resource abundance and is consumed by an herbivore, H. The second plant, P_2, is resistant to herbivory, but does have an intrinsic background mortality rate. We explored a number of formulations of the resource growth function, including exponential, logistic, and semichemostat growth. We can write our system of differential equations as

$$\dot{R} = f(R) - c_1 r_1 P_1 R - c_2 r_1 P_2 R$$

$$\dot{P_1} = r_1 P_1 R - c_3 r_3 H P_1 - m_1 P_1$$

$$\dot{P_2} = r_2 P_2 R - m_2 P_2$$

$$\dot{H} = r_3 H P_1 - m_3 H$$

where c_j are conversion efficiency parameters that relate the loss in prey or resource to the gain in the consumer species, r_j are intrinsic rates of growth, m_j are mortality rates and the resource supply, $f(R)$, is equal to one of the following:

$$f(R) = r_0 R$$

$$f(R) = r_0 (R_{max} - R)$$

$$f(R) = r_0 R \left(1 - \frac{R}{R_{max}} \right).$$

We can also add a predator, T, to this system and assume that the only source of mortality for the herbivore, H, is the predator. This gives the system of equations

(box continued)

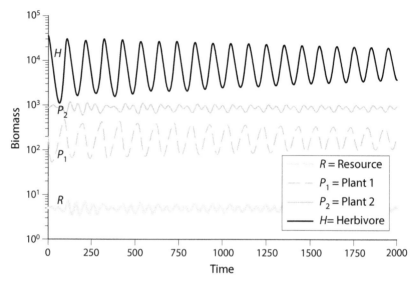

Food web dynamics for one limiting plant resource, two plant species and an herbivore when the resource, R, has logistic growth.

$$\dot{R} = f(R) - c_1 r_1 P_1 R - c_2 r_2 P_2 R$$

$$\dot{P_1} = r_1 P_1 R - c_3 r_3 H P_1 - m_1 P_1$$

$$\dot{P_2} = r_2 P_2 R - m_2 P_2$$

$$\dot{H} = r_3 H P_1 - c_4 r_4 T H - m_3 H$$

$$\dot{T} = r_4 T H - m_4 T.$$

We can simulate dynamics for both models and compare the stability and persistence of all elements of the food web as a function of whether there is a top predator, or what the form of the resource dynamics at the base of the system is.

As an example, the figure within this box shows the outcomes without a top predator under logistic resource growth.

Model Variants

We also explored various model variants that add nonlinearities to the simple structure of the model in the previous section; however, the same general

(box continued)

patterns emerged from our simulations. For example, a function response can be incorporated that limits the growth rate that can be attained for high prey biomass levels. This saturation can arise from a handling time, where each prey item requires time to process. With this, we need to introduce handling time or half-saturation constants, k_i for each interaction that govern the rate at which growth reaches its maximum rate as the resource increases.

$$\dot{R} = f(R) - c_1 r_1 \frac{RP_1}{R + k_1} - c_2 r_2 \frac{RP_2}{R + k_2}$$

$$\dot{P}_1 = r_1 \frac{RP_1}{R + k_1} - q_2 \frac{P_1 H}{P_1 + k_3} - m_1 P_1$$

$$\dot{P}_2 = r_2 \frac{RP_2}{R + k_2} - m_2 P_2$$

$$\dot{H} = r_3 \frac{P_1 H}{P_1 + k_3} - m_3 H$$

Most aspects of the assembly of this simple food web were intuitive, such as the fact that each consumer could survive only if its resource—R, P_1, or H for P_1, H, and T, respectively—was present. Further, given that the conditions for persistence of the two plants and their herbivore are met, then invasion by one of the species into a food web that is lacking it will be successful, and not lead to the extinction of any of the species.

However, we also observed unexpected behavior. Successful invasion by a second plant species drove the top predator extinct. Similarly, for slightly different model parameters, successful invasion by the top predator drove a plant species extinct. In particular, when herbivore-resistant P_2 had traits that let it invade into and persist in a system in which R, P_1, H, and T stably coexisted, the top consumer, T, was displaced by P_2's invasion (Fig. 1). Similarly, when the top predator, T, had traits that allowed it to invade a stable system with R, P_1, P_2, and H, the herbivore-resistant plant P_2 was displaced (Fig. 2). For the suites of traits considered, there was no region of the parameter space where all four species exhibited stable persistence either in this model, or a similar model in which all consumer-resource interactions had Michaelis–Menten-like dynamics.

Moreover, for T to persist, its equilibrial requirement for H (i.e., its H^*) had to be lower than the equilibrium abundance of H in the absence of the top predator, T, but in the presence of P_2. However, when T invaded,

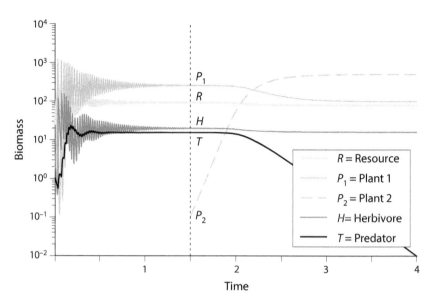

FIGURE 1. A food web consisting of a plant resource, a plant, its herbivore and a top predator grew to equilibrium. At time = 1.5, a few individuals of an herbivory-resistant plant were introduced. When the herbivory-resistant plant has traits that allow it to invade this system, the top predator is invariably driven to extinction given this model (Box 1).

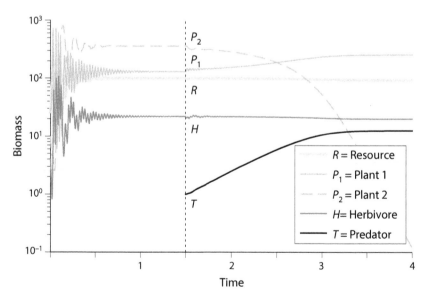

FIGURE 2. A food web containing a plant resource, two plant species, and an herbivore that consumes only plant 1 went to equilibrium. At time = 1.5, a predator that had traits that allowed it to invade this system was introduced. As the predator increased in abundance, it reduced the abundance of the herbivore, increasing the abundance of plant 1, and decreasing the resource. This caused the extinction of the herbivory-resistant plant.

it decreased the abundance of the herbivore, H, causing the abundance of P_1, the plant susceptible to herbivory, to increase sufficiently that it drove the resource low enough to exclude the herbivore-resistant plant, P_2.

This case illustrates how, even in a simple model, an invader can disrupt currently coexisting species in ways rarely seen in nature. We do not mean to imply that low-dimensional models such as our simple one-resource model can describe high-diversity ecosystems. Rather, our point is that there is, at present, no general theory of the multitrophic level tradeoffs that likely explain the multispecies, multitrophic-level persistence that we consistently see in food webs both before and after invasion or speciation.

Universal Food Web Tradeoffs?

How might we explore and quantify food web tradeoffs? As a first step, we might choose traits that are linked to fitness differences among individuals within a species and that similarly determine the outcomes of interactions among species in food webs. Given such traits, consider an n-dimensional trait space where each species can be classified by n quantitative traits, and these traits are parameters for an explicit model of species dynamics in a food web. The n traits of a species give a vector $V \in R^n$ of trait values for each of n traits. Although in theory a species that has the best value for each trait could outcompete all other species, in reality, physical, energetic, and evolutionary constraints and tradeoffs would restrict the feasible sets of traits that are attainable by any species. For instance, protein allocated to create larger and more water-efficient kidneys cannot simultaneously be used to create either additional musculature for faster locomotion or to produce more offspring. Carbon and nitrogen allocated by a plant to create more roots cannot also create more leaves, or more secondary compounds or structures that protect leaves from herbivores, or more seeds. Greater time spent foraging for food could reduce time spent attracting mates or increase the likelihood of encountering a predator. Moreover, traits that increase body size within a species have predictable costs, such as lower maximal per capita net reproductive rates and lower carrying capacities (e.g., see Brown 1984, Brown et al. 2004). Tradeoffs seem ubiquitous.

Constraints and tradeoffs result in a restricted subset of trait combinations, A, that are attainable via evolution, so $A \subset R^n$. However, the vast majority of these attainable trait combinations will have low fitness relative to some other attainable combinations, and thus would be reduced or eliminated from a species by natural selection, leaving a smaller subset of successful trait combinations, S_U, each of which represents a species that could arise via natural selection but that cannot be driven to extinction in a multispecies food web for at least some set of physical environmental conditions, that is to say, in some habitat types.

We hypothesize that the net result of 3 billion years of natural selection may be that the suite of species on Earth is composed of those species with traits that occur on $S_U \subset A$, which represents a single, universal, subset of those possible trait combinations that are successful in a food web. S_U is the hypothesized universal food web tradeoff surface on which, we suggest, all of life on Earth may have evolved. A particular portion of S_U represent the range of traits of groups of species that live in a given biome, such as those plants that live in deserts, or, for a different portion, those that live in Mediterranean-type ecosystems, or for a different portion, tropical rainforests. For each biome there may be multiple clusters of species in trait space corresponding to different functional groups such as herbivores, predators, or plants.

An alternative perspective should also be considered that posits realm-specific tradeoffs surfaces, S_R. S_R could depend on the evolutionary interplay of phylogeny and the selective forces that led to biome-specific adaptations during radiative speciation. Unless there were truly convergent evolution of organisms independent of their phylogenetic origins, selection among the particular taxonomic groups present in a geographic realm would seem likely to increase the potential for multispecies persistence to depend on unique local interspecific tradeoffs. For instance, consider the plant traits, S_R, for a suite of plant species that evolved in a realm that lacked mammalian herbivores, versus the traits, $S_{R'}$, for plants that evolved in a realm with such herbivores. Or, similarly consider the traits, S_R, of birds that evolved large body size and loss of the ability to fly on islands that lacked mammalian predators, versus those that evolved on islands with mammalian predators, $S_{R'}$. For both cases, S_R and $S_{R'}$ seem likely to differ, causing, as observed on some oceanic islands, extinction rather than persistence after invasion by taxa from a different realm.

Conclusions

Persistence following speciation and migration is paradoxical precisely because, over millions of years, pre-existing phylogenetic differences, mutations, and recombination would be expected to contribute to the evolution in one realm of species with traits that would, by chance, be superior to those that evolved in similar species in some other realm. The emergence of species with superior traits should have led to the exclusion and extinction of resident species in realms invaded by the species from a superior realm (Tilman 1999, 2011). However, the fossil record for at least the past ~400 million years seems to overwhelmingly show persistence, not extinction, after both speciation and transrealm migrations.

We suggest that one potential driver of such food web persistence may be unavoidable allocation-based tradeoffs that individuals, and thus species, face during evolution. But, in what way are such tradeoffs manifested? Will tradeoffs influence food web connectance patterns (e.g., May 1973; Martinez 1992; Dunne et al. 2002) or lead to weak trophic interactions combined with each species having density-dependent growth (McCann et al. 1998; McPeek 2012), or spatial coupling of small subwebs (McCann et al. 2005), or Janzen–Connell effects? Each of these possibilities can contribute to food web persistence. Or, will allocation-based tradeoffs somehow give rise to consumer–resource–food web species parameters that assure persistence (Leibold 1989, Chase et al. 2001, Schmitz 1994, Polis and Strong 1996, McPeek 1996)? Or will the study of species functional traits, and their relationships (Wright et al 2004, Reich 2014) reveal patterns that explain food web persistence?

Ecological scaling relationships might also provide answers. For instance, both r values (maximum per capita population growth rates) and abundances are lower for species with greater body mass and both are also temperature dependent (Brown et al. 2004, Enquist et al. 1998). Abundances and spatial distributions are linked (Brown 1984), and species richness scales geographically with latitude, temperature, and various measures of habitat harshness (Allen et al. 2004, Marks et al. 2016). For structural reasons, the mass of the stem of a plant scales with its height, and height is a critically important determinant of competitive ability for light (Givnish 1982). Root-to-shoot ratios change predictably along productivity gradients, as does plant height. Each such relationship embodies a component of the tradeoffs species face. Their synthesis could be insightful.

We suggest that the solution to the food web persistence paradox may require an even broader synthesis, one that tests the potential ability of evolutionarily unavoidable allocation-based tradeoffs to better explain adaptive radiations and the traits of new taxa relative to existing taxa. Speciation occurs in the ecological context of communities and ecosystems. The effects of species and of the physical environment seem likely to be major forces shaping the traits of emerging species, as would be any factors that constrain traits in ways that "improvements" in one trait are linked to "costs" to one or more other traits. The hypothesis of a universal tradeoff surface might be a useful framework for better understanding the evolutionary ecology of speciation, and thus provide a stronger evolutionary basis for community and ecosystem ecology.

References

Allen, A. P., J. H. Brown, and J. F. Gillooly. 2002. Global biodiversity, biochemical kinetics, and the energetic-equivalence rule. Science 297:1545–1448.

Barabás, G., M. J. Michalska-Smith, and S. Allesina. 2017. Self-regulation and the stability of large ecological networks. Nature, Ecology & Evolution 1:1870.

Benton, M. J. 1995. Diversification and extinction in the history of life. Science 268:52–58.

Benton, M. J. 1996. On the nonprevalence of competitive replacement in the evolution of tetrapods. In: D. Jablowski, D. H. Erwin, and J. H. Lipps, eds. Evolutionary paleobiology. Chicago: University of Chicago Press, 185–210.

Brown, J. H. 1984. On the relationship between abundance and distribution of species, The American Naturalist 124:255–279.

Brown, J. H., J. F. Gillooly, A. P. Allen, V. M. Savage, and G. B. West. 2004. Toward a metabolic theory of ecology. Ecology 85:1771–1789

Chase, J., M. A. Leibold, and E. Simms. 2001. Plant tolerance and resistance in food webs: Community-level predictions and evolutionary implications. Evolutionary Ecology 14:289–314.

Chesson, P., and N. Huntley. 1997. The roles of harsh and fluctuating conditions in the dynamics of ecological communities. The American Naturalist 150:519–553.

Chesson, P., and J. J. Kuang. 2008. The interaction between predation and competition. Nature 456:235–238.

Davis, M. A. 2003. Biotic globalization: does competition from introduced species threaten biodiversity? BioScience 53:481–489.

Dunne, J. A., R. J. Williams, and N. D. Martinez. 2002. Food-web structure and network theory: the role of connectance and size. Proceedings of the National Academy of Sciences of the United States of America 99:12917–12922.

Enquist, B. J., J. H. Brown, and G. B. West. 1998. Allometric scaling of plant energetics and population density. Nature 395:163–165.

Flannery, T. 2001. The eternal frontier. Melbourne: The Text Publishing Company.

Flynn, L. J., R. H. Tedford, and Q. Zhanxiang. 1991. Enrichment and stability in the Pliocene mammalian fauna of North China. Paleobiology 17:246–265.

Givnish, T. J. 1982. On the adaptive significance of leaf height in forest trees. The American Naturalist 120:353–381.

Goldblatt, P. 1993. Biological relationships between Africa and South America: An overview. In: P. Goldblatt, ed. Biological relationships between Africa and South America. New Haven: Yale University, 3–14

Gould S. J., and C. B. Calloway. 1980. Clams and brachiopods: Ships that pass in the night. Paleobiology 6:383–396.

Grant, P. R. 1986. Ecology and evolution of Darwin's finches. Princeton, N.J.: Princeton University Press.

Grant, P. R., and B. R. Grant. 2006. Evolution of character displacement in Darwin's finches. Science 313:224–226.

Holt, R. D., J. Grover, and D. Tilman. 1994. Simple rules for interspecific dominance in systems with exploitative and apparent competition. The American Naturalist 144:741–771.

Kneitel, J. M., and J. M. Chase. 2003. Trade-offs in community ecology: linking spatial scales and species persistence. Ecology Letters 7:69–80.

Leibold, M. A. 1996. A graphical model of keystone predators in food webs: trophic regulation of abundance, incidence, and diversity patters in communities. The American Naturalist 147:784–812.

Lindberg, D. R. 1991. Marine biotic interchange between the northern and southern hemispheres. Paleobiology 17:308–324.

MacArthur, R. H. 1972. Geographical ecology: Patterns in the distribution of species. New York: Harper and Row.

Marks, C., H. Muller-Landau, and D. Tilman. 2016. Tree diversity, tree height and environmental harshness in eastern and western North America. Ecology Letters 19:743–751.

Marshall, L. G., S. D. Webb, J. J. Sepkowki, Jr., D. M. Raup. 1982. Mammalian evolution and the Great American Interchange. Science 215:1351–57.

Martinez, N. D. 1992. Constant connectance in community food webs. The American Naturalist 139:1208–1218.

May, R. M. 1973. Stability and complexity in model ecosystems. Princeton, N.J.: Princeton University.

McCann, K., S., A. Hastings, and G. R. Huxel. 1998. Weak trophic interactions and the balance of nature. Nature 395:794–798.

McCann, K. S., J. B. Rasmussen, and J. Umbanhowar. 2005. The dynamics of spatially coupled food webs. Ecology Letters 8:513–523.

McPeek, M. A. 1996. Food web structure, and the persistence of habitat specialists and generalists. The American Naturalist 148:S124–S138.

McPeek, M. A. 2012. Intraspecific density dependence and a guild of consumers coexisting on one resource. Ecology 93:2728–2735.

Morley, R. J. 2003. Interplate dispersal paths for megathermal angiosperms. Perspectives in plant ecology, Evolution and Systematics 6:5–20.

Myers, J. A., and K. E. Harms. 2009. Seed arrival, ecological filters, and plant species richness: a meta-analysis. Ecology Letters 12:1250–1260.

Patzkowski, M. E., and S. M. Holland. 2007. Diversity partitioning of a Late Ordovician marine biotic invasion: controls on diversity in regional ecosystems. Paleobiology 33:295–309.

Pennington, R. T., and C. W. Dick. 2004. The role of immigrants in the assembly of the South American rainforest tree flora. Philosophical Transactions of the Royal Society of London B: Biological Sciences 359:1611–1622.

Raup, D. M., and J. J. Sepkoski. 1986. Periodic extinctions of families and genera. Science 231:833–836.

Reich, P. B. 2014. The worldwide 'fast–slow' plant economics spectrum: A traits manifesto. Journal of Ecology 102:275–301.

Sax, D. F., S. D. Gaines, and J. H. Brown. 2002. Species invasions exceed extinctions on islands worldwide: a comparative study of plants and birds. The American Naturalist 160:766–783.

Tilman, D. 1982. Resource competition and community structure. Princeton, N.J.: Princeton University Press.

Tilman, D. 1990. Mechanisms of plant competition for nutrients: The elements of a predictive theory of competition. In: J. Grace and D. Tilman, eds. Perspectives on plant competition. New York: Academic Press, 117–141.

Tilman, D. 1997. Community invasibility, recruitment limitation, and grassland biodiversity. Ecology 78:81–92.

Tilman, D. 1999. The ecological consequences of changes in biodiversity: A search for general principles. The Robert H. MacArthur Award Lecture. Ecology 80:1455–1474.

Tilman, D. 2011. Diversification, biotic interchange and the Universal Tradeoff Hypothesis. The American Naturalist 178:355–371.

Tilman, D., and S. Pacala. 1993. The maintenance of species richness in plant communities. In: R. E. Ricklefs and D. Schluter, eds. Species diversity in ecological communities. Chicago: University of Chicago Press.

Turnbull, L. A., M. J. Crawley, M. Rees. 2000. Are plant populations seed-limited? A review of seed sowing experiments. Oikos 88:225–238.

Vermeij, G. J. 1989. Invasion and extinction: the last three million years of North Sea pelecypod history. Conservation Biology 3:274–281.

Vermeij, G. J. 1991. When biotas meet: Understanding biotic interchange. Science 253:1099–1104.

Vos, M., A. M. Verschoor, B. W. Kooi, F. L. Wäckers, D. L. DeAngelis, and W. M. Mooij. 2004. Ecology 85:2783–2794.

Vrba, E. S. 1992. Mammals as a key to evolutionary theory. Journal of Mammalogy 73:1–28.

Webb, S. D. 1991. Ecogeography and the Great American Interchange. Paleobiology 17:266–280.

Webb, S. D. 2006. The great American biotic interchange: patterns and processes. Annals of the Missouri Botanical Garden 93:245–257.

Wright, I. J., P. B. Reich, M. Westoby, et al. 2004. The worldwide leaf economics spectrum. Nature 428:821–827.

What Is the Species Richness Distribution?

Pablo A. Marquet, Mauricio Tejo, and Rolando Rebolledo

Variety and change are quintessential to life as we know it. This variety is usually quantified as the number of entities in a given place and time. If the entities under analysis are species, then the resulting quantification will be one of the local or alpha diversity of the place. Although this is a common practice, the answer to why we find the number of species we do is by far more complex than the simplicity of measuring it. Indeed, this is the famous diversity problem or, paraphrasing Hutchinson (1959), why are there so many kinds of species? The answer to this question is still a matter of contention as we know that there may not be a single explanation for why the number of species changes across gradients in latitude, altitude, depth, and why and how it is influenced by variables such as temperature and productivity (e.g., Brown 1998, Gaston 2000, Allen et al. 2002, Fine 2015, Zhou et al. 2016), as well as by the area sampled (Arrhenius 1921) or the area of the landmass containing the community under analysis (MacArthur and Wilson 1963, Rosenzweig 1995). Our focus in this chapter is rather different; we are not interested in "Why so many?" but on the much simpler, albeit related question: "What is the probability of finding s species in a particular community, to begin with?" Interestingly, this question, which is logically previous, but related (see Fig. 1), to the question of "what is the probability of finding s species with k individuals?" or, the species abundance distribution (SAD), has not been asked often, and although there exist some theoretical answers it remains as an empirical challenge. To answer the question of what is the probability of finding s species in a given community is equivalent to know what we call the species richness distribution (SRD). To contextualize the SRD, it is instructive to analyze in some detail what we know about SADs as both distributions should be related.

The study of macroecological patterns in the distribution of individuals among species, or SAD, puzzled theoretical and empirical ecologists during the last 80 years (Fisher 1943, Preston 1948, McArthur 1960, May 1975), this time was fertile for the proliferation of alternative models and explanations, rather than for synthesis and unification (Marquet et al. 2003, McGill et al. 2007). Interestingly, the renewed interest upon

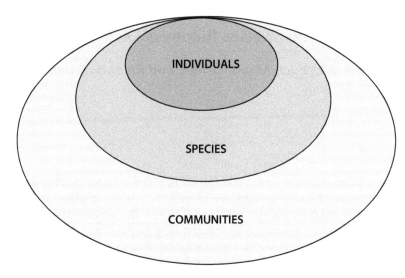

FIGURE 1. Nested relationship between individuals, species, and communities that illustrate the link between the species abundance distribution (SAD) or the probability of finding s species with n individuals and the species richness distribution (SRD), or the probability of finding c communities with s species. Any theory that predicts either the number of individuals, species, or their incidence across communities could potentially say something about SADs and SRDs.

first-principles theories (Marquet et al. 2014, 2015) and particularly in symmetrical or neutral theories grounded on the first principles of stochastic processes, have radically changed the study of SADs, leading to important advances by introducing simple stochastic models (e.g., Hubbell 2001, Volkov et al. 2003, Alonso and McKane 2004, Etienne and Alonso 2005) originally developed in population genetics (Leigh 2007) and based upon the interplay between dispersal limitation, speciation, and density dependence in birth and death processes (Hubbell 2001; Volkov et al. 2003, 2005; Marquet et al. 2017). The problem of what processes underlay a given SAD has been solved in general, although some auxiliary hypotheses on the form of density dependence and dispersal may be needed for different types of systems and organisms (Volkov et al. 2007; but see Marquet et al. 2017). Further, it should not go unnoticed how remarkable it is that we can we can find an explanation at all for the SAD, considering that the abundance of any given species can be affected by a large number of factors such as resource availability, environmental fluctuations, and by the number and kind of interacting species within metacommunities. The same is true in the case of the SRD, as we know that many factors affect the number of species found in a local community.

The key to solving this problem is rather simple. If the complexity of the problem under analysis is large, reduce it to what is known on the basis of first principles. This, in many cases, amounts to developing neutral or symmetric approaches that serve the purpose of providing a null hypothesis of what is to be expected if everything else is not playing a role in generating the pattern, but the simplest processes we know are at play (see Marquet et al. 2014, 2015).

The issue we confront in this chapter is what is the species richness distribution (SRD) or the probability of observing s species in a given local community (i.e., $P(s)$). This, as we show further on, has a long and distinguished history in ecology but is largely still an open and unanswered question, at least from an empirical point of view.

History of the Pattern

One of the first attempts to derive the probability of observing s species in a local community was by Barton and David (1959) in the context of a simple urn model. Indeed, they focused on the changes in the number of coexisting species as a process akin to the allocation of balls of k different colors among N identical boxes. Although their main aim was to derive a test for species association, they also derived the expression for the expected occurrence of s species under the assumption of independence.

In an ecological context, different colors correspond to different species and the different boxes to different local communities or samples. Under the assumption that all species behave independently of each other and have the same abundance (i.e., they have the same probability of becoming part of a given community), $P(s)$ follows a binomial distribution:

$$P(s) = \binom{K}{s} p^s (1-p)^{K-s}$$

(1)

$$s = 0, 1, \ldots, K,$$

where K corresponds to the pool of species available for colonizing a local community, p is the probability of finding any given species in a local community, and s is the local community richness. Further, they show that the binomial is a good approximation provided that the value of p does not vary too much (see Barton and David 1959, Pielou 1975). If it does, as a consequence of, for example, species having different abundances, then it can be approximated by a binomial distribution with probability generating function:

$$(Q + P)^\theta \text{ with } P = \bar{p} + \frac{var(p)}{\bar{p}}, Q = 1 - P, \text{ and } \theta = \frac{K}{1 + \frac{var(p)}{\bar{p}^2}}.$$

where $var(p)$ is the variance and $\bar{p}$ is the mean.

These expressions have been commonly used to assess the expected number of samples or communities with different number of species to assess community-wide positive or negative association among species (e.g., Pielou and Pielou 1967, Pielou 1975, Strong 1982); that is, an unexpectedly large number of samples or communities containing more or fewer species than expected, respectively.

The second attempt at this question came from MacArthur and Wilson (1963, 1967) under the theory of insular biogeography (TIB), one of the first neutral theories in community ecology. This time, however, the interest was not in providing a statistical description of the expectation of finding s species in a given community but on understanding how s varies depending upon the processes of colonization and extinction. These authors discuss the change in the number of species found on a single island and showed that this can be studied as a simple birth/death process. To do this they defined $P_s(t)$ as the probability that at time t a focal island contains s species λ_s, as the rate of immigration of new species onto the island (either from the pool or other islands) when s are present, and μ_s the rate of extinction of species on the island when s are present. The functional form of λ_s and μ_s correspond to the intersecting curves of TIB's now-classical graphical model; that is, λ_s is a decreasing function of s while μ_s increases with s. These two rates are assumed to be the same for any species. Thus, the probability of observing s species on an island, or set of identical islands, follows the *master equation*

$$\frac{dP_s(t)}{dt} = P_{s+1}(t)\mu_{s+1} + P_{s-1}(t)\lambda_{s-1} - P_s(t)(\lambda_s + \mu_s) \tag{2}$$

where s is the number of species, λ_s is the rate of colonization or speciation that increases s to $s+1$ and μ_s is the rate of extinction decreasing s to $s-1$.

Equation 2 leads to a steady state or equilibrium condition that satisfies $P_s = P_0 \Pi_{i=1}^s \frac{\lambda_{i-1}}{\mu_i}$. The invariant distribution of the process, if it exists, will depend upon the functional form of λ_s and μ_s. For example, the neutral theory of ecology derived the SAD using the following birth/death process

(Volkov et al. 2003) that expresses the probability that the kth species contains n individuals:

$$\frac{dP_{n,k}(t)}{dt} = P_{n+1,k}(t)d_{n+1,k} + P_{n-1,k}(t)b_{n-1,k} - P_{n,k}(t)(b_{n,k} + d_{n,k}) \qquad (3)$$

where n is the number of individuals, b_n is the birth rate that increases n to $n+1$ and d_n is the death rate that decreases n to $n-1$. If we assume that all species have the same birth and death rates and that these are density independent (i.e., $b_n = bn$ and $d_n = dn$), one obtains that the equilibrium distribution is Fisher's log series (Kendall 1948, Volkov et al. 2003). Notice that Equation 3 is identical in structure to Equation 2.

Interestingly, MacArthur and Wilson (1963, 1967) were not interested in solving for the invariant distribution, which would be our SRD and neither was Hubbell's neutral theory, although he was aware of the problem of connecting abundance to the number of species in a given community (Hubbell 2001 and personal communication). The invariant distribution was obtained by Goel and Richter-Dyn (1974). To solve Equation 2 one needs to specify an initial condition P_0 and some boundary conditions, which in this case can be deduced by noticing that when there are no species present in the island, $\mu_0 = 0$ and $\lambda_0 \neq 0$, and when the island is saturated with species, that is, the number of species is equal to the number of species in the pool K, then $\mu_K \neq 0$ and $\lambda_K = 0$. Thus, the stochastic process associated with the number of species is confined between the two reflecting states 0 and K. Once these conditions are established the crucial step is to define the functional form of the extinction and colonization rates.

Goel and Richter-Dyn (1974) considered two scenarios for these rates: (1) $\lambda_s = \lambda(K - s)$ and $\mu_s = \mu s$ and (2) $\lambda_s = \lambda(K - s)^2$ and $\mu_s = \mu s^2$. The first scenario reasons that because the area and hence the amount of resources present on the island are fixed, as the number of species in the island increases, the average population size of any given species decreases, hence the probability of extinction of a species can be hypothesized to monotonically increase with the number of species present in the island. Similarly, it is reasonable to assume that the probability of a new species to become present on the island depends upon the species already present, because the more species that become established on the island the lower the chances that a new immigrant individual will belong to a new species, hence the immigration rate should decrease monotonically as the number of species established on the island increases, and becomes $\lambda_K = 0$ when the number of species present on the island is equal to the richness in the pool. Scenario 2 considers nonlinear dependencies in both colonization and extinction appealing to the fact that at small s the number of species colonizing the island is likely to be higher as good dispersing species will likely arrive very fast causing a larger initial decrease in the immigration rate, which later levels off as s increases. Similarly, extinction may not be the same for each species as assumed in Scenario 1, but likely it will increase as the number of species increases due to negative interspecific interactions (see also MacArthur and Wilson 1963, 1967).

As shown by Goel and Richter-Dyn (1974) as $t \to \infty$ under scenario 1, P_s tends to

$$P_s^* = \binom{K}{s}\left(\frac{\lambda}{\lambda+\mu}\right)^s\left(\frac{\mu}{\lambda+\mu}\right)^{K-s},$$

$$s=0,1,\ldots,K$$

(4)

Our Equation 4 is the same as Equation 1 where the probability of success p corresponds to the probability that an immigration event occurs and $1-p$ corresponds to the probability of an extinction event. For the case of Scenario 2 the equilibrium or steady-state distribution is

$$P_s^* = \frac{\binom{K}{s}^2\left(\frac{\lambda}{\mu}\right)^s}{\sum_{s=0}^{K}\binom{K}{s}^2\left(\frac{\lambda}{\mu}\right)^s}$$

(5)

In Figure 2, we illustrate the theoretical shapes of P_s^* under Scenarios 1 and 2. The problem, however, is with testing what is the most appropriate model for the SRD as this would require finding identical islands in terms of area and isolation, variables known to affect λ and μ. Further, as Equation 2 also applies to sites within a continuous landmass one could construct the SRD from a large number of equal-area samples from communities, and assume that sites are similar in all other relevant variables, such as latitude, altitude, temperature, and precipitation, among others. Once this is done it could be compared to Equations 4 or 5 but this would require knowing K or being able to estimate it. Empirically, the form of the SRD is rarely analyzed, and eventually impossible to construct. Thus, the question of what the SRD is seems im-

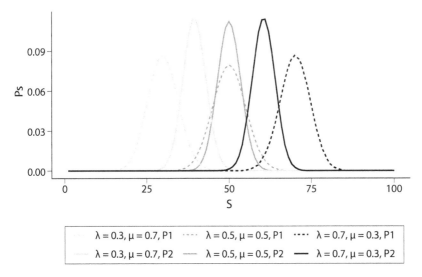

FIGURE 2. Graphical representation of the probability mass functions associated to scenario 1 (P1) as in Equation 4 and scenario 2 (P2) as in Equation 5 for different values of λ and μ and for $K = 100$.

possible to answer from an empirical perspective, unless we assume a different kind of symmetry. In addition to the neutrality in species we will need to assume neutrality in places; i.e., all sites are equal, which we know they are not.

From Individual to Species Dynamics

As we noted in Figure 1 and previously here, the SAD and the SRD are not independent but share many similarities let alone the fact that both can be modeled under the assumption of neutrality. But more importantly, they should be compatible, from a mathematical point of view. The question is then are the neutral derivations of SAD and SRD compatible? To examine this question we can proceed in the following way: Let us represent the number of species living in an island as $k = 1, 2, \ldots, K$ and for any $t \in R_+$, let $(N_1, \ldots, N_K)$ (t) be the abundance vector of the local community; i.e., $N_k(t)_{t \in R_+}$ is the abundance of species k at time t, which in turn is associated with the stochastic process described by the master equation for the abundance of species k in the focal island (local community) given by Equation 3, where $P_{n,k}(t) = P(N_k(t) = n)$, and $b_{n',k}$ and $d_{n',k}$ are the birth and death rates, respectively, when the number of individuals of species k present in the focal island is n'. As mentioned above Equations 2 and 3 have a similar structure but different meaning: The former looks at the interspecies dynamics and the latter looks at the intraspecies dynamics, within the island community. They are, however, related as Equation 2 is a coarse-grained version of Equation 3, to the extent that the presence of a species on an island implies that it has at least one individual and similarly its extinction corresponds to the death of the last individual on the island.

Mathematically, however, these two approaches are not compatible, as they cannot be true at the same time because if Equation 3 holds true, the distribution of the number of species in the local community is no longer Markovian as assumed by these models. This is one case where a neutral model could be rejected for making strong assumptions not related to the assumption of neutrality (Haegeman and Etienne 2008). To see this, let $\{S(t)_{t \in R_+}\}$ be the stochastic process accounting for the number of species present in a focal island A, say, and consider K=pool of species. Then, $S(t) = \Sigma 1_{[1,\infty)}(N_k(t))$, where $1_{[1,\infty)}$ denotes the indicator function of $[1,\infty)$, describes the number of species inhabiting island A. Then, the event $\{s \text{ species at time } t\}$ now depends upon a large number of possibilities or arrangement of the abundance vector $(N_1, \ldots, N_K)(t)$:

$$\{S(t) = s\} = \bigcup_{k_1, \ldots, k_s} \{(N_{k_1}(t), \ldots, N_{k_s}(t)) \in [1, \infty)^s\}$$

so that times between transitions $S \to S \pm 1$ are not longer exponentially distributed. This implies that $S(t)$ is a function of a Markov process, but it is not Markov itself. Thus, to derive the SRD we either resort to non-Markovian approaches, use a different formalism (e.g., diffusion approximation) or we treat richness as an observable on the state variable abundance. In what follows we will explore this latter alternative.

SRD as an Observable on SAD

Let $\{N_k(t)_{k \in K}\}$ be the species abundance of K (pool) species at time t. From the dynamics of the master equations governing its probability distributions (Equation 3), we can notice that the species do not interact, and then, they are statistically independent. Now, we can construct the SRD from this model under a neutral assumption. First, note that the species richness is actually an observable of the species abundance. Indeed, for a given local community we say that "the species k is present at time t" if $N_k(t) \geq 1$ Then, the number of species at time t is given by $S(t) = \Sigma_{k=1}^{K} 1_{[1,\infty)}(N_k(t))$. The event "$s$ *species are present in the local community at time* t", denoted as $\{S(t)=s\}$, for $0 \leq s \leq K$, is described by all the combinations of s elements of the K elements $(N_1(t), \ldots, N_K(t))$ which are in $[1, \infty)^s$ and the remaining in $\{0\}^{K-s}$. When all the species have the same probability distribution (a neutrality assumption), the probability of the previous event is given by:

$$P(S(t) = s) = \binom{K}{s} P(N(t) \geq 1)^s \left[1 - P(N(t) \geq 1)\right]^{K-s},$$

where $N(\cdot)$ is a process sharing the same probability distribution of the $N_k(\cdot)$s. Again, the binomial feature of the SRD is obtained. Additionally, if we consider a large pool and that the probability of finding a single species decreases inversely proportional to the pool size (i.e., $P(N(t) \geq 1)$ is $O(1/K)$), we can get the following approximation of the probability distribution of $S(t)$ as $K \to \infty$:

$$P(S(t)=s) \underset{K\uparrow\infty}{\to} \frac{\lambda(t)^s}{s!} e^{\{-\lambda(t)\}}$$

Where $\lambda(t) = \lim_{K\uparrow\infty} K \, P(N(t) \geq 1)$. This is the Poisson approximation of the previous binomial distribution due to the law of rare events, which in our context means that as long as the pool increases the species become increasingly scarce in the local community. In this way, it may be interpreted as a zero-sum assumption, and as the pool is infinity, the limit SRD is now a pure-birth process.

Final Remarks

This chapter aimed at presenting a pattern that has remained virtually unknown in ecology and which represents an unanswered question at least empirically. In doing so, however, we identified some other important concepts and problems. First, the importance of distinguishing between states and observables. States are properties of a system that are required to implement a given theory, but whose determination lies outside of it; that is, they cannot be predicted from first principles (Harte et al. 2008, Harte 2011). On the other hand, observables are associated with calculations or operations done on the states. We show here that, from the perspective of stochastic processes, richness and abundance cannot be states of the system at the same time (as they are in MaxEnt for example; Harte 2011), but richness should be defined as an observable upon abundance. The alternative is to use a different formulation for the problem as a diffusion process (as done in Marquet et al. 2017 for SADs) or derive non-Markovian approaches. A common feature of the master equations used to model abundance and richness (Equations 2 and 3) is that they describe Markov-type processes, as future states of the system under analysis depend only on the present state, but not on the past history that led to this present state (e.g., Karlin 1968). Although this is a common assumption in ecological and evolutionary models, a large body of experimental data and analyses shows the importance of history (or memory) in affecting current states at the level of individuals, populations, and lineages (e.g., Boyer 1976, Agur 1985, Losos 1995, Peterson 2002, Ogle 2015). In this context it seems desirable to develop non-Markovian models in macroecology (that is, models with memory; Feller 1968) specially as an alternative to neutral models for the distribution of species richness; life is a historical process and the explicit consideration of history may be the simplest way of breaking the symmetry of neutrality.

Acknowledgments

The authors acknowledge support from project FONDECYT 1161023, AFB-170008, and Proyecto Redes 180018.

References

Agur, Z. 1985. Randomness, synchrony and population persistence. JTheoretical Biology. 112:677–693.

Allen, A. P., J. H. Brown, and J. F. Gillooly. 2002. Global biodiversity, biochemical kinetics, and the energetic-equivalence rule. Science 297:1545–1548.

Arrhenius, O. 1921. Species and area. Journal of Ecology 9:95–99.

Barton, D. E., and David, F. N. 1959. The dispersion of a number of species. Journal of the Royal Statistical Society: Series B 21190–21194.

Boyer, J. F. 1976. The effects of prior environments on *Tribolium castaneum*. Journal of Animal Ecology 45:865–874.

Brown, J. H., and M. V. Lomolino. 1998. Biogeography. Sunderland, Mass.: Sinauer.

Feller, W. 1968. An introduction to probability theory and its applications. Vol. 1. New York: Wiley.

Fine, P. V. 2015. Ecological and evolutionary drivers of geographic variation in species diversity. Annual Review of Ecology, Evolution, and Systematics 46:369–392.

Fisher, R. A., A. S. Corbet, and C. B. Williams. 1943. The relation between the number of species and the number of individuals in a random sample of an animal population. Journal of Animal Ecology 12:42–58.

Gaston, K. J. 2000. Global patterns in biodiversity. Nature 405:220–227.

Goel, N. S., and N. Richter-Dyn. 2016. Stochastic models in biology. New York: Academic Press.

Haegeman, B., and R. S. Etienne. 2008. Relaxing the zero-sum assumption in neutral biodiversity theory. Journal of Theoretical Biology 252:288–294.

Harte, J., T. Zillio, E. Conlisk, and A. B. Smith. 2008. Maximum entropy and the statevariable approach to macroecology. Ecology 89:2700–2711.

Harte, J. 2011. Maximum entropy and ecology: A theory of abundance, distribution, and energetics. Oxford: Oxford University Press.

Hubbell, S. P. 2001. The unified neutral theory of biodiversity and biogeography. Princeton, N.J.: Princeton University Press.

Hutchinson, G. E. 1959. Homage to Santa Rosalia or why are there so many kinds of animals? The American Naturalist 93:145–159.

Karlin, S. 1968. A first course in stochastic processes. New York: Academic Press.

Kendall, D. G. 1948. On some modes of population growth leading to RA Fisher's logarithmic series distribution. Biometrika 35:6–15.

Leigh, E. G. 2007. Neutral theory: A historical perspective. Journal of Evolutionary Biology 20:2075–2091.

Losos, J. B., and F. R. Adler. 1995. Stumped by trees? A generalized null model for patterns of organismal diversity. American Naturalist 145:329–342.

MacArthur, R. H. 1960. On the relative abundance of species. American Naturalist 94:25–36.

MacArthur, R. H., and E. O. Wilson. 1963. An equilibrium theory of insular zoogeography. Evolution 17:373–387.

MacArthur, R.H., and E. O. Wilson. 1967. The theory of island biogeography. Vol. 1. Princeton, N.J.: Princeton University Press.

Marquet, P. A., J. Keymer, and H. Cofre. 2003. Breaking the stick in space. In: T. M. Blackburn and K. Gaston, eds. Macroecology: Concepts and consequences. Hoboken: Blackwell, 64–84.

Marquet, P. A., A. P. Allen, J. H. Brown, et al. 2014. On theory in ecology. BioScience 64:701–710.

Marquet, P. A., A. P. Allen, J. H. Brown, et al. 2015. On the importance of first principles in ecological theory development. BioScience 65:342–343.

Marquet, P. A., G. Espinoza, S. R. Abades, A. Ganz, and R. Rebolledo. 2017. On the proportional abundance of species: Integrating population genetics and community ecology. Scientific Reports 7:16815.

May, R. M. 1975. Patterns of species abundance and diversity. In: M. L. Cody and J. M. Diamond, eds. Ecology and evolution of communities. Cambridge, Mass.: Harvard University Press, 81–120.

McGill, B. J., et al. 2007. Species abundance distributions: moving beyond single prediction theories to integration within an ecological framework. Ecology Letters 10:995–1015.

Ogle, K., J. J. Barber, G. A. Barron-Gafford, L. P. Bentley, J. M. Young, T. E. Huxman, M. E. Loik, and D. T. Tissue. 2015. Quantifying ecological memory in plant and ecosystem processes. Ecology Letters 18:221–235.

Peterson, G. D. 2002. Contagious disturbance, ecological memory, and the emergence of landscape pattern. Ecosystems 5:329–338.

Pielou, D. P., and E. C. Pielou. 1967. The detection of different degrees of coexistence. Journal of Theoretical Biology 16:427–437.

Pielou, E. C. 1975. Ecological diversity. New York: Wiley.

Preston, F. W. 1948. The commonness and rarity of species. Ecology 29:254–283.

Rosenzweig, M. L. 1995. Species diversity in space and time. Cambridge: Cambridge University Press.

Strong, D. R., Jr. 1982. Harmonious coexistence of hispine beetles on *Heliconia* in experimental and natural communities. Ecology 63:1039–1049.

Volkov, I., J. R. Banavar, F. He, S. P. Hubbell, and A. Maritan. 2005. Density dependence explains tree species abundance and diversity in tropical forests. Nature 438:658–661.

Volkov, I., J. Banavar, S. P. Hubbell, and A. Maritan. 2003. Neutral theory and relative species abundance in ecology. Nature 28:1035–1037.

Volkov, I., J. R. Banavar, S. P. Hubbell, A. Maritan. 2007. Patterns of relative species abundance in rainforests and coral reefs. Nature 450:45–49.

Zhou, J., Y. Deng, L. Shen, C. Wen, Q. Yan, D. Ning, et al. 2016. Temperature mediates continental-scale diversity of microbes in forest soils. Nature Communications 7:12083

Two Sides of the Same Coin

High Non-Neutral Diversity and High-Dimensional Trait Space in Pathogen Populations and Ecological Communities

Mercedes Pascual

It is underappreciated that central questions on diversity in community ecology are similar to those in infectious disease immunology and ecology (Holt and Dobson 2005, Seabloom et al. 2015). "What processes shape biodiversity and its structure, and what are the implications of such structure for the resilience of ecosystems broadly defined?" is analogous to asking, "what processes shape genetic diversity of pathogens and how does this impact control and subsequent exposure to infection?" Both these sets of questions concern the fundamental connection between "microscopic" processes at the level of individuals and "macroscopic" patterns at the level of populations and communities, as well as their consequences for ecosystem function and conservation on the one hand, and for transmission and control of infection on the other. There is a vast ecological literature on this former connection that continues to motivate active research on food webs, island biogeography and, more generally, the interface of community ecology and evolution. There is also a convergence of rapid developments based on theory, computational models, and statistical inference for complex adaptive systems (CAS). Although these fields are almost completely isolated, they have strong conceptual overlap once differences of temporal, spatial, and organizational scales are taken into account: these range from pathogen phylodynamics and strain theory (e.g., Koelle et al. 2006, Buckee et al. 2008, Lemey et al. 2009, Volz et al. 2013, Bedford et al. 2015), through island biogeography and community phylogenetics (e.g., Webb et al. 2002, Kembel and Hubbell 2006, Cavender-Bares 2009, O'Dwyer et al. 2015, Valente et al. 2015, Cadotte et al. 2016), to macroevolution (e.g., Rabosky 2013, Rosindell et al. 2015).

Here, I focus on the open question of non-neutral forces stabilizing coexistence in ecological systems of high diversity. Such non-neutral forces refer to ecological interactions that specifically depend on trait differences between individuals and species, so that diversity and the structure of this diversity do not simply reflect stochastic colonization and extinction events. A more general agenda of this chapter is to further promote the dialogue between disease and community ecologists on the structure of biodiversity emerging from eco-evolutionary assembly dynamics (Holt and Dobson 2005, Seabloom et al. 2015). I present some speculative ideas on coexistence of free-living species, born from thinking about the vast genetic diversity underlying antigenic variation in *Plasmodium falciparum* malaria in hyperendemic regions, and the more limited genetic diversity in seasonal influenza globally. In particular, I will rely on the recent evidence for a non-neutral structure of the extensive genetic variation that influences competitive interactions between *P. falciparum* parasites for human hosts.

As a fundamental null model, Hubbell's neutral theory revealed the limited information on underlying microscopic processes contained in traditional macroscopic patterns used by ecologists to study frequently occurring patterns in communities of species (e.g., rank–abundance curves) (Hubbell 2001). The success of this theory in explaining widespread and long-studied empirical regularities in ecology, motivated the search for patterns that could effectively differentiate between neutral and non-neutral hypotheses, including signatures of an important role of ecological interactions related to niche partitioning and niche formation. Efforts at the interface of ecology and evolution developed within the new area of community phylogenetics (Webb et al. 2002, Cavender-Bares 2009). The object of study has been the link between underlying diversity processes and the topology of phylogenies, as well as the distribution of traits on these phylogenies, for species coexisting in a given geographical region. Until recently, community phylogenetics has relied to a large extent on quantitative indicators and predictions from conceptual models, rather than dynamical assembly models, to interrogate data on underlying non-neutral processes. Success to date has been mixed.

In an almost "parallel universe," disease ecologists developed the area now known as *pathogen phylodynamics* to address the interplay of population dynamics and genetic/antigenic diversity (e.g., Ferguson et al. 2003, Grenfell et al. 2004) . Here, phylogenies and antigenic variation on these phylogenies are considered *within* pathogen species, so that the scope of these studies has been at a lower taxonomic scale than that of community phylogenetics. (Experimental studies of community assembly and niche differentiation above the single-species level have been conducted for viruses of plants (Seabloom et al. 2013), and it is intriguing to think about the possibility of phylodynamic models at that level). A focus on dynamics, computational models, and statistical inference has been prevalent, facilitated by the feasibility of sampling pathogen systems over evolutionary time scales. Consideration of spatial variation and connectivity has also been important (e.g., Bedford et al. 2015).

Theory on the phylodynamics of seasonal H3N2 influenza has shown the delicate sensitivity of tree topology to the underlying rates of ecological and evolutionary processes (Zinder et al. 2013, Bedford et al. 2015). The shape of the tree and the distribution of antigenic variation (the phenotype of relevance to competition for hosts), depend closely on parameter values and not just on the processes themselves. This means that without a dynamical understanding based on numerical simulation of these systems and without the recourse of associated computational methods of statistical inference it would be impossible to confront different hypotheses on underlying processes. A pattern of coexisting branches whose phenotypes are further apart than expected in a neutral model (where all genetic variation leads to viruses that are seen as equivalent by the immune system) only occurs in phylodynamic models with a sufficiently fast, intermediate speed of antigenic evolution, and not just on the basis of the intensity of competition (Zinder et al. 2013). The same processes that could give rise to this pattern of coexisting branches and niche differentiation result in H3N2 influenza in the successive replacement of antigenic types if the antigenic mutation rate is slower. Thus, the rate of antigenic evolution is not fast enough to allow for the exploration and establishment of different niches (Zinder et al. 2013). More generally, we can expect that there are no simple phylogenetic patterns that we can identify as definite signatures of niche differentiation without relying on a combination of quantitative theory and statistical inference. The expected "aggregated" pattern of niche differentiation in community phylogenetics is probably too simple. (Note that the I am using this term *aggregated* in the way parasite and plant ecologists apply it to describe distributions that are mean aggregated and whose variance is much larger than the mean. The use of overdispersion versus clustering happens to be the opposite in community phylogenetics.)

It is worth pausing here to make some aspects of these analogies clearer. Pathogen populations are competition systems where infected individuals are the consumers and susceptible individuals, who have not yet acquired immunity, are the resources. Competition for hosts occurs through the acquisition of immunity and cross immunity. Depending on their individual history of exposure, hosts can be available or unavailable to infection by individual antigenic variants of a given pathogen. The traits that mediate competition are therefore the antigens that the adaptive immune system recognizes, and antigenic variation determines the outcome of competition for hosts. We can map this kind of variation to the so-called niche differences used by ecologists, conferring an advantage to the rare and a disadvantage to the common in a frequency-dependent manner (Chesson 2000).

Antigenic Diversity in Malaria Populations

Early mathematical models of frequency-dependent competition within pathogen populations were developed as part of a strain theory for *P. falciparum* malaria, with possible extensions to other pathogens including influenza, and more recently, the explicit consideration of evolution (Gupta et al. 1994, 1996, Recker et al. 2007, Zinder et al. 2013, Artzy-Randrup et al. 2012). *P. falciparum* is one of several pathogens whose antigenic variation is known to be extremely large even at the local scale of small human populations in high transmission regions, with this variation encoded by multigene families that undergo frequent recombination (Deitsch 2009). Another remarkable example is *Trypanosoma brucei*. In malaria, the major antigen of the blood stage of infection, the surface protein Pf erythrocyte membrane protein 1 (PfEMP1), is encoded by the multigene family known as *var* and exported to the surface of infected erythrocytes causing adhesion to the microvasculature and influencing parasite clearance. Cumulative diversity curves for the number of genetic variants in local African endemic populations can reach thousands to tens of thousands (when considering a threshold of 96% genetic divergence), compared to a few hundred in South America where transmission is much lower (Day et al. 2017). Moreover, each parasite contains 50 to 60 *var* genes that are sequentially expressed in the different waves of parasitemia within a single infection. Thus, the combinatorial complexity of the possible arrangements or *repertoires* that constitute individual parasite phenotypes is potentially astronomical.

We can ask whether there is structure to this huge level of genetic diversity at the population level. That is, are local parasite populations just random assemblages of these building blocks or are they alternatively shaped by ecological/immunological interaction? Frequency-dependent (or stabilizing) competition is the ecological interaction of interest here, also referred to in the literature as immune selection. This question is of interest as it extends from the structure of biodiversity to a better understanding of the resilience of the malaria system to control efforts in hyperendemic regions. Clinical cases in such regions are only the tip of the iceberg, with most of the human population carrying the parasite asymptomatically and constituting a large reservoir of infection (Tiedje et al. 2017).

Randomness may be expected from the large pool of genetic variation and from *var* repertoires undergoing high rates of recombination in the sexual stage of the life cycle in the mosquito vector. Recent deep sampling of all children in villages of Gabon has revealed the opposite. *P. falciparum* isolates from any two children's blood exhibit extremely low overlap of their *var* genetic composition, with no two sets of genes the same, and lower average overlap than expected from the random sampling of the large gene pool (Day et al. 2017).

Strain theory has needed significant extensions to keep pace with rapidly increasing understanding of the complex biology of the *P. falciparum* malaria system. Besides explicit extension to multicopy genes (Artzy-Randrup 2012), two important additional model developments have involved the formulation of neutral models that can be used as alternative null hypotheses for stabilizing competition (immune selection), and computational systems that can reach the combinatorial complexity observed in nature (size of the gene pool and length of gene repertoires in individual parasites). For the former, note that the randomizations applied earlier to examine the similarity (or equivalently, the overlap) of any two individual isolates, provide a very specific null model that only controls for the frequency of gene types and isolate length. This null model is not dynamic and does not consider processes, as such; it cannot tell us whether similar patterns could arise dynamically from the demography of the system in the complete absence of specific immunity. Here, demography refers to the processes of transmission and recovery from infection, and even the acquisition of immunity, but immunity that does not depend on the specific identity of the parasite. That is, these neutral models include the processes of colonization and extinction of individual infections, but not specific memory of what variants hosts have been previously exposed. The latter extension involved making existing computational models open to evolutionary innovation to achieve the diversity levels observed in nature, with consideration of gene recombination as a major source of new gene variants. The dynamical neutral models of He et al. (2018) were applied not only to interrogate field molecular data about the importance of immune selection, but more fundamentally to even ask what aspects of structure would be informative to answer this question. He et al. (2018) showed that networks of genetic similarity and an ensemble of their properties can differentiate immune selection from neutrality. When applied to data from the deep sampling of another hyperendemic population in West Africa, network analyses provided unequivocal evidence for an important role of frequency-dependent competition in structuring the parasite population by selecting against recombinants (He et al. 2018, Rorick et al. 2018).

Networks provide the counterparts of phylogenetic trees for pathogen systems with recombination; their edges represent genetic similarity and nodes corresponding to *var* repertoires. The overall size of the pool of gene

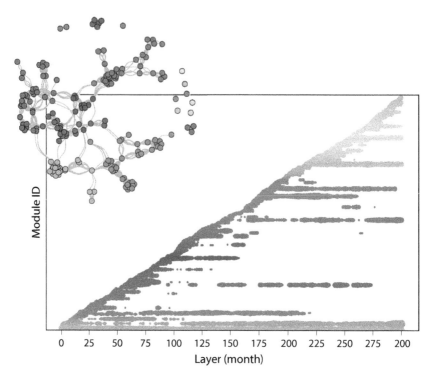

Module ID

0   25   50   75   100 125 150 175 200 225 250 275 200
Layer (month)

FIGURE 1. Networks of genetic similarity can be applied instead of phylogenetic trees, to identify signatures of nonneutral processes in systems such as *P. falciparum* malaria with reticulate evolution as the result of recombination. A node in the network represents a given parasite genome containing a specific combination of genes encoding different variants of a surface antigen. This repertoire is akin to a set of traits defining ecological interactions between individuals. The edges linking two nodes depict the similarity of these combinations. For the unweighted network illustrated in the inset, a threshold has been applied to the originally weighted links to keep only those with the highest values, reflecting the strongest competitive interactions between parasites for hosts based on their genetic/antigenic overlap. Modularity can be studied in these networks to identify sets of parasites with higher similarity within, rather than between, these groups. The extension to a multilayer formalism considers similarity links both within a given time slice and between successive time slices. This representation allows one to identify modules that are defined over time, indicated with a given identification in the main graph (Pilosof et al., 2019). Here, the birth and death of modules is illustrated through time. The long branches indicate persistence of groups of parasites that we can consider strains since they share similar "niches" in host immune space (they can infect groups of hosts with similar immune memory). Such persistence characterizes nonneutral dynamics resulting from stabilizing (frequency-dependent) competition of parasites for hosts. Long persistence is a signature of balancing selection (stabilizing competition), which manifests itself at two different levels of organization, corresponding respectively to genes and combinations of genes. (Figure courtesy of Shai Pilosof.)

variants profoundly influences the network topology of repertoires. In particular, the expected pattern of niche partitioning or limiting similarity, with clear strain clusters, only occurs in the models with medium pool sizes that are below those of empirical systems in West Africa. At the limit of large gene pools, limiting similarity forms a more complex structure that is statistically distinguishable from neutral assemblages (He et al. 2018). Repertoire similarity networks can also be considered to include the temporal dimension by using a multilayer formalism, with each layer representing a given time slice. Here, modules or niches become more evident (Fig. 1) and exhibit another clear distinction with those formed under neutrality; in particular, they persist for much longer (Pilosof et al. 2019). This brings me back to the main point of this long digression into *P. falciparum*.

High-Dimensional Trait Spaces and Non-Neutral Assembly

Immune selection is a form of frequency-dependent competition acting on the niche differences created by antigenic variation among individuals. It is therefore a form of balancing selection. This role has been recognized at the level of the *var* genes themselves. In fact, the pool of variation in these building blocks has evolved over deeper time scales than those of ecology/epidemiology, also under balancing selection, presumably to escape the immune system. The long persistence of *var* genes themselves has been documented in humans and going back to primates (Larremore et al. 2015). Thus, balancing selection operates at the different levels of organization of genes and gene repertoires constituting strains. Where we have a hyperdiverse strain population, we also have a hyperdiverse gene pool. The coexistence of a large number of strains (and genes) in local populations is enabled by the same force that creates such a large pool over longer temporal scales and larger regions.

Hyperdiverse communities of species such as the rain forest and coral reefs are also the result of an ongoing and dynamical assembly process involving the interplay of ecology and evolution. A myriad of frequency-dependent type interactions is likely to be at play, as postulated by Janzen and Connell (Janzen 1970, Connell 1970). Such interactions have been experimentally demonstrated for coexisting trees in the rainforest via negative impacts on the recruitment of conspecifics mediated by soil biota (Mangan et al., Nature 2010). A global latitudinal comparison of temperate and tropical regions provides further evidence for their importance in limiting the abundance of common plant species and stabilizing the persistence of rare ones in diverse rainforests (LeManna et al. 2017).

Under environmental conditions favoring abundant growth and leading to limitation of the general resources fueling such growth, the same ecological forces that ultimately shape species coexistence today in ecological time, are likely to have shaped the genetic variation involved in such interactions over deeper evolutionary time. The kinds of specific interactions that would be at play would involve interactions with natural enemies, mainly between hosts (trees) and (fungal) pathogens, but also between prey and predators/parasitoids (and possibly also mutualistic interactions involving microorganisms such as fungi and pollinators, although their outcome in evolutionary assembly models would have to be examined). The result might be in analogy to the malaria system (the complete opposite to pure neutrality), a rich structure of vast complexity, where ecological interactions of a frequency-dependent type operate in a large-dimensional trait space. Such structure would exist but be extremely difficult to characterize, as limiting similarity at such a limit would not look like simple clusters or niches. The questions of how to characterize such structure and how to infer underlying dominant processes remain open. The importance of high-dimensional individual variation to species coexistence in diverse forest ecosystems has been argued by Clark (2010) on a different basis.

Finally, one could argue that this kind of high dimensionality might not be relevant as it would effectively become neutral-like. A profound distinction would exist, however, in aspects of diversity structure that concern persistence at different levels of organization. This matters, because losing species or strains would mean losing much more than diversity itself—namely, the intricate product of assembly over times that are much longer than those of human impacts. In the case of malaria, a better understanding of such loss might help us monitor and perhaps even achieve elimination. In the case of species communities and conservation, it would provide a powerful argument for the profound intrinsic value of the "entangled bank," beyond issues of resilience and ecosystem services.

Acknowledgments

I thank Qixin He and Shai Pilosof for stimulating discussions and insights on the *var* system, and for their research on the computational models and network analyses, Karen Day for an ongoing collaboration on this fascinating system, and Bob Holt and Andy Dobson for comments on an earlier version of this chapter. I also thank Peter Raven for listening to me and discussing these ideas in incipient form at a meeting of the James S. McDonnell

Foundation. I am grateful for the support of the Fogarty International Center at the National Institutes of Health, Ecology and Evolution of Infectious Diseases (award: R01 TW009670), and of the joint NIH-NSF-NIFA Ecology and Evolution of Infectious Disease (award R01 AI149779).

References

Artzy-Randrup, Y., M. M. Rorick, K. Day, D. Chen, A. Dobson, and N. Pascual. 2012. Population structuring of multi-copy, antigen-encoding genes in *Plasmodium falciparum*. eLife 1:e00093.

Bedford, T., S. Riley, I. G. Barr, S. Broor, M. Chadha, N. J. Cox, et al. 2015. Global circulation patterns of seasonal influenza viruses vary with antigenic drift. Nature 523:217–220.

Buckee, C. O., K. A. Jolley, M. Recker, B. Penman, P. Kriz, S. Gupta, and M.C.J. Maiden. 2008. Role of selection in the emergence of lineages and the evolution of virulence in *Neisseria meningitidis*. Proceedings of the National Academy of Sciences of the United States of America 105:15082–15082.

Cadotte, M. W., and T. J. Davies. 2016. Phylogenies in ecology: A guide to concepts and methods. Princeton N.J.: Princeton University Press.

Cavender-Bares, J., K. H. Kozak, P.V.A. Fine, and S. W. Kembel. 2009. The merging of community ecology and phylogenetic biology. Ecology Letters 12:693–715.

Chesson, P. 2000. Mechanisms of maintenance of species diversity. Annual Review of Ecology and Systematics 31:343–366.

Clark, J. S. 2010. Individuals and the variation needed for high species diversity in forest trees. Science 327:1129–1132.

Connell, J. H. 1970. On the role of natural enemies in preventing competitive exclusion in some marine animals and in rain forest trees. In: P. J. Den Boer and G. R. Gradwell. Dynamics of population. Wageningen: Centre for Agricultural Publishing and Documentation (PUDOC).

Day, K., Y. Artzy, K. Tiedje, V. Rougeron, D. Chen, T. Rask, and M. Pascual. 2017. Evidence of strain structure in *Plasmodium falciparum* var gene repertoires in children from Gabon, West Africa. Proceedings of the National Academy of Sciences of the United States of America 114:E4103–E4111.

Deitsch, K. W., S. A. Lukehart, and J. R. Stringer. 2009. Common strategies for antigenic variation by bacterial, fungal and protozoan pathogens. Nature Reviews: Microbioly 7:493–503.

Ferguson, N. M., A. P. Galvani, and R. M. Bush. 2003. Ecological and immunological determinants of influenza evolution. Nature 422:428–433.

Grenfell, B. T., O. G. Pybus, J. R. Gog, J. L. N. Wood, J. M. Daly, J. A. Mumford, and E. C. Holmes. 2004. Unifying the epidemiological and evolutionary dynamics of pathogens. Science 303:327–332.

Gupta, S., K. Trenholme, R. M. Anderson, and K. P. Day. 1994. Antigenic diversity and the transmission dynamics of *Plasmodium falciparum*. Science 263:961–963.

Gupta, S., M.C.J. Maiden, I. M. Feavers, S. Nee, R. M. May, and R. M. Anderson. 1996. The maintenance of strain structure in populations of recombining infectious agents. Nature Medicine 2:437–442.

He, Q., S. Pilosof, K. E. Tiedje, S. Ruybal-Pesantez, Y. Artzy-Randrup, E. B. Baskerville, K. Day, and M. Pascual. 2018. Networks of genetic similarity reveal non-neutral processes shape strain structure in *Plasmodium falciparum*. Nature Communications 9:1817.

Holt, B., and A. P. Dobson. 2005. Extending the principles of community ecology to address the epidemiology of host-pathogen systems. In: S. K. Collinge and C. Ray, eds. Ecology of emerging infectious diseases. Oxford: Oxford University Press, 6–27.

Hubbell, S. 2001. The unified neutral theory of biodiversity and biogeography. Princeton, N.J.: Princeton University University.

Janzen, D. H. 1970. Herbivores and the number of tree species in tropical forests. The American Naturalist 104:940.

Kembel, S., and S. P. Hubbell. 2006. The phylogenetic structure of a neotropical forest tree community. Ecology 87:86–99.

Koelle, K., S. Cobey, B. Grenfell, and M. Pascual. 2006. Epochal evolution shapes the phylodynamics of interpandemic influenza A (H3N2) in humans. Science 314:1898–1903.

LaManna, J. A., S. A. Mangan, A. Alonso, N. A. Bourg, W. Y. Brockelman, S. Bunyavejchewin, et al. Plant diversity increases with the strength of negative density dependence at the global scale. Science 356 (6345):1389–1392.

Larremore, D. B., S. A. Sundararaman, W. Liu, W. R. Proto, A. Clauset, D. E. Loy, et al. 2015. Ape parasite origins of human malaria virulence genes. Nature Communications 6:8368.

Lemey, P., A. Rambaut, A. J. Drummond, and M. A. Suchard. 2009. Bayesian phylogeography finds its roots. PLOS Computational Biology 5:e1000520.

Mangan, S. A., S. A. Schnitzer, E. A. Herre, K. M. L. Mack, M. C. Valencia, E. I. Sanchez, and J. D. Bever. 2010. Negative plant–soil feedback predicts tree-species relative abundance in a tropical forest. Nature 466:752–755.

O'Dwyer, J. P., S. W. Kembel, and T. J. Sharpton. 2015. Backbones of evolutionary history test biodiversity theory for microbes. Proceedings of the National Academy of Sciences of the United States of America 112(27):8356–8361

Pilosof, S., Q. He, K. Tiedje, S. Ruybal-Pesantez, K. Day, and M. Pascual. 2019. Competition for hosts modulates vast antigenic diversity to generate persistent strain structure in *Plasmodium falciparum*. PLOS Biology 17(6):e3000336.

Rabosky, D. L. 2013. Diversity-dependence, ecological speciation, and the role of competition in macroevolution. Annual Review of Ecology, Evolution, and Systematics. 44:481–502

Recker, M., O. G. Pybus, S. Nee, and S. Gupta. 2007. The generation of influenza outbreaks by a network of host immune responses against a limited set of antigenic types. Proceedings of the National Academy of Sciences of the United States of America 104:7711–6.

Rorick, M. M., Y. Artzy-Randrup, S. Ruybal-Pesántez, K. E. Tiedje, T. S. Rask, A. Oduro, A. Ghansah, K. Koram, K. P. Day, and M. Pascual. 2018. Signatures of competition and strain structure within the major blood-stage antigen of *P. falciparum* in a local community in Ghana. Ecology and Evolution 8(7):3574-3588.

Rosindell, J., L. J. Harmon, and R. S. Etienne. 2015. Unifying ecology and macroevolution with individual-based theory. Ecology Letters 18(5):472–482.

Seabloom, E. W., E. T. Borer, C. Lacroix, C. E. Mitchell, and A. G. Power. 2013. Richness and composition of niche-assembled viral pathogen communities. PLOS One 8(2):e55675.

Seabloom, E. W., E. T. Borer, K. Gross, A. E. Kendig, C. Lacroix, C. E. Mitchell, E. A. Mordecai, and A. G. Power. 2015. The community ecology of pathogens: Coinfection, coexistence and community composition. Ecology Letters 18(4):401–415

Tiedje, K. E., A. Oduro, G. Agongo, T. Anyorigiya, D. Azongo, T. Awine, and K. P. Day. 2017. Seasonal variation in the epidemiology of asymptomatic *Plasmodium falciparum* infections across two catchment areas in Bongo District, Ghana. The American Journal of Tropical Medicine and Hygiene, 97(1):199–212.

Valente, L. M., A. B. Phillimore, and R. S. Etienne. 2015. Equilibrium and non-equilibrium dynamics simultaneously operate in the Galapagos islands. Ecology Letters 18(8): 844–852.

Volz, E. M., K. Koelle, and T. Bedford. 2013. Viral phylodynamics. PLOS Computational Biology 9:e1002947.

Webb, C. O., D. D. Ackerly, M. A. McPeek, and M. J. Donoghue. 2002. Phylogenies and community ecology. Annual Review of Ecology, Evolution, and Systematics 33:475–505.

Zinder, D., T. Bedford, S. Gupta, and M. Pascual. 2013. The roles of competition and mutation in shaping antigenic and genetic diversity in influenza. PLOS Pathogens 9(1):e1003104.

PART IV

ECOLOGICAL COMMUNITIES AND ECOSYSTEMS

What Regulates Growth across Levels of Organization?

Ian Hatton

Ever since Thomas Malthus claimed the "passion between the sexes" remains constant, biologists have assumed geometric or exponential growth potential in living matter at all levels, from the replication of cells and individuals, to the productivity of populations and communities. This assumption has led to biology's most profound insights, as sure as "the struggle for existence is the doctrine of Malthus" (Darwin 1859), but also an enduring question: What regulates growth? A more general answer to this question at all levels would have sweeping applied implications, from cancer and disease dynamics, to juvenile development, and from fisheries and forest productivity to global carbon cycling. Many of ecology's most contentious dichotomies and paradoxes, going back decades, are also fundamentally related to the question of what regulates growth, so that any general understanding could have enormous theoretical significance. This is admittedly a broad question that applies across scales, but research has begun to identify critical properties of growth and regulation that recur in very different kinds of systems, suggesting that a partial, but far-reaching answer may be close at hand.

In this chapter, we summarize why this question remains among the most important unsolved problems in ecology, but that recent evidence has revealed deep symmetries in growth across scales, pointing at the possibility that a more comprehensive quantitative understanding of growth and regulation may be within reach. We show that despite the huge diversity of biological structures, their growth trajectories are often surprisingly similar across taxa and levels of organization. These dynamic similarities become even more evident by considering the maximum growth plotted against mass across different systems, which follows a power law with remarkably similar $\sim\!\frac{3}{4}$ exponents.

The fact that the exponent of these scaling relations is significantly less than one indicates that growth at multiple levels is subexponential, which contrasts markedly with the "Malthusian law." Subexponential growth has strong stabilizing properties for both consumer–resource and competitive dynamics, with implications for several ecological dichotomies and para-doxes. The origin of this growth law, particularly since the publication of the "Metabolic Theory of Ecology" (Brown et. al. 2004), has been widely assumed to follow from basal metabolic ~¾ body mass scaling, given the essential role of energy for growth and reproduction. Theories proposed to explain ~¾ scaling are thus based on structural or geometric con-straints that limit the supply or dissipation of energy, (Brown et al. 2004, Kooijman 2000, West et al. 1997, Hou et al. 2008). Only recently has evi-dence accumulated to suggest this order of causality is mistaken. Instead, metabolic architecture exhibits considerable scope to adjust to growth and may evolve and adapt to efficiently fuel growth (Glazier 2015, Hat-ton et al., 2019). Different lines of evidence now suggest that ~¾ scaling appears to be fundamentally a problem of relative growth, urging a re-orientation of theory to consider dynamical processes (rather than struc-tural constraints) for understanding biological scaling laws. As of yet however, this growth law, representing among the most ubiquitous pat-terns in all of biology, is almost completely unstudied, and so it is not surprising that the problem remains unsolved.

Dichotomy and Paradox

Growth and regulation together, are a simply coupled positive and negative feedback, but the topic is a prolific parent of controversy, spawning more questions than answers. If the regulation of growth is intrinsic, how may it have evolved, and by what behavioral or physiological means does it operate? If extrinsic, what biotic or abiotic components are directly or in-directly responsible, and by what regulatory pathways? Below are listed five well-known dichotomies and two paradoxes with rich literatures that we cannot hope to adequately summarize but are not so entirely unrelated as they might at first appear.

Dichotomies

1. *Density dependence versus independence.* Are groups and popula-tions primarily regulated by intrinsic density-dependent factors or extrinsic environmental or interspecific factors, independent of their density?
2. *Group versus individual selection.* Can groups and populations possess regulatory traits that benefit the group at the expense of the individual, or does individual fitness always dominate under selection?
3. *Top-down versus bottom-up control.* Are populations and trophic communities extrinsically regulated by predators and

pathogens (top-down) or by the supply of their resource (bottom-up)?

4. *The ecosystem concept.* Are ecosystems functional units with intrinsic regulatory capacity, or is "ecosystem" merely a convenient designation for the sum of all parts?

5. *The Gaia hypothesis.* Are the biosphere and its abiotic components capable of large-scale negative feedbacks that resemble individual homeostasis, acting to maintain conditions suitable for life, or are feedbacks in global cycles largely decoupled from the action of living systems?

One thing that differentiates these dichotomies is scale. Whereas the first and second dichotomies focus on family groups and populations, the third dichotomy focuses on populations and communities, the fourth on whole ecosystems, and the fifth on the entire biosphere. One thing that unites these dichotomies, on the other hand, is that they are all to varying degrees concerned with the origin and agency of negative feedbacks on growth. With the exception of density dependence, which is now widely observed, these topics all continue to be debated, but rarely within the wider framework of growth and regulation. The regulation of growth is also at the basis of the stability and persistence of consumer–resource and competitive interactions, and thus related to two much-studied paradoxes.

Paradoxes

1. *The paradox of enrichment.* Populations and communities can persist over large gradients of environmental enrichment, such as rainfall or nutrient gradients, translating into orders of magnitude changes in biomass of all trophic communities. Early consumer–resource models, however, predict that increases in resource productivity are transferred entirely to consumers, giving an increasingly top-heavy ecological pyramid until the system is destabilized.

2. *The paradox of the plankton.* A great multitude of species with different growth rates can persist on a limited number of resources, as exemplified in plankton communities. Early competition models, however, predict increasing diversity causes exclusion and extinction of less-fit competitors.

Each paradox originates from ecology's simplest mathematical models that predict instability or extinction where observation suggests there should not be. This has led to the study of factors specific to different interactions, including heterogeneity in space and time, coevolution and Red Queen dynamics, physical constraints and trade-offs, niche separation, and refugia, among a long list of others. But these paradoxes may stem more from how we model intrinsic growth and regulation, rather than a failure to account for additional factors. In the next sections we show some of the striking similarities in growth across taxa and levels of organization that suggest a quantitative form of growth that has not been considered in these models. In the final section we show how incorporating this form of growth into these models, can potentially resolve the paradoxes, and cast light on some of these dichotomies.

The Universality of the Growth Curve

From birth to maturity, the growth curves followed by mammals as different as shrew and whale are astonishingly similar (Fig. 1). This S-shaped growth curve describes increasing mass (m) in time (t) by a compounding growth term and a saturating loss or regulatory term. Different animals can be compared by rescaling absolute units of mass and time to the relative fraction of adult mass reached in each relative fraction of time to maturity. This results in a universal growth curve (Fig. 1B).

Three common ways to describe the growth curve are as follows (a and b are positive constants that are not the same in each model),

Logistic: $dm/dt = am - bm^2$ (1)

Bertalanffy: $dm/dt = am^k - bm$ for $k < 1$ (2)

Gompertz: $dm/dt = am - bm \log m$ (3)

These models all have a growth term with a weaker mass dependency than that of the loss term, so that the growth curve eventually plateaus. Each model tends to be used at different levels of organization as follows, with representative examples shown in Figure 2.

1. *Logistic* (Equation 1) is often used in population biology to model population growth up to carrying capacity. It assumes that the growth term (am) is exponential, but the loss term (bm^2) follows mass action where individuals negatively interact as the square of their density until the population stops growing at carrying capacity (a/b in Equation 1). Logistic has also been introduced in the prey growth term of classic Lotka–Volterra consumer–resource models and is the basis for Lotka–Volterra competition models, from which we obtain the two paradoxes listed previously (May 1972).

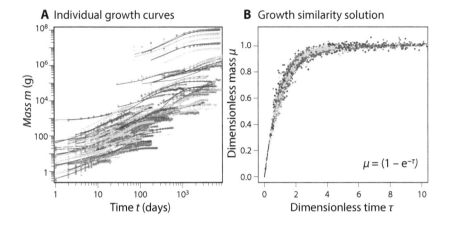

A Individual growth curves

B Growth similarity solution

$\mu = (1 - e^{-\tau})$

Growth equation: $\dfrac{dm}{dt} = am^k - bm$

Solution: $m(t) = \left[\dfrac{a}{b} - \left(\dfrac{a}{b} - m_0^{1-k} \right) e^{-bt(1-k)} \right]^{\frac{1}{1-k}}$

Dimensionless mass: $\mu = \left(\dfrac{m}{m_\infty} \right)^{1-k}$

Dimensionless time: $\tau = (1-k)bt - \log\left[1 - \left(\dfrac{m_0}{m_\infty} \right)^{1-k} \right]$

FIGURE 1. Mammalian individual growth curves and their similarity solution. (A) Growth curves through ontogeny (birth to maturity) spanning shrew to whale (150 species) plotted on logarithmic axes. Points are actual measurements of mass and time, while lines are best fits from the Bertalanffy model (Equation 2) setting $k = \frac{3}{4}$. (B) Similarity solution to the Bertalanffy model which collapses all growth curves in (A) to a universal curve. The values of the constants a and b from fits in A, can be used to rescale mass m and time t to dimensionless mass and time, as shown above (m_0 is birth mass and m_∞ is adult mass), following West et al. 2001.

2. *Bertalanffy* (Equation 2) is often used in physiology to model individual growth up to maturity. It assumes that the growth term is subexponential ($k < 1$), so that growth rate declines with mass. This has been thought to follow metabolic scaling, be it via constraints on heat dissipation over a surface ($k = \frac{2}{3}$), or on energetic supply through a space-filling network ($k = \frac{3}{4}$)

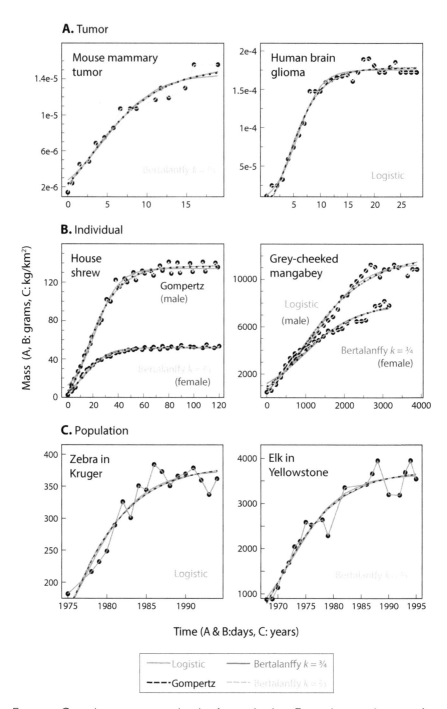

A. Tumor

Mouse mammary tumor

Human brain glioma

Bertalanffy $k = \frac{3}{4}$

Logistic

B. Individual

Mass (A, B: grams, C: kg/km²)

House shrew

Gompertz (male)

Bertalanffy $k = \frac{3}{4}$ (female)

Grey-cheeked mangabey

Logistic (male)

Bertalanffy $k = \frac{3}{4}$ (female)

C. Population

Zebra in Kruger

Logistic

Elk in Yellowstone

Bertalanffy $k = \frac{3}{4}$

Time (A & B:days, C: years)

———Logistic ——— Bertalanffy $k = \frac{3}{4}$

----Gompertz ---- Bertalanffy $k = \frac{2}{3}$

FIGURE 2. Growth curves across levels of organization. Example growth curves for (A) malignant tumors, (B) individual males and females (taken from Fig. 1A), and (C) populations during periods of growth. These can all be described almost identically with different models (equations 1 to 3). Logistic, Gompertz, and Bertalanffy (for $k = \frac{3}{4}$ and $k = \frac{2}{3}$) were fit by nonlinear least squares. The best-fit model is stated in each plot, but is only marginally better than any other. There is also no consistent best model within any given level of organization for these and other data.

(Bertalanffy 1957, West et al. 2001). As discussed previously, however, evidence has accumulated to suggest that metabolism instead adjusts to growth, so that this growth term is not yet mechanistically understood (Hatton et. al. 2019). The loss term is rationalized as catabolic breakdown or maintenance requirements, which acts in proportion to the body as a whole and so has an exponent of one.

3. *Gompertz* (Equation 3) is often used in oncology to model tumor growth and schedule cancer treatments (Rodriguez-Brenes et al. 2013). It assumes that growth is initially exponential, but that the growth rate itself declines exponentially, in contrast to the logistic model, where growth rate declines linearly. Whereas logistic regulation can be rationalized as an extrinsic carrying capacity, Gompertz assumes an intrinsic relative loss of growth potential.

The rationale for the use of any of these models is often vague and can nearly always apply to other levels of organization. Moreover, models are largely indistinguishable in their ability to fit empirical growth curves across these levels (examples in Fig. 2; Rodriguez-Benes 2013; Ricklefs 2003; Vrana et al. 2019). The fact that different research fields commonly favor a particular model appears to be based more on historical precedent than on mechanistic or empirical/statistical grounds. This means that a similarity solution to any of the three growth models (Equations 1–3) for data of tumors, individuals, or populations would likely reveal an analogous dimensionless growth pattern to that shown in Figure 1B.

Despite the dynamic resemblance observed across levels of organization, the similarities are even more compelling if one considers the power-law scaling of maximum growth versus mass. This research has revealed exceptional regularities that span all major taxa and that recur across distinct levels of organization.

Similar Scaling from Embryos to Ecosystems

Rather than considering the entire growth curve as in Figures 1 and 2, we can instead measure the maximum growth at the inflection point of the growth curve and the maximum or asymptotic mass. If we do this for very different sized organisms, plotting maximum growth against mass on logarithmic axes, we observe different species arrayed along a straight line, called a power law. Growth–mass power laws can be written as,

$$dm/dt = cm^k, \tag{4}$$

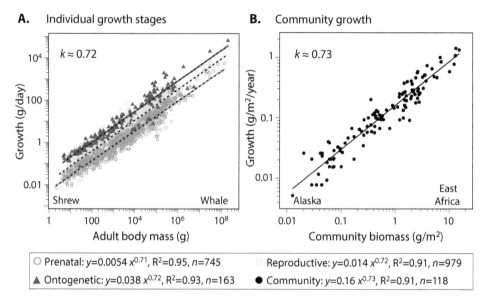

A. Individual growth stages

B. Community growth

○ Prenatal: $y=0.0054\,x^{0.71}$, $R^2=0.95$, $n=745$ Reproductive: $y=0.014\,x^{0.72}$, $R^2=0.91$, $n=979$

▲ Ontogenetic: $y=0.038\,x^{0.72}$, $R^2=0.93$, $n=163$ ● Community: $y=0.16\,x^{0.73}$, $R^2=0.91$, $n=118$

FIGURE 3. Mammal growth-mass scaling from embryos to ecosystems. (A) Mammal prenatal, ontogenetic and reproductive growth vs. adult body mass exhibit near identical growth-mass scaling ($k \approx 0.72$). Data derive from Hatton et al. 2019. (B) Mammal whole trophic community productivity to biomass aggregated cross large herbivore (5–500 kg) populations in Africa, southeast Asia and North America over a combined area of >200,000 km². Each point is a separate protected ecosystem and derives from the censusing of some 1850 mammal populations. Growth calculations follow Appendix S2 of Hatton et al. 2015.

where c is a constant coefficient (elevation on logarithmic axes) and k is a dimensionless scaling exponent (slope on logarithmic axes; Charnov et al. 2007). In contrast with Equations 1–3, which model an entire individual growth curve, the growth–mass power-law equation [Equation 4] is the locus of inflection points from different growth curves, though can also describe an individual growth curve if a suitable loss term is included, as in Equation 2. All of the previously cited growth models (1 through 3) fit to individuals of very different sizes will give similar ∼¾ scaling, following Equation 4.

Recent findings show highly regular growth–mass scaling, spanning the tree of life, and recurring across distinct levels of organization. Below we list several levels of organization where these relations have been observed, all of which tend to converge on similar exponents near $k = ¾$, with examples for mammals shown in Figure 3.

1. *Prenatal growth versus adult body or birth mass.* This relation, observed in mammals and birds, is obtained from maximum fetal growth rates or calculated as birth mass divided by gestation time, and plotted against birth mass or adult body mass, for different species. The relation reveals ~¾ mass scaling across species, and often through much of the gestation period of single individuals (Ricklefs 2010).
2. *Individual ontogenetic growth versus body mass.* This relation, observed across vertebrates, is obtained from the maximum growth of the ontogenetic growth curve (e.g., Fig. 2B) and has similar ~¾ exponents within many major taxa, but different coefficients, *c*, between groups, such as endotherms and ectotherms (Case 1978).
3. *Reproductive growth versus body mass.* This relation is obtained from population time series during a growth period (e.g., Fig. 2C), or calculated from life-history traits, such as age of maturity, gestation, litter size, and lifespan for different species (Hamilton et al. 2011). This relation is among the most universal scaling laws in biology, spanning all eukaryotes, and exhibiting both ~¾ scaling within and across major taxa (Brown et al. 2004, Hatton et al. 2015, 2019).
4. *Social insect colony growth versus whole colony mass.* This relation represents the aggregate growth and mass of all individuals in colonies of wasps, bees, termites, and ants, and is measured from the growth curves of whole colonies grown in the lab. These relations scale at ~¾ and appear to fall directly on the individual growth–mass relation, so that social insects appear to be superorganisms, not only in various qualitative characteristics, but also in their quantitative dynamics (Hou et al. 2010).
5. *Community production versus community biomass.* This relation represents a higher level of organization, aggregating the growth and mass of a community of different species in space, and has been observed across terrestrial and aquatic ecosystems, including forests, grasslands, lakes, and oceans (Hatton et al. 2015). Maximum community growth is measured by estimating biomass per unit area at different periods in time, across many different ecosystems along an environmental gradient that typically span whole biomes of the world. These relations also scale near ¾, and constitute among the largest-scale regularities known in biology.

The ecosystem scaling laws (number 5) have only begun to be explored. Scaling at this level is exemplified in Figure 3B, showing separate protected areas for large mammals from the barrens of Alaska to the fertile plains of east Africa. Across this three-order-of-magnitude gradient in biomass density, the mean body mass of all the large mammals in the community is nearly constant (Hatton et al. 2015). In other words, the relative frequencies of small and large mammals are largely constant, and only their numerical density changes across the biomass gradient. This suggests that community-level ~¾ scaling (number 5) arises independently from individual-level scaling (numbers 1 to 3). A constant mean body mass across biomass gradients also suggests that each population in the community is likely to be following this same subexponential ($k < 1$) growth pattern. Subexponential growth of a population or community means that individual reproduction declines with increased crowding from the maximum shown in Figure 3A (diamonds). This form of density dependence acts in the same relative way at all scales, assuming external conditions are not limiting, but is a relatively weak form of self-regulation. Subexponential growth implies that the population never actually saturates, contrasting with the growth models listed previously (Equations 1–3). But because the growth form is scale-free, it can have powerful stabilizing properties for consumer–resource and competitive interactions across very different assemblages.

The Stability of Subexponential Growth

Consumer–resource and competitive interactions are two of the most fundamental interactions in ecology, and yet two questions prevail. How do consumers keep from overexploiting their resource or causing destabilization, especially at high densities, as per the paradox of enrichment, and how do diverse competitors coexist on a shared resource, as per the paradox of the plankton? In trying to answer these questions, ecologists have considered many other factors that may contribute to the balance of ecosystems, but it is hard to see how any of these factors could generally apply to the diversity of life. Subexponential growth-mass scaling across populations and communities (number 5, previous section), on the other hand, implies a generic intrinsic, albeit weak, form of self-regulation that can have robust stabilizing properties. Here, we consider subexponential growth in the Lotka–Volterra equations, which were the first equations to describe the dynamics of consumer–resource and competitive interactions, and on which most simple contemporary models are based.

Consumer-resource

We can consider a consumer (C)–resource (B) system as follows,

$$dC/dt = gqBC - mC,$$

$$dB/dt = rB^k - qBC \tag{5}$$

This system could describe single predator and prey populations or whole trophic communities. When $k=1$, this model is equivalent to the Lotka–Volterra predator–prey equations, with an exponential resource growth term (rB). The resource loss term (qBC) is a linear (mass action) functional response, where q is interaction strength. Consumers grow depending on their growth efficiency g in converting consumption (qBC) into offspring, and are lost at rate m. This model ($k=1$) is well-known to give neutrally stable cycling whose amplitude depends, unrealistically, on starting densities (Fig. 4A). When $k<1$, however, the model is asymptotically stable, so that for any parameters and initial conditions, any perturbation returns to equilibrium (Fig. 4B). Subexponential resource growth also tends to be stable for nonlinear functional response curves such as the saturating type II, donor-control, or ratio dependence (Hatton et al. 2015).

When k is set near ¾, this model has also been shown to yield realistic equilibrium trophic structure between carnivores and herbivores across the rainfall gradient in African savanna, leading to a predator–prey power law in biomass densities ($C^* = {}^{rg}/_m B^{*k}$; Hatton et al. 2015). Subexponential prey growth means that if prey are driven down to lower densities, their reproduction increases, whereas at higher densities, their reproduction declines. This intrinsic negative feedback on growth suffices to regulate consumer–resource interactions in the simplest models in a way that is general and robust.

Competition

Next, we can consider a model for competition among n species, x_i, as follows,

$$dx_i / dt = r_i x_i^k - r_i x_i / \kappa_i \Sigma^n_{j=1} \alpha_{ij} x_j^k \tag{6}$$

Again, when $k=1$, this model is equivalent to the Lotka–Volterra competition equations (e.g., Fig. 4C for $n=2$ species), or with a transformation of variables, to the replicator equation used in evolutionary theory (Page and Nowak 2003). The competition coefficient, α_{ij}, measures the

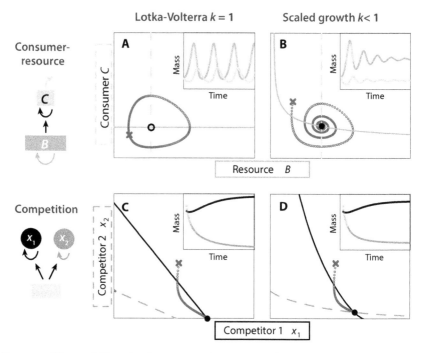

FIGURE 4. The stability of sub-exponential growth. Exponential ($k=1$) and sub-exponential ($k<1$) growth terms are compared in Lotka–Volterra models [Equations 5 and 6]. When $k<1$, it causes convex curvature to zero growth lines (isoclines) in phase space, which dampens neutral oscillations in consumer-resource interactions (A, B), and tends to lead to stable coexistence where exclusion is expected for competitors (C, D). The "×" is an arbitrary initial condition for the time trajectory and the black circle is an asymptotically stable equilibrium point.

effect of species j on species i, and κ_i is the species carrying capacity. This model ($k=1$) can exhibit a range of behaviors depending on parameters and initial conditions, but increasing diversity, n, tends to beget instability, leading to extinction of one or more interactions (May 1972, Barabas et al. 2017). When $k<1$, however, the system is often stabilized (e.g., Fig. 4D for $n=2$ species), although the model can exhibit competitive exclusion with very different population carrying capacities (κ_i).

A simpler alternative that ignores different carrying capacities (κ_i) is to consider species that grow with different rate constants r_i, and that are lost from the system in proportion to their densities, such that the total density of all species is kept constant. This model is obtained by setting $\alpha_{ij}=r_j/r_i$ and $\alpha_{ii}=\kappa_i=1$. Such a system could describe different autocatalytic template replicators in a flow reactor, for which the model was initially devised (Szathmáry 1991), or a community of herbivores all targeted equally by consumers that have no prey preference. This model has been shown to be globally stable by the existence of a Lyapunov function for $k < 1$, implying any number of species will coexist regardless of their growth constants (r_i; Varga and Szathmáry 1998). Subexponential growth means that at higher density, reproduction declines, whereas at lower density it increases, so that a species can always invade at low density and there is no competitive exclusion. This result turns prevailing theory on its head and the task is then "to elucidate the devious strategies" which make for competitive exclusion in otherwise stable and enduring natural systems (May 1972).

These models seek to highlight the potent stability properties of subexponential growth in the simplest models, without the need to invoke additional factors. The consumer–resource model (Equation 5, with $k < 1$) is consistent with predictions for both top-down and bottom-up regulation, and overcomes the "paradox of enrichment," as even at very-high densities (modeled by lowering q) the system is always stable. The competition model (Equation 6, with $k < 1$), and particularly the simpler variant of it described previously, overcomes the "paradox of the plankton," as any diversity of species can coexist on a limited variety of resource. Although we have only begun to explore the theoretical consequences of these growth patterns, understanding the basis for subexponential population growth could have vital repercussions at the heart of ecology.

Recapitulation

The question "what regulates growth?" resounds from cell biology to macroecology and is at the foundation of several dichotomies and paradoxes that continue to be debated. It underpins how we model the dynamics of cell aggregates, individuals, populations, and communities, and has both pure and applied importance that cuts across fields of life science. But a general answer to such a broad question bearing on such a diverse assortment of processes might seem overwhelming. Indeed, the loss or regulatory terms of the growth equations we have considered here alone include carrying capacity (Equation 1), catabolism (Equation 2), loss of growth potential (Equation 3), consumption (Equation 5), and competition (Equation 6); these have barely scratched the surface of the ways we might approach the question. And yet, amazingly, there are aspects of growth dynamics that are remarkably conserved across living systems.

Both the dynamic trajectory and the scaling of maximum growth rate appear universal across major taxa and distinct levels of organization. Very different species growth trajectories can be rescaled to collapse onto a singular dimensionless growth curve (Fig. 1; West et. al. 2001). Different sigmoid growth models also give indistinguishable growth curves across levels of organization (Fig. 2). Finally, very similar growth–mass scaling laws recur across life stages and levels of organization, including fetal, ontogenetic and reproductive growth, and population and community productivity in space (Fig. 3). Such growth–mass scaling implies a weak but systematic form of self-regulation that acts in the same relative way at all scales. It also implies that a partial, but wide-ranging answer to our question may be possible if we could understand the origin of this regularity.

The ubiquity of approximately ¾ growth scaling from algae to elephants and from embryos to ecosystems challenges what is arguably biology's oldest quantitative assumption. Subexponential growth has profound pure and applied implications that have only begun to be explored. Although the problem is found in every area of biology, part of the answer to what regulates growth may be the same across very different kinds of systems, urging new theory that transcends different fields. This simple scaling law is at the dynamical basis of living systems, it spans the tree of life, and recapitulates across levels of organization. As yet, it seems, there are few scientists studying what should be regarded as a very core problem. The fact that it remains unsolved, therefore, may have more to do with neglect than intractability. Understanding where this growth pattern comes from, could tell us a great deal about how growth is regulated across scales.

References

Barabás, G., M. J. Michalska-Smith, and S. Allesina. 2017. Self-regulation and the stability of large ecological networks. Nature Ecology & Evolution 1:1870.

Bertalanffy, L. V. 1957. Quantitative laws in metabolism and growth. The Quarterly Review of Biology 32:217–231.

Brown, J. H., J. F. Gillooly, A. P. Allen, V. M. Savage, and G. B. West. 2004. Toward a metabolic theory of ecology. Ecology 85:1771–1789.

Case, T. J. 1978. On the evolution and adaptive significance of postnatal growth rates in the terrestrial vertebrates. Quarterly Review of Biology 53:243-282.

Charnov, E. L., R. Warne, and M. Moses. 2007. Lifetime reproductive effort. The American Naturalist. 170:E129–E142.

Darwin, C. 1859. On the origin of species by means of natural selection, or the preservation of favoured races in the struggle for life. New York: D. Appleton.

Glazier, D. S. 2015. Is metabolic rate a universal 'pacemaker' for biological processes? Biological Reviews 90:377–407.

Hamilton, M. J., A. D. Davidson, R. M. Sibly, and J. H. Brown. 2011. Universal scaling of production rates across mammalian lineages. Proceedings of the Royal Society B: Biological Sciences 278:560–566.

Hatton, I. A., A. P. Dobson, D. Storch, E. Galbraith, and M. Loreau. 2019. Linking scaling laws across eukaryotes. Proceedings of the National Academy of Sciences of the United States of America 116, 21616–21622.

Hatton, I. A., K. S. McCann, J. M. Fryxell, T. J. Davies, M. Smerlak, A.R.E. Sinclair, and M. Loreau. 2015. The predator-prey power law: Biomass scaling across terrestrial and aquatic biomes. Science 349:aac6284.

Hou, C., M. Kaspari, H.B.V. Zanden, and J. F. Gillooly. 2010. Energetic basis of colonial living in social insects. Proceedings of the National Academy of Sciences of the United States of America 107:3634–3638.

Kooijman, S.A.L.M. 2000. Dynamic energy and mass budgets in biological systems. Cambridge: Cambridge University Press.

May, R. M. 1972. Stability and complexity in model ecosystems. Princeton, N.J.: Princeton University Press.

Page, K. M., and M. A. Nowak. 2002. Unifying evolutionary dynamics. Journal of Theoretical Biology 219:93-98.

Peters, R. H. 1983. The ecological implications of body size. 1st ed. Cambridge: Cambridge University Press.

Ricklefs, R. E. 2003. Is rate of ontogenetic growth constrained by resource supply or tissue growth potential? A comment on West et al.'s model. Functional Ecology 17:384–393.

Ricklefs, R. E. 2010. Embryo growth rates in birds and mammals. Functional Ecology 24:588–596.

Rodriguez-Brenes, I. A., N. L. Komarova, and D. Wodarz. 2013. Tumor growth dynamics: insights into evolutionary processes. Trends in Ecology & Evolution 28:597–604.

Szathmáry, E. 1991. Simple growth laws and selection consequences. Trends in Ecology & Evolution 6:366–370.

Varga, Z., and E. Szathmáry. 1997. An extremum principle for parabolic competition. Bulletin of Mathematical Biology 59:1145–1154.

Vrána, J., V. Remeš, B. Matysioková, K. M. Tjørve, and E. Tjørve. 2019. Choosing the right sigmoid growth function using the unified-models approach. Ibis 161:13–26.

West, G. B., J. H. Brown, and B. J. Enquist. 1997. A general model for the origin of allometric scaling laws in biology. Science 276:122–126.

West, G. B., J. H. Brown, and B. J. Enquist. 2001. A general model for ontogenetic growth. Nature 413:628–631.

The Ecosystem

Superorganism, or Collection of Individuals?

Michel Loreau

The nature of the ecosystem has been a matter of debate since its inception. When Tansley (1935) first defined the ecosystem concept, he did so in opposition to the then prevailing view of Clements (1916), who conceived plant communities as superorganisms; i.e., as higher-level biological entities that have properties of functional organization similar to those of individual organisms (Wilson and Sober 1989). According to Clements, plant communities develop regularly to a climax, just as do individual organisms to their adult stage during ontogeny. Although Tansley rejected Clements's superorganismic view, he did recognize a significant amount of organization in communities and ecosystems; he went as far as saying that "mature well-integrated plant communities . . . had enough of the characters of organisms to be considered as *quasi-organisms*, in the same way that human societies are habitually so considered [emphasis in the original]" (Tansley 1935, pp. 289–290). In contrast, Gleason developed an explicitly individualistic view of plant associations, which he regarded as the mere product of "the coincidence of environmental selection and migration over an area of recognizable extent" (Gleason 1926, p. 26).

Admittedly, the ecosystem concept has evolved substantially since the time of Clements, Gleason, and Tansley. In particular, the initial focus of ecosystem ecology on patterns of energy flow in closed systems has given way to a more dynamic view of ecosystems as open, hierarchical, spatially heterogeneous, and temporally variable complex adaptive systems (Levin 1998, O'Neill 2001). But the tension between the superorganismic and individualistic viewpoints of Clements and Gleason has nonetheless persisted to this day, and has resurfaced in various disguises throughout the history of ecology and its sister sciences.

The longstanding controversy over group selection (Wilson and Wilson 2007) is one manifestation of this tension in evolutionary biology. Indeed, pure group selection leads logically to the emergence of superorganisms (Wilson and Sober 1989). The rejection of group selection theories by many evolutionary biologists is rooted in an individualistic view of natural selection (Williams 1966), which echoes Gleason's view of plant associations.

The controversy over Lovelock's (1979) Gaia hypothesis is another manifestation of this tension. Lovelock, a geochemist, proposed that the entire Earth system behaves as a sort of superorganism, Gaia, in which organisms collectively contribute to self-regulating feedback mechanisms that keep Earth's surface environment stable and habitable for life. Evolutionary biologists such as Dawkins (1982) opposed this hypothesis based on the argument that the Earth system is not a unit of selection, and hence there is no reason why evolution should lead to a planetary environment that is favorable for life. The debate that ensued (Lenton 1998, Free and Barton 2007) is a vivid example of a dialogue where the parties have been talking past each other, which has often characterized the relationship between some branches of ecosystem ecology and biogeochemistry, which have recurrently leaned towards a superorganismic viewpoint, and a hard core of evolutionary biology, which has upheld a strict individualistic viewpoint against all odds.

The disturbing aspect of these debates is that they seem to recur without showing any sign of resolution. Yet there is ample evidence now that neither a strict superorganismic viewpoint nor a strict individualistic viewpoint hold good. These extremes distract attention from the real challenges involved in assessing the degree of integration of ecosystems and in understanding its consequences for ecosystem functioning, stability, and services in a rapidly changing world.

At one extreme, the individualistic viewpoint fails to take into account the manifold interactions that bind individual organisms to their biotic and abiotic environment and that define much of their ecology. Each living organism requires abiotic or biotic resources to stay alive, grow, and reproduce. Resource consumption inevitably leads to competition for resources among individuals, both within and between species. Consumers are themselves resources to higher-level consumers such as predators, parasites, and diseases and any leftover is recycled into inorganic nutrients by decomposers. Many organisms cooperate with other organisms from the same and other species to facilitate their access to resources, enhance their breeding success, or avoid predation. There is increasing evidence that organisms also actively modify their physical environment to meet their needs. These myriad interactions generate complex ecological networks in ecosystems, in which all organisms are inextricably embedded (Olff et al. 2009). As a consequence, each organism contributes to shaping the environment, and hence also the fitness, of other organisms, thereby becoming an actor in their ecology and evolution (Loreau 2010). Thus, there is no ground for a purely individualistic view suggesting that individual organisms behave as independent particles in some sort of inert medium.

At the other extreme, the superorganismic viewpoint fails to take into account the lability of communities and ecosystems, which generally lack clear-cut physical boundaries and show constant changes in at least part of their species composition. Ecosystem-level selection of ecosystem properties requires rather stringent conditions, in particular the existence of long-lasting and localized interactions between ecosystem components (Loreau 2010). Although some small-scale ecosystem properties may approach this ideal situation, no ecosystem in nature is so fully integrated and localized as to bypass any influence of individual selection. Therefore, it is reasonable to expect a combination of individual- and ecosystem-level selection to operate under natural conditions, with individual selection probably often prevailing as many ecological interactions are not strongly localized. As different levels of selection generally drive evolution in different directions, ecosystems are expected to evolve suboptimal properties that result from a compromise between individual- and ecosystem-level selection (Loreau 2010). As a logical consequence, ecosystems cannot be fully-fledged superorganisms.

Thus, it seems perfectly reasonable, based on existing theoretical and empirical evidence, to reject the two extreme views that ecosystems are either superorganisms or mere coincidental collections of individuals. Note, however, that even this fairly obvious conclusion still seems hard to accept for some evolutionary and ecosystem ecologists. If we accept this conclusion, the question then becomes, where do ecosystems lie along the continuum between these two extremes? To what extent are ecosystems integrated units of organization? This is a much more difficult question, to which there is probably no universal answer. Nevertheless, we might expect ecology to have accumulated enough knowledge to be able to narrow down uncertainty to some confidence interval bounded away from the two extreme viewpoints, for at least some ecosystems. The unfortunate truth is that we have no such confidence interval—which probably explains why the two extreme viewpoints resurface periodically.

There are several reasons why assessing the degree of integration of ecosystems has proved so difficult. First, the ecosystem concept is broad enough to be applied to a wide range of different systems and scales, from minute microcosms to the entire biosphere. These widely different systems are likely to show substantial differences in their functional organization and integration. Second, there was a vigorous backlash against the ecosystem concept after the initial enthusiasm for systems analysis, which views complex systems as cybernetic systems stabilized by feedback loops around a relatively constant equilibrium (Patten and Odum 1981, O'Neill 2001). This backlash resulted in a loss of interest in ecosystem ecology for some time, especially from theoretical ecologists, who, by and large, followed the new trends toward complex nonlinear dynamics and complex networks in physics and other sciences. Third, ecosystems are "medium-number systems" (O'Neill et al. 1986) that present considerable methodological difficulties. On the one hand, they are too complex to be fully accounted for by simple dynamical models (although these models have proved extremely useful to study some of their properties). On the other hand, individual organisms—the elementary particles of ecology—belong to a myriad of different species that occupy different niches, and their numbers are much smaller than those of physical particles, which precludes straightforward application of statistical approaches borrowed from thermodynamics and statistical mechanics (Loreau 2010). Therefore, there is no simple, universal approach to study and model ecosystems.

These difficulties have contributed to fuel the individualistic viewpoint, which prevails in many areas of ecology. The consequences of this state of affairs are profound, and extend way beyond a mere philosophical issue. The prevalence of the individualistic viewpoint may even be an obstacle to the discovery of ecosystem-level patterns and processes. As an example, we recently discovered general power-law relationships between prey biomass and predator biomass and between prey biomass and prey production with exponents consistently near ¾ at the scale of whole ecosystems across a wide range of terrestrial and aquatic biomes (Hatton et al. 2015). This discovery came as a surprise even to us as these robust large-scale patterns suggest that ecosystems are more tightly constrained and integrated than previously believed. Perhaps the most intriguing aspect of this finding is that ecosystem production follows the same near-¾ scaling law with biomass as does individual growth with body mass (Brown et al. 2004). Ecosystem production scaling emerges over large numbers of individuals and size structure is often near constant, indicating that similar growth dynamics at the ecosystem and individual levels arise independently (Hatton et al. 2015). Thus, similar basic processes and patterns may re-emerge across systems and levels of organization.

The mechanistic basis of these ecosystem-level patterns is still unclear, but so is the much-debated mechanistic basis of the corresponding individual-level patterns (Glazier 2010, Glazier 2015). Yet, the lack of a convincing explanation for individual-level allometries has not prevented the flourishing of a wealth of empirical and theoretical studies on this topic. One can hardly say the same for ecosystem-level patterns, which, in comparison, remain very poorly studied. One possible mechanism for the re-emergence of sublinear scaling relationships between production and biomass across levels of organization is a form of system-level density dependence in biological activity. Although density dependence has traditionally been studied as a mechanism of population regulation, interactions between individuals from different species that use shared resources might lead to stronger regulation of entire trophic levels than of their component populations, thus generating more robust patterns at the ecosystem level than at the population level. Theoretical and empirical research on this issue would be particularly exciting.

By contrast, in other areas of research, individual- and population-centered approaches have fostered the emergence of novel perspectives and understanding that might reinvigorate ecosystem ecology. For instance, they have contributed to the clarification and the resolution in large part of the long-standing controversy over the relationship between the diversity and stability of ecosystems (McCann 2000). The idea that stability emerges at the ecosystem level as a result of the diversity of population-level processes is deeply anchored in ecosystem ecology and has been central to the cybernetic approach to ecosystems (Odum 1953, Patten 1975). This tenet was overturned in the 1970s by May (1972, 1973) and others, who showed theoretically that, everything else being equal, diversity and complexity should beget instability, not stability. The fact that the stability of an ecosystem's aggregate properties may be very different from that of its component populations (May 1974) seems to have gone virtually unnoticed at that time. Only when large-scale experiments manipulating plant diversity revealed contrasting effects of species diversity on the temporal variability of population- and ecosystem-level properties such as plant biomass and production did it become clear that diversity–stability relationships were qualitatively different at the population and ecosystem levels (Tilman 1996, Tilman et al. 2006, Hector et al. 2010).

A large body of theory has subsequently been developed to identify the mechanisms by which biodiversity stabilizes ecosystem properties while at the same time often destabilizing population dynamics. A striking feature of this theory is that it uses population dynamical models to derive predictions on ecosystem stability (de Mazancourt et al. 2013, Loreau and de Mazancourt 2013), thereby establishing an explicit link between population- and ecosystem-level dynamical properties. Thus, paradoxically, population ecology has been instrumental in laying the theoretical foundations of one of the core principles of ecosystem ecology, that ecosystem stability emerges from the diversity of its component populations. This theory shows that asynchronous population fluctuations that arise from differences in the way species respond to changes in their biotic and abiotic environment are key to the stabilizing effect of biodiversity on ecosystem properties. Interestingly, deterministic differences between species' response niches provide a population-level mechanistic basis for the statistical averaging of aggregate ecosystem properties, thus linking deterministic and statistical behaviors across levels of organization (Loreau 2010). This theory has since been expanded to provide novel predictions on the spatial scaling of ecosystem stability and of its relationship with biodiversity, thus opening up a promising new area of research in ecosystem ecology (Wang and Loreau 2014, 2016).

These and other recent advances at the interface between population and ecosystem ecology show that the tension between population- and ecosystem-centered approaches in ecology can be extremely fertile if it is not congealed in opposed extreme viewpoints. Ecosystems are neither superorganisms nor mere collections of individuals; they are dynamic entities in which ecosystem-level constraints and individual-level variability interact to shape ecological patterns and processes across scales and levels of organization. The relative importance of ecosystem-level constraints and individual-level variability, however, is still a largely unsolved question. The fact that ecosystems are complex medium-number systems puts important theoretical and methodological obstacles in the way of providing a general answer to this question. Recent advances suggest that both top-down (studying whole-ecosystem patterns and processes and searching for underlying mechanisms) and bottom-up (examining the properties that emerge from the aggregation of component populations) approaches will contribute to this goal. Hopefully, narrowing the gap between these two approaches will pave the way for more integrative ecological approaches that fully account for the dynamic interplay between individuals and ecosystems.

Acknowledgments

I gratefully acknowledge support by the TULIP Laboratory of Excellence (ANR-10-LABX-41).

References

Brown, J. H., J. F. Gillooly, A. P. Allen, V. M. Savage, and G. B. West. 2004. Toward a metabolic theory of ecology. Ecology 85:1771–1789.

Clements, F. E. 1916. Plant succession: An analysis of the development of vegetation. Washington, D.C.: Carnegie Institution of Washington.

Dawkins, R. 1982. The extended phenotype: The gene as the unit of selection. Oxford: Freeman.

Mazancourt, C., F. Isbell, A. Larocque, F. Berendse, E. De Luca, J. B. Grace, et al. 2013. Predicting ecosystem stability from community composition and biodiversity. Ecology Letters 16:617–625.

Free, A., and N. H. Barton. 2007. Do evolution and ecology need the Gaia hypothesis? Trends in Ecology & Evolution 22:611–619.

Glazier, D. S. 2010. A unifying explanation for diverse metabolic scaling in animals and plants. Biological Reviews 85:111–138.

Glazier, D. S. 2015. Is metabolic rate a universal 'pacemaker' for biological processes? Biological Reviews 90:377–407.

Gleason, H. A. 1926. The individualistic concept of the plant association. Bulletin of the Torrey Botanical Club 53:7–26.

Hatton, I. A., K. S. McCann, J. M. Fryxell, T. J. Davies, M. Smerlak, A.R.E. Sinclair, and M. Loreau. 2015. The predator-prey power law: Biomass scaling across terrestrial and aquatic biomes. Science 349:aac6284.

Hector, A., Y. Hautier, P. Saner, L. Wacker, R. Bagchi, J. Joshi, M. Scherer-Lorenzen, et al. 2010. General stabilizing effects of plant diversity on grassland productivity through population asynchrony and overyielding. Ecology 91:2213–2220.

Lenton, T. M. 1998. Gaia and natural selection. Nature 394:439–447.

Levin, S. A. 1998. Ecosystems and the biosphere as complex adaptive systems. Ecosystems 1:431–436.

Loreau, M. 2010. From populations to ecosystems: Theoretical foundations for a new ecological synthesis. Princeton, N.J.: Princeton University Press.

Loreau, M., and C. de Mazancourt. 2013. Biodiversity and ecosystem stability: A synthesis of underlying mechanisms. Ecology Letters 16 (S1):106–115.

Lovelock, J. 1979. Gaia: A new look at life on Earth. Oxford: Oxford University Press.

May, R. M. 1972. Will a large complex system be stable? Nature 238:413–414.

May, R. M. 1973. Stability and complexity in model ecosystems. Princeton, N.J.: Princeton University Press.

May, R. M. 1974. Ecosystem patterns in randomly fluctuating environments. In R. Rosen and F. M. Snell, eds. Progress in Theoretical Biology. New York: Academic Press, 1–52.

McCann, K. S. 2000. The diversity-stability debate. Nature 405:228–233.

O'Neill, R. V. 2001. Is it time to bury the ecosystem concept? (With full military honors, of course!). Ecology 82:3275–3284.

O'Neill, R. V., D. L. DeAngelis, J. B. Waide, and T.F.H. Allen. 1986. A hierarchical concept of ecosystems. Princeton, N.J.: Princeton University Press.

Odum, E. P. 1953. Fundamentals of ecology. Philadelphia: Saunders.

Olff, H., D. Alonso, M. P. Berg, B. K. Eriksson, M. Loreau, T. Piersma, and N. Rooney. 2009. Parallel ecological networks in ecosystems. Philosophical Transactions of the Royal Society of London B: Biological Sciences 364:1755–1779.

Patten, B. C. 1975. Ecosystem linearization: An evolutionary design problem. American Naturalist 109:529–539.

Patten, B. C., and E. P. Odum. 1981. The cybernetic nature of ecosystems. American Naturalist 118:886–895.

Tansley, A. G. 1935. The use and abuse of vegetational concepts and terms. Ecology 16:284–307.

Tilman, D. 1996. Biodiversity: Population versus ecosystem stability. Ecology 77:350–363.

Tilman, D., P. B. Reich, and J.M.H. Knops. 2006. Biodiversity and ecosystem stability in a decade-long grassland experiment. Nature 441:629–632.

Wang, S., and M. Loreau. 2014. Ecosystem stability in space: Alpha, beta and gamma variability. Ecology Letters 17:891–901.

Wang, S., and M. Loreau. 2016. Biodiversity and ecosystem stability across scales in metacommunities. Ecology Letters 19:510–518.

Williams, G. C. 1966. Adaptation and natural selection: A critique of some current evolutionary thought. Princeton, N.J.: Princeton University Press.

Wilson, D. S., and E. Sober. 1989. Reviving the superorganism. Journal of Theoretical Biology 136:337–356.

Wilson, D. S., and E. O. Wilson. 2007. Rethinking the theoretical foundation of sociobiology. Quarterly Review of Biology 82:327–348.

Untangling Food Webs

Robert M. Pringle

Just as the ecology of an organism is defined in large part by what it eats and what eats it, the properties of a community emerge largely from the network of trophic interactions among its members. Consequently, food webs are central to almost all ecological research, if not as the direct object of study then as the context in which species interactions and other processes are situated (Paine 1966, May 1983, Polis et al. 2004, McCann 2012).

But although food webs are fundamental to our understanding of ecology, we do not yet understand their most fundamental feature—the basic architecture of nodes and links that comprise the network. In vanishingly few cases and with inordinate effort, investigators have compiled something roughly approaching a complete map of trophic interactions for the set of macroscopic consumer and producer populations present at a site (Cohen et al. 2003, Brown and Gillooly 2003). But even the most finely resolved networks have missing pieces (and gaping holes if we include parasites and microbes (Lafferty et al. 2008)) and are merely static averages of what are inherently dynamic systems (Cohen et al. 2003).

Barriers to Knowing What Wild Consumers Actually Eat

Visitors to a zoo, standing in front of some big mammal from some exotic place, might field a basic question from a curious child: "What does it eat?" Although the informational placard provides only the vaguest of information ("plants"), the parents may assume that scientists know the answer. But with rare exceptions, they would be wrong. Zoo directors may appreciate the depths of our ignorance on this count better than anyone. As Mike Jordan of the Chester Zoo put it, "detailed information about the diet of the majority of free-ranging mammals and birds does not exist and often only the most generalized approximation of food items consumed is known" (Jordan 2005).

There are two major problems with the quality of empirical data used to construct food webs. First, they are poorly resolved taxonomically, with food items often lumped at the level of genus, family, or order, or else categorized into broad functional groups (large versus small, animal versus plant, grass versus shrub, foliage versus fruit) (Paine 1988, Solow and Beet 1998, Winemiller 2007). Such coarse data may be severely mismatched with the precision of consumers' foraging decisions: The distinctions among resource types that are most readily perceived and quantified by ecologists may or may not be those that are salient to the animals. Second, dietary data are poorly resolved in space and time, often being drawn either from a single population and averaged across time (as is common in field studies) or from individuals sampled at many points in space and time and averaged across both (as is common in studies of museum specimens). Consequently, we know little about individual dietary variation within populations, about dietary differences between populations across environmental and geographic gradients (i.e., dietary beta diversity), or about dietary shifts in response to changing seasonal or climatic conditions.

The simple reason for these problems is that it is extremely difficult to accurately and representatively characterize the diet of most free-ranging consumers. This statement is coincidentally illustrated by something happening nearby as I write this. About ten meters away from me, a habituated warthog (*Phacochoerus africanus*) is grazing the unusually homogeneous lawn of the Chitengo Camp in Gorongosa National Park, Mozambique. I know that warthogs in this park eat mostly grass (Pansu et al. 2019), and that the dominant grass in this particular lawn is *Urochloa mosambicensis* (Hack.) Dandy. But although the lawn is unusually homogeneous, it nonetheless has multiple species of grasses and forbs coexisting at small spatial scales, and I cannot see—even through binoculars—exactly which species are being eaten. And although this pig is unusually tame, it is not tame enough that I could reach into its mouth, remove the food, and sort the foliage by species. The best I could do instead is walk up to the place where the animal is grazing and try to identify which plants have been bitten or not (Kleynhans et al. 2011). But now the warthog has walked off—how long would I have to follow it before I had a complete list of the plant species in its diet? Would its foraging decisions be altered by my following it around? Probably. And before long, I would come across a plant species that I could not identify, perhaps one that only a few people in the world could identify, perhaps even one that has no name because it has never been scientifically described. The biota of Mozambique, like that of many African countries, is understudied and poorly understood.

I face all of these problems just for the tamest of warthogs; never mind trying to follow one of Gorongosa's shier or more lethal large herbivores at close range. And forget about shooting large numbers of them and sifting through their guts—a once-preferred method of diet analysis for large

African herbivores (Field 1972). In addition to being legally prohibited and ethically outrageous, stomach-contents analysis would not solve the problem. Buss (1961) shot dozens of Ugandan elephants in 1959, and although he was able to identify some forage items to the species level—71 elephant stomachs collectively contained 266 kg of *Combretum collinum* and 59 kg of *Vitex doniana*—the overwhelmingly dominant food types were identified only as "mature grass" and "young grass" (3793 and 479 kg respectively). Even some of the nongrass species in Buss' elephant stomachs could only be identified to the genus or family level, and others could not be identified at all ("unidentified woody materials" weighed in at 27 kg, making it the seventh-most-abundant food type). Microhistologic examination of feces to visually match undigested plant fragments with reference specimens is an ethically uncomplicated nonlethal alternative, but it is extremely laborious and tends to yield low-quality data (Newmaster et al. 2013). Stable-isotope analysis is a profoundly important tool for inferring many food-web properties (Layman et al. 2012), but it provides only coarse-grained insights into the taxonomic composition of a consumer's diet.

Similar limitations pertain to other traditional methods of diet assessment. Expert opinion (Stier et al. 2016) is unreliable. Cafeteria choice experiments (Ford 2014) are unwieldy. Gastric lavage and allied techniques (Holechek and Pieper 1982) require capture and can harm animals. Some study species are more tractable than others. Sea otters conveniently consume all their prey at the ocean's surface where people can see them, which facilitates the study of individual- and population-level dietary variation (Estes et al. 2003, Tinker et al. 2008). Sea stars (Paine 1966) and caterpillars (Hebert et al. 2004, Janzen et al. 2017) conveniently sit on their foods for a long time while consuming them. But for the vast diversity of animals that lack such agreeable traits, conventional approaches are insufficient to thoroughly and accurately identify food types to species, and our knowledge of diet composition and food-web structure remains spotty at best. These difficulties are compounded for species-rich food types that are not easily identified from a distance or as partially digested fragments. That category includes essentially all arthropods and herbaceous plants, especially in the tropics.

Cryptic Diversity and Taxonomic Imprecision Compound the Challenges of Food-Web Analysis

Identifying many organisms to species level is a serious challenge even for a taxonomic specialist with a high-quality specimen in hand. Many ecologists underestimate the difficulties associated with identifying and distinguishing species and overestimate their own ability to make determinations by consulting published field guides, keys, and reference collections. Many community-level studies proceed on the grounds that similar-looking species are probably ecologically "close enough," even if not quite the same thing. This is the implicit premise underpinning the reliance of many field studies upon supraspecific lumpings or morphospecies determinations by nonspecialists (Oliver and Beattie 1996). It is tempting to view taxonomic imprecision as functionally inconsequential—unlikely to bias our conceptual understanding of food-web organization, and perhaps even necessary to achieve theoretical clarity.

To what extent is that true? That is an unsolved question that needs answering. More precisely: When is what degree of taxonomic approximation sufficient to capture the mechanistic essence of trophic interactions and predict the outcomes of ecological processes? More simply: How good is good enough?

A growing body of evidence indicates that even fairly narrow approximations are not good enough to understand the structure and dynamics of food webs. Over the past 15 years, we have learned that cryptic species are commonplace, and that accounting for them can dramatically alter our understanding of consumers' dietary niches and hence food-web architecture (Hebert et al. 2004; Janzen et al. 2017; Smith et al. 2006, 2007, 2008). In a now-famous example, the neotropical skipper butterfly *Astraptes fulgerator,* thought since 1775 to be a single wide-ranging species, was discovered to be "a complex of ≥10 food plant specialists with differing ecological attributes" in northwestern Costa Rica alone (Hebert et al. 2004). Farther up the food chain, a braconid wasp parasitoid of Costa Rican skipper caterpillars, *Apanteles leucostigmus*, formerly considered a generalist consumer of 32 skipper species, was found to comprise at least 36 species, each of which eats only "one or a very few closely related species of caterpillars." Examination of the six microgastrine braconid genera of northwestern Costa Rica revealed more than 300 provisional species, 95% of which were undescribed and 90% of which "attack only 1 or 2 species of caterpillars" (Smith et al. 2008). Cryptic diversity and cryptic host specificity were likewise found within the tachinid fly parasitoids of this region: 16 presumed generalist species were found to represent 73 mitochondrial lineages, of which only 9 were true generalists (Smith et al. 2007).

In short, what appeared to be a fairly generalized plant–herbivore–parasitoid food web resolved, on more rigorous inspection, into a series of far more specialized food chains. One obvious general lesson is that fine-grained taxonomic distinctions are not ecologically trivial. Heaps of closely related butterfly species, identical enough to pass for one another in plain sight for hundreds of years, are ecologically and trophically disparate. Any theory of food webs that elided such distinctions in the search for "useful generalizations" (Lawton 1999) would be anti-progress, because the mischaracterization of network architecture, diet breadth, and niche overlap would preclude any reliable inferences about the eco-evolutionary processes that produced those attributes. The quest for a predictive conceptual framework of ecological specialization (Poisot et al. 2011, Vamosi et al. 2014) is doomed if we are routinely and unknowingly mistaking specialists for generalists in nature.

This kind of problem is not confined to megadiverse tropical-forest food webs of tiny wasps and flies and caterpillars. In a semiarid African savanna ecosystem in Kenya, we have found similar patterns of cryptic diversity and dietary specificity among large mammalian herbivores. Since 2008, Jake Goheen, Todd Palmer, and I have maintained the UHURU study—a network of 1-ha experimental plots where we simulate size-biased extinction by selectively excluding first the megaherbivores (elephants and giraffes), followed by successively smaller sets of species (meso-herbivores, then dwarf antelopes), until only hares and rodents remain (Pringle 2012, Goheen et al. 2013, Kartzinel et al. 2014, Goheen et al. 2018). To assess the ecological impacts of removing larger species, we regularly monitor plants, trap small mammals, and survey other animal populations and ecosystem processes (Coverdale et al. 2016, 2018, 2019; Ford et al. 2014, 2015; Long et al. 2017; Louthan et al. 2013, 2014, 2018; Ngatia et al. 2014; Pringle et al. 2011, 2014, 2016; Titcomb et al. 2017; Young et al. 2013, 2015, 2017). Among the few-dozen small-mammal species in this region, there are two genera, *Mus* and *Crocidura*, that each contain multiple species that we cannot distinguish in the field (Goheen et al. 2013, Young et al. 2015). A 'species' known to us for the first several years of the study as *Gerbilliscus robustus* was later revealed to be two species from different genera, *G. robustus* and *Taterillus harringtoni* (Goheen et al. 2013). Two species of hares (*Lepus* spp.) that occur in the plots can be distinguished based on mitochondrial DNA, but our attempt to identify these two haplotypes based on the reference DNA sequences available in GenBank produced hopelessly confusing results (Kartzinel et al. 2019); we are forced to refer to them as Hare A and Hare B. Notably, fecal DNA analysis reveals that Hare A and Hare B—whoever they are—have different diets (Kartzinel et al. 2019). For plants, our initial list of 105 morphotaxa in the experimental plots has been painstakingly refined and expanded over the past decade with the assistance of taxonomic experts and DNA barcoding, currently numbering 189 (Goheen et al. 2013; Kartzinel et al. 2014, 2015; Gill et al. 2019).

Just as in Costa Rica, these fine-grained taxonomic distinctions have implications for our understanding of ecological specialization and food-web architecture. African savanna herbivores are often classified as grazers, browsers, or mixed-feeders—a taxonomically coarse typology that refers to the proportion of monocots (primarily grasses, family Poaceae) versus all other plant lineages ("browse") in the diet (du Toit and Olff 2014). The most common large-herbivore species at our Kenyan site include three pairs of species that consume roughly equivalent proportions of grass and browse (as inferred from carbon stable-isotope analysis), yet strikingly partition different plant species within those categories (as inferred from fecal DNA metabarcoding): plains and Grevy's zebra (*Equus quagga* and *E. grevyi*, respectively), Cape buffalo and domestic cattle (*Syncerus caffer* and *Bos indicus*, respectively), and elephant and impala (*Loxodonta africana* and *Aepyceros melampus*, respectively) (Kartzinel et al. 2015). Thus, depending on the taxonomic resolution with which diet is assessed, one could conclude that diet composition is highly redundant within these pairs (comprising similar proportions of grass and browse), or that each species is relatively specialized and distinct (consuming different amounts of particular grass and browse species). The as-yet unanswered question is to what extent these subtler distinctions influence broader system-level processes and properties—competition, coexistence, productivity, stability—and thus to what extent we must account for all nodes and edges before we can have a functional understanding of a food-web network.

Old Wine in New Bottles

These perspectives throw modern light onto a longstanding problem. Concerns about the inadequacy of empirical data to resolve food-web structure and dynamics go way back. Ecologists' interest in "the structure of food webs" (May 1983, Pimm 1979) intensified in the late 1970s, with hopes that general rules of community organization could be distilled (Cohen 1977, Briand and Cohen 1984). May (1983) outlined the contours of an emerging field: "Although a good deal of scattered information about individual food webs has been available for some time, it is only in the last 10 years or so that people have begun a systematic attempt to understand what factors determine the structure of food webs."

However, the accuracy and resolution of the empirical food webs being used to guide theoretical development left much to be desired. This was especially true for the small and inconspicuous species at low trophic levels. Whereas large vertebrates were rarely overlooked and were often resolved to the species level, basal consumers and resources were often lumped into coarse taxonomic or functional groups (not always the correct ones) or else omitted entirely. Paine (1988) argued that existing empirical food webs provided only an approximate "road map" of interactions in a community:

These qualitative descriptions were never intended to be data, to serve as grist for the theoretician's mill. I do not believe that clever theory can overcome this handicap and generate testable, interesting predictions about web structure and dynamics. Profitable theory can be done, and often is, for theory's sake. However, when theory is developed in concert with data, the partnership should be more or less equal. This has not been the case with food webs, where theory seems far ahead of the data, often to the theory's detriment. I know of no one who, having assembled a data set on feeding relationships, considers those data to constitute much more than an incomplete preliminary description. I believe a fresh start is called for.

Paine argued that food webs must at least be subjected to "common sense" scrutiny as to whether they represent "a biologically realistic representation," and that "whenever possible, species should be identified rather than aggregated so that individual roles can be identified and ties to mainstream ecological mathematics facilitated." He also suggested that it might be more profitable to shift focus away from patterns of connectance in complex whole-community food webs as a basis for theory, and towards the use of interaction strength as a currency to generate more easily testable predictions.

The latter recommendation foreshadowed many advances in community ecology throughout the 1990s. By focusing on simplified bi- and tri-trophic modules of strongly interacting species, it was possible to demonstrate the importance of indirect effects in shaping communities and their responses to perturbations (Power 1990, Strauss 1991, Polis 1994, Wootton 1994, Menge 1995, Holt and Polis 1997, Schmitz et al. 2000). The ability to experimentally exclude or add individual species in natural communities or mesocosms facilitated the bridging of theory and empiricism that had been lacking in whole-community connectedness-web approaches. Dynamic models of interactions within these modules could be parameterized and tested in ways that complex food-web networks could not.

Via judicious simplification, these developments sidestepped the logistical challenges of fully characterizing food-web architecture. Yet these two approaches are not substitutable: The value of understanding food-web modules underscores the importance of resolving food-web architecture. Attempts to construct food-web theory piecewise by linking modules will struggle to reproduce the emergent properties that arise at successively higher levels of organization, which are difficult if not impossible to predict based on their modular subsystems. In other words, "food webs are more than the sum of their tri-trophic parts," just as a cell is more than a bag of molecules (Cohen et al. 2009). Although it may ultimately prove possible to build upwards towards predictive and dynamic food-web models (McCann 2012), doing so will minimally require a precise understanding of how the constituent pieces fit together and how each modifies the others—as any home-furniture assembler would attest.

Meanwhile, the development and testing of complex network models continues to be hindered by the scarcity of good data. New-and-improved theory, computers, and numerical techniques mean that we can now do for complex networks what could not be done in the 1970s. This has rein-vigorated efforts to characterize ecological communities and their dynamics using information about network architecture (Strogatz 2001, Dunne et al. 2002, Jordano et al. 2003, Tylianakis et al. 2008, Allesina et al. 2008, Stouffer and Bascompte 2011). Yet the theoretical advances in this area have not been matched by improvements in the datasets necessary to pa-rameterize and test predictive models. Brown and Gillooly's (2003) ob-servation that "theoretical progress has been hampered by lack of ade-quate data" echoes the concerns previously voiced by May (1983) and Paine (1988) and subsequently voiced by others (Lafferty et al. 2008). Yet there has been little concerted effort over the past four decades to rectify this situation. In 1983, May had counted 62 empirical food webs. Thirty-two years later, Cirtwill et al. (2015) found 196 webs of sufficient quality to be usable; of these 196, only 31 were from terrestrial ecosystems, and those 31 were drawn from a mere 19 primary sources with a mean age of more than 50 years (as of 2016). Cohen et al. (2009) found a total of three webs (all aquatic, and two from the same lake in different years) that in-cluded information on both link structure and the average body mass and population density of each taxon in the network.

Ecological Forensics: Inroads Using Molecular Methods

One bright spot in the landscape painted above is that recently devel-oped molecular and bioinformatics techniques such as DNA barcoding and metabarcoding can vastly facilitate both reliable taxonomic assigna-tion and delineation (Hebert et al. 2004, Janzen et al. 2017) and dietary analysis (Taberlet et al. 2007; Pompanon et al. 2011; Wirta et al. 2014; Craine et al. 2015; Kartzinel et al. 2015, 2019; Kartzinel and Pringle 2015; Evans et al. 2016; Atkins et al. 2019; Pringle et al. 2019). Allied techniques are enabling us to approach the microbial component of food webs for the first time (Henderson et al. 2015, Reese et al. 2018, Kartzi-nel et al. 2019). Integration of DNA-based methods with complementary approaches such as stable-isotope and fatty-acid analyses can compen-sate for the limitations of each method in isolation (Traugott et al. 2013, Nielsen et al. 2018).

These approaches, if creatively harnessed to the conceptual frameworks that community ecologists have been honing for decades, have revolutionary potential. Lawton's (1999) lament that "community ecology is a mess" grew from the perception that system-specific contingencies thwart theoretical predictions and prevent ecologists from scaling up. This diagnosis resonated with many community ecologists, and the prescription—a retreat from mechanism, "reductionism, and experimental manipulation" in favor of the search for large-scale, "detail-free statistical patterns"—set an enduring tone for the field. But "contingency" simply means that our working model is incorrect, incomplete, or both. Contingency has mechanistic underpinnings, and they too can be untangled. Judicious simplification and abstraction will continue to be valuable tools for coping with ecological complexity. But it is increasingly unnecessary to reflexively shy away from complexity, as the reach and power of our tools grow more and more to scale with that complexity. And it is increasingly easy to envision a near future in which Lawton's (1999) "overwhelmingly complicated . . . intermediate scales" cease to seem quite so overwhelming.

References

Allesina, S., D. Alonso, and M. Pascual. 2008. A general model for food web structure. Science 320:658–661.

Atkins, J. L., R. A. Long, J. Pansu, J. H. Daskin, A. B. Potter, M. E. Stalmans, C. E. Tarnita, and R. M. Pringle. 2019. Cascading impacts of large-carnivore extirpation in an African ecosystem. Science 364:173–177.

Briand, F., and J. E. Cohen. 1984. Community food webs have scale-invariant structure. Nature 307:264–267.

Brown, J. H., and J. F. Gillooly. 2003. Ecological food webs: high-quality data facilitate theoretical unification. Proceedings of the National Academy of Sciences of the United States of America 100:1467–1468.

Buss, I. O. 1961. Some observations on food habits and behavior of the African elephant. Journal of Wildlife Management 25:131–148.

Cirtwill, A. R., D. B. Stouffer, and T. N. Romanuk. 2015. Latitudinal gradients in biotic niche breadth vary across ecosystem types. Proceedings of the Royal Society B: Biological Sciences 282:20151589.

Cohen, J. E. 1977. Ratio of prey to predators in community food webs. Nature 270:165–167.

Cohen, J. E., T. Jonsson, and S. R. Carpenter. 2003. Ecological community description using the food web, species abundance, and body size. Proceedings of the National Academy of Sciences of the United States of America 100:1781–1786.

Cohen, J. E., D. N. Schittler, D. G. Raffaelli, and D. C. Reuman. 2009. Food webs are more than the sum of their tritrophic parts. Proceedings of the National Academy of Sciences of the United States of America 106:22335–22340.

Coverdale, T. C., J. R. Goheen, T. M. Palmer, and R. M. Pringle. 2018. Good neighbors make good defenses: Associational refuges reduce defense investment in African savanna plants. Ecology 99:1724–1736.

Coverdale, T. C., T. R. Kartzinel, K. L. Grabowski, R. K. Shriver, A. A. Hassan, J. R. Goheen, T. M. Palmer, and R. M. Pringle. 2016. Elephants in the understory: Opposing direct and indirect effects of consumption and ecosystem engineering by megaherbivores. Ecology 97:3219–3230.

Coverdale, T. C., I. J. McGeary, R. D. O'Connell, T. M. Palmer, J. R. Goheen, M. Sankaran, D. J. Augustine, A. T. Ford, and R. M. Pringle. 2019. Strong but opposing effects of associational resistance and susceptibility on defense phenotype in an African savanna plant. Oikos 128:1772–1782.

Craine, J. M., E. G. Towne, M. Miller, and N. Fierer. 2015. Climatic warming and the future of bison as grazers. Scientific Reports 5:16738.

Dunne, J. A., R. J. Williams, and N. D. Martinez. 2002. Food-web structure and network theory: The role of connectance and size. Proceedings of the National Academy of Sciences of the United States of America 99:12917–12922.

Estes, J. A., M. L. Riedman, M. M. Staedler, M. T. Tinker, and B. E. Lyon. 2003. Individual variation in prey selection by sea otters: patterns, causes and implications. Journal of Animal Ecology 72:144–155.

Evans, D. M., J.J.N. Kitson, D. H. Lunt, N. A. Straw, and M.J.O. Pocock. 2016. Merging DNA metabarcoding and ecological network analysis to understand and build resilient terrestrial ecosystems. Functional Ecology 30:1904–1916.

Field, C. R. 1972. The food habits of wild ungulates in Uganda by analyses of stomach contents. African Journal of Ecology 10:17–42.

Ford A. T., J. R. Goheen, D. J. Augustine, M. F. Kinnaird, T. G. O'Brien, T. M. Palmer, R. M. Pringle, and R. Woodroffe. 2015. Recovery of African wild dogs suppresses prey but does not trigger a trophic cascade. Ecology 96:2705–2714.

Ford, A. T., J. R. Goheen, T. O. Otieno, L. Bidner, L. A. Isbell, T. M. Palmer, D. Ward, R. Woodroffe, and R. M. Pringle. 2014. Large carnivores make savanna tree communities less thorny. Science 346:346–349.

Gill, B. A., P. M. Musili, S. Kurukura, A. A. Hassan, J. R. Goheen, W. J. Kress, M. Kuzmina, R. M. Pringle, and T. R. Kartzinel. 2019. Plant DNA-barcode library and community phylogeny for a semi-arid East African savanna. Molecular Ecology Resources 19:838–846.

Goheen, J. R., D. J. Augustine, K. E. Veblen, D. M. Kimuyu, T. M. Palmer, L. M. Porensky, et al. 2018. Conservation lessons from large-mammal manipulations in East African savannas: The KLEE, UHURU, and GLADE experiments. Annals of the New York Academy of Sciences 1429:31–49.

Goheen, J. R., T. M. Palmer, G. K. Charles, K. M. Helgen, S. N. Kinyua, J. E. Maclean, B. L. Turner, H. S. Young, and R. M. Pringle. 2013. Piecewise disassembly of a large-herbivore community across a rainfall gradient: the UHURU experiment. PLOS One 8:e55192.

Hebert, P. D. N., E. H. Penton, J. M. Burns, D. H. Janzen, and W. Hallwachs. 2004. Ten species in one: DNA barcoding reveals cryptic species in the Neotropical skipper butterfly *Astraptes fulgerator*. Proceedings of the National Academy of Sciences of the United States of America 101:14812–14817.

Henderson, G., F. Cox, S. Ganesh, A. Jonker, W. Young, Global Rumen Census Collaborators, and P. H. Janssen. 2015. Rumen microbial community composition varies with diet and host, but a core microbiome is found across a wide geographical range. Scientific Reports 5:14567.

Holechek, J. L., M. Vavra, and R. D. Pieper. 1982. Botanical composition determination of range herbivore diets: A review. Journal of Range Management 35:309–315.

Holt, R. D., and G. A. Polis. 1997. A theoretical framework for intraguild predation. American Naturalist 149:745–764.

Janzen, D. H., J. M. Burns, Q. Cong, W. Hallwachs, T. Dapkey, R. Manjunath, M. Hajibabaei, P.D.N. Hebert, and N. V. Grishin. 2017. Nuclear genomes distinguish cryptic species suggested by their DNA barcodes and ecology. Proceedings of the National Academy of Sciences of the United States of America 114:8313–8318.

Jordan, M.J.R. 2005. Dietary analysis for mammals and birds: a review of field techniques and animal-management applications. International Zoo Yearbook 39:108–116.

Jordano, P., J. Bascompte, and J. M. Olesen. 2003. Invariant properties in coevolutionary networks of plant–animal interactions. Ecology Letters 6:69–81.

Kartzinel, T. R., P. A. Chen, T. C. Coverdale, D. L. Erickson, W. J. Kress, M. L. Kuzmina, D. I. Rubenstein, W. Wang, and R. M. Pringle. 2015. DNA metabarcoding illuminates dietary niche partitioning by African large herbivores. Proceedings of the National Academy of Sciences of the United States of America 112:8019–8024.

Kartzinel, T. R., J. R. Goheen, G. K. Charles, E. DeFranco, J. E. Maclean, T. O. Otieno, T. M. Palmer, and R. M. Pringle. 2014. Plant and small-mammal responses to large-herbivore exclusion in an African savanna: five years of the UHURU experiment. Ecology 95: 787–787.

Kartzinel, T. R., J. C. Hsing, P. M. Musili, B.R.P. Brown, and R. M. Pringle. 2019. Covariation of diet and gut microbiome in African megafauna. Proceedings of the National Academy of Sciences of the United States of America 116:23588–23593.

Kartzinel, T. R., and R. M. Pringle. 2015. Molecular detection of invertebrate prey in vertebrate diets: Trophic ecology of Caribbean island lizards. Molecular Ecology Resources 15:903–914.

Kleynhans, E. J., A. E. Jolles, M.R.E. Bos, and H. Olff. 2011. Resource partitioning along multiple niche dimensions in differently sized African savanna grazers. Oikos 120: 591–600.

Lafferty, K. D., S. Allesina, M. Arim, C. J. Briggs, G. De Leo, A. P. Dobson, et al. 2008. Parasites in food webs: The ultimate missing links. Ecology Letters 11:533–546.

Lawton, J. H. 1999. Are there general laws in ecology? Oikos 84:177–192.

Layman, C. A., M. S. Araujo, R. Boucek, C. M. Hammerschlag-Peyer, E. Harrison, Z. R. Jud, et al. 2012. Applying stable isotopes to examine food-web structure: an overview of analytical tools. Biological Reviews 87:545–562.

Long, R. A., A. Wambua, J. R. Goheen, T. M. Palmer, and R. M. Pringle. 2017. Climatic variation modulates the indirect effects of large herbivores on small-mammal habitat use. Journal of Animal Ecology 86:739–748.

Louthan, A. M., D. F. Doak, J. R. Goheen, T. M. Palmer, and R. M. Pringle. 2013. Climatic stress mediates the impacts of herbivory on plant population structure and components of individual fitness. Journal of Ecology 101:1074–1083.

Louthan, A. M., D. F. Doak, J. R. Goheen, T. M. Palmer, and R. M. Pringle. 2014. Mechanisms of plant–plant interactions: concealment from herbivores is more important than abiotic-stress mediation in an African savannah. Proceedings of the Royal Society B: Biological Sciences 281:20132647.

Louthan, A. M., R. M. Pringle, J. R. Goheen, T. M. Palmer, W. F. Morris, and D. F. Doak. 2018. Aridity weakens population-level effects of multiple species interactions on *Hibiscus meyeri*. Proceedings of the National Academy of Sciences of the United States of America 115:543–548.

May, R. M. 1983. The structure of food webs. Nature 301:566–568.

McCann, K. S. 2012. Food webs. Princeton, N.J.: Princeton University Press.

Menge, B. A. 1995. Indirect effects in marine rocky intertidal interaction webs: Patterns and importance. Ecological Monographs 65:21–74.

Newmaster, S. G., I. D. Thompson, R. A. D. Steeves, A. R. Rodgers, A. J. Fazekas, J. R. Maloles, R. T. McMullin, and J. M. Fryxell. 2013. Examination of two new technologies to assess the diet of woodland caribou: video recorders attached to collars and DNA barcoding. Canadian Journal of Forest Research 43:897–900.

Ngatia, L. W., K. Ramesh Reddy, P. K. Ramachandran Nair, R. M. Pringle, T. M. Palmer, and B. L. Turner. 2014. Seasonal patterns in decomposition and nutrient release from East African savanna grasses grown under contrasting nutrient conditions. Agriculture, Ecosystems and Environment 188:12–19.

Nielsen, J. M., E. L. Clare, B. Hayden, M. T. Brett, and P. Kratina. 2018. Diet tracing in ecology: Method comparison and selection. Methods in Ecology and Evolution 9:278–291.

Oliver, I., and A. J. Beattie. 1996. Invertebrate morphospecies as surrogates for species: A case study. Conservation Biology 19:99–109.

Paine, R. T. 1988. Food webs: Road maps of interactions or grist for theoretical development? Ecology 69:1648–1654.

Paine, R. T. 1966. Food web complexity and species diversity. American Naturalist 100:65–75.

Pansu, J., J. A. Guyton, A. B. Potter, J. L. Atkins, J. H. Daskin, B. Wursten, T. R. Kartzinel, and R. M. Pringle. 2019. Trophic ecology of large herbivores in a reassembling African ecosystem. Journal of Ecology 107:1355–1376

Pimm, S. L. 1979. The structure of food webs. Theoretical Population Biology 16:144–158.

Poisot, T., J. D. Bever, A. Nemri, P. H. Thrall, and M. E. Hochberg. 2011. A conceptual framework for the evolution of ecological specialisation. Ecology Letters 14:841–851.

Polis, G. A. 1994. Food webs, trophic cascades, and community structure. Australian Journal of Ecology 19:121–136.

Polis, G. A., M. E. Power, and G. R. Huxel. 2004. Food webs at the landscape level. Chicago: University of Chicago Press.

Pompanon, F., B. E. Deagle, W.O.C. Symondson, D. S. Brown, S. N. Jarman, and P. Taberlet. 2011. Who is eating what: Diet assessment using next generation sequencing. Molecular Ecology 21:1931–1950.

Power, M. E. 1990. Effects of fish in river food webs. Science 250:811-814.

Pringle, R. M. 2012. How to be manipulative: Intelligent tinkering is key to understanding ecology and rehabilitating ecosystems. American Scientist 100:30–37.

Pringle, R. M., J. R. Goheen, T. M. Palmer, G. K. Charles, E. DeFranco, R. Hohbein, A. T. Ford, and C. E. Tarnita. 2014. Low functional redundancy among mammalian browsers in regulating an encroaching shrub (Solanum campylacanthum) in African savannah. Proceedings of the Royal Society B: Biological Sciences 281:20140390.

Pringle, R. M., T. R. Kartzinel, T. M. Palmer, T. J. Thurman, K. Fox-Dobbs, C. C. Y. Xu, et al. 2019. Predator-induced collapse of niche structure and species coexistence. Nature 570:58–64.

Pringle, R. M., T. M. Palmer, J. R. Goheen, D. J. McCauley, and F. Keesing. 2011. Ecological importance of large herbivores in the Ewaso ecosystem. In: N. Georgiadis, ed. Smithsonian Contributions to Zoology (Conserving Wildlife in African Landscapes: Kenya's Ewaso Ecosystem), no. 632. Washington, D.C.: Smithsonian Institution Scholarly Press. 43–54.

Pringle, R. M., K. M. Prior, T. M. Palmer, T. P. Young, and J. R. Goheen. 2016. Large herbivores promote habitat specialization and beta diversity of African savanna trees. Ecology 97:2640–2657.

Reese, A. T., F. C. Pereira, A. Schintlmeister, D. Berry, M. Wagner, L. P. Hale, et al. 2018. Microbial nitrogen limitation in the mammalian large intestine. Nature Microbiology 3:1441–1450.

Schmitz, O. J., P. A. Hambäck, and A. P. Beckerman. 2000. Trophic cascades in terrestrial systems: A review of the effects of carnivore removals on plants. American Naturalist 155:141–153.

Smith, M. A., J. J. Rodriguez, J. B. Whitfield, A. R. Deans, D. H. Janzen, W. Hallwachs, and P.D.N. Hebert. 2008. Extreme diversity of tropical parasitoid wasps exposed by iterative integration of natural history, DNA barcoding, morphology, and collections.

Proceedings of the National Academy of Sciences of the United States of America 105:12359–12364.

Smith, M. A., D. M. Wood, D. H. Janzen, W. Hallwachs, and P.D.N. Hebert. 2007. DNA barcodes affirm that 16 species of apparently generalist tropical parasitoid flies (Diptera, Tachinidae) are not all generalists. Proceedings of the National Academy of Sciences of the United States of America 104:4967–4972.

Smith, M. A., N. E. Woodley, D. H. Janzen, W. Hallwachs, and P.D.N. Hebert. 2006. DNA barcodes reveal cryptic host-specificity within the presumed polyphagous members of a genus of parasitoid flies (Diptera: Tachinidae). Proceedings of the National Academy of Sciences of the United States of America 103:3657–3662.

Solow, A. R., and A. R. Beet. 1998. On lumping species in food webs. Ecology.

Stier, A. C., J. F. Samhouri, S. Gray, R. G. Martone, M. E. Mach, B. S. Halpern, C. V. Kappel, C. Scarborough, and P. S. Levin. 2017. Integrating expert perceptions into food web conservation and management. Conservation Letters 10:67–76.

Stouffer, D. B., J. Bascompte. 2011. Compartmentalization increases food-web persistence. Proceedings of the National Academy of Sciences of the United States of America 108:3648–3652.

Strauss, S. Y. 1991. Indirect effects in community ecology: Their definition, study and importance. Trends in Ecology & Evolution 6:206–210.

Strogatz, S. H. 2001. Exploring complex networks. Nature 410:268–276.

Taberlet, P., E. Coissac, F. Pompanon, L. Gielly, C. Miquel, A. Valentini, et al. 2007. Power and limitations of the chloroplast trnL (UAA) intron for plant DNA barcoding. Nucleic Acids Research 35:e14.

Tinker, M. T., and G. Bentall, J. A. Estes. 2008. Food limitation leads to behavioral diversification and dietary specialization in sea otters. Proceedings of the National Academy of Sciences of the United States of America 105:560–565.

Titcomb, G., B. F. Allan, T. Ainsworth, L. Henson, T. Hedlund, R. M. Pringle, et al. 2017. Interacting effects of wildlife loss and climate on ticks and tick-borne disease. Proceedings of the Royal Society B: Biological Sciences 284:20170475.

du Toit, J. T., and H. Olff. 2014. Generalities in grazing and browsing ecology: using across-guild comparisons to control contingencies. Oecologia 174:1075–1083.

Traugott, M., S. Kamenova, L. Ruess, J. Seeber, and M. Plantegenest. 2013. Empirically characterising trophic networks: what emerging DNA-based methods, stable isotope and fatty acid analyses can offer. Advances in Ecological Research 49:177–224.

Tylianakis, J. M., R. K. Didham, J. Bascompte, and D. A. Wardle. 2008. Global change and species interactions in terrestrial ecosystems. Ecology Letters 11:1351–1363.

Vamosi, J. C., W. S. Armbruster, and S. S. Renner. 2014. Evolutionary ecology of specialization: insights from phylogenetic analysis. Proceedings of the Royal Society B: Biological Sciences 281:20142004.

Winemiller, K. O. 2007. Interplay between scale, resolution, life history and food web properties. In: N. Rooney, K. S. McCann, and D.L.G. Noakes, eds. From energetics to ecosystems: The dynamics and structure of ecological systems. Dordrecht, Netherlands: Springer, 101–126.

Wirta, H. K., P.D.N. Hebert, R. Kaartinen, S. W. Prosser, G. Várkonyi, and T. Roslin. 2014. Complementary molecular information changes our perception of food web structure. Proceedings of the National Academy of Sciences of the United States of America 111:1885–1890.

Wootton, J. T. 1994. Putting the pieces together: Testing the independence of interactions among organisms. Ecology 75:1544–1551.

Young, H. S., D. J. McCauley, R. Dirzo, J. R. Goheen, B. Agwanda, C. Brook, et al. 2015. Context-dependent effects of large-wildlife declines on small-mammal communities in central Kenya. Ecological Applications 25:348–360.

Young, H. S., D. J. McCauley, R. Dirzo, C. L. Nunn, M. G. Campana, B. Agwanda, et al. 2017. Interacting effects of land-use and climate on rodent-borne disease in central Kenya. Philosophical Transactions of the Royal Society B: Biological Sciences 372:20160116.

Young, H. S., D. J. McCauley, K. M. Helgen, J. R. Goheen, E. Otárola-Castillo, T. M. Palmer, et al. 2013. Effects of mammalian herbivore declines on plant communities: Observations and experiments in an African savanna. Journal of Ecology 101:1030–1041.

What Determines the Abundance of Lianas and Vines?

Helene C. Muller-Landau and Stephen W. Pacala

Climbing plants—a group that includes both lianas (woody) and vines (nonwoody)—are found in a majority of the world's forests, at widely varying abundances. Climbing plants are more abundant in tropical than in temperate forests, in dry tropical forests than in wet tropical forests, and in younger forests and forest edges (Schnitzer 2005, DeWalt et al. 2015). The abundance of climbers is critically important to determining forest carbon stores and cycling (van der Heijden et al. 2015, Schnitzer 2018). Where there are more lianas and vines, trees grow more slowly, forests store less carbon, and forest structure is altered, with implications for the diversity and abundance of other plants and animals within the forest (Schnitzer et al. 2015). Further, the effects of lianas differ among tree species, and thus the relative performance of tree species depends on liana abundance (Muller-Landau and Visser 2019). Thus, the question of what determines the abundance of lianas and vines within a forest is central in forest ecology. Yet surprisingly, we have not yet answered the question of what determines the abundance of climbing plants in any forest, much less what explains variation in their abundance across forests and among tree species within forests.

Climbing plants have inherently lower investment in structural support than self-supporting woody plants (trees and shrubs). Trees invest a substantial part of their carbon and other resources in constructing and maintaining trunks and branches to compete for light. Lianas and vines, as structural parasites on trees, invest far less (Darwin 1865). This then poses a conundrum: Given that climbing plants receive similar benefits (light) at lower cost (stems), why aren't they more common relative to trees? After all, common though they are, lianas and vines are much less abundant than trees in the vast majority of forests and are completely absent from other forests. There are parallels here to the classic question "why is the world green?" (Hairston et al. 1960). In both cases, we need to step back and recognize that the world as we are used to seeing it is not the only way it could be, and that the current state requires explanation.

A first answer to the question of why lianas and vines aren't more common might be that, as parasites, they depend on trees to be their hosts, and cannot themselves dominate forests. However, local patches in which lianas or vines dominate and the canopy is very low (e.g., liana-choked gaps) are found in many temperate and tropical forests, although they generally occupy only a small fraction of the landscape (Terramura et al. 1991, Foster et al. 2008). Further, there are so-called "liana forests" in French Guiana, Bolivia, Brazil, and elsewhere in the tropics, in which lianas dominate over large areas (e.g., Pérez-Salicrup et al. 2001, Foster et al. 2008, Tymen et al. 2016).

In this essay, we consider what mechanisms regulate and limit the abundance of climbing plants at different scales, and how these might contribute to explaining variation in their prevalence within and among forests. We say a mechanism *regulates* climber abundance if it introduces negative density-dependence that prevents climbers from increasing in abundance without bounds (Turchin 1995). By *limiting* factors, we mean both density-dependent and density-independent factors that affect the abundances at which climbers are regulated. For simplicity, we refer in the remainder of the text simply to lianas; however, almost all our arguments apply equally to vines.

We propose that liana abundance is regulated and limited at three distinct and interacting scales: in the proportion of trees infested with lianas; in the liana load within the crowns of individual host trees; and in the proportion of the landscape that is in a liana-dominated, low-canopy state. We suggest that a disease ecology or host–parasite ecology framework offers useful insights for understanding the first two scales. Thus, the proportion infested can be understood as a function of the rates at which uninfested trees are colonized and infested, the rates at which infested host trees lose lianas, and the demographic rates of infested and uninfested hosts. We suggest that the biomass of lianas within a tree crown (parasite load) can be understood in terms of selection for the highest reproductive number, expressed as new hosts infested per infested host (Anderson and May 1982), which necessarily must balance the benefits of higher liana load and associated resource pre-emption against the cost to the parasite of negative impacts this load imposes on its host (Ichihashi and Tateno 2011). Finally, the proportion of the landscape that is in a liana-dominated, low-canopy state doesn't fit within a disease ecology or host–parasite framework; it is as though a parasite had a free-living alternative lifestyle that competed directly with its host. In the remainder of this essay, we discuss what regulates and limits liana abundance at each of these scales in turn.

Proportion of Trees Infested with Lianas

We can explain the proportion of trees infested with lianas in fundamentally the same way that we explain the prevalence of a disease or parasite in a host population. The proportion of infested trees must depend on (1) the *liana colonization rate*, the rate at which liana-free trees become infested (analogous to disease transmission); (2) the *liana loss rate*, the rate at which infested trees lose lianas (analogous to recovery from an infectious disease); and (3) the *host demographic rates*, specifically mortality and recruitment of infested and uninfested hosts (Fig. 1A). If these rates and their dependence upon the density and frequency of infested and liana-free trees are known, then we can solve the corresponding differential equations for the expected stable equilibrium proportion of trees infested with lianas, and the solution will illuminate exactly how the proportion infested relates to the parameters determining liana colonization, liana loss, and host demographic rates.

It is useful to frame this discussion with reference to a simplified model. This enables us to discuss factors in terms of their influences on parameters of the simplified model, or deviations from the simplified model forms. Let I be the number of liana-infested trees, and S the number of liana-free trees susceptible to liana colonization. In the simplest formulation, all liana-free trees are susceptible, and thus $S=N-I$, where N is the total number of trees, assumed constant for simplicity. A simple frequency-dependent model for the liana colonization rate (liana-free trees infested per unit of time) is $\beta SI/N=\beta I(N-I)/N$, where β is a constant that reflects the liana colonization pressure per infested tree (Fig. 1B). The simplest model of the liana loss rate (liana-infested trees losing lianas per time) is that it equals υI, where υ is a constant for the rate at which liana-infested trees lose lianas. And the simplest model for mortality of infested hosts is αI, where α is a constant infested tree mortality rate (Fig. 1C). This leads to the following differential equation for the number of infested individuals:

$$\frac{dI}{dt} = \beta I\left(\frac{N-I}{N}\right) - \upsilon I - \alpha I$$

Here, the proportion of trees infested by lianas is *regulated* by the negative density-dependence of colonization (the $N-I$ term): The number of trees newly colonized by lianas eventually declines as the availability of susceptible trees goes down. Under this model, the proportion of trees infested with lianas tends to a stable equilibrium,

$$\frac{\overline{I}}{N} = 1 - \frac{\upsilon + \alpha}{\beta}.$$

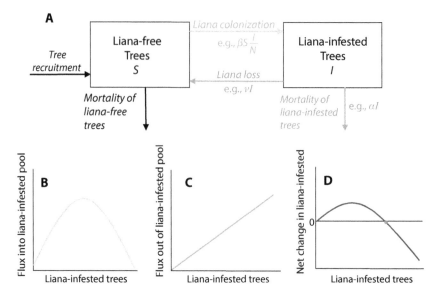

FIGURE 1. (*Caption on facing page.*)

The proportion infested increases with the liana colonization rate β, and declines with the liana loss rate v and the infested tree mortality rate α; thus, all three parameters can be said to *limit* the proportion infested.

We now discuss each of the three sets of underlying rates—liana colonization, liana loss, and host demography—and its controls. We thereby link the simplified model above to natural history knowledge about these processes.

Liana Colonization

In general, we expect the number of new trees infested in a forest stand to be a unimodal function of the total proportion infested, with an initial increase due to increased availability of lianas to do the colonizing, and an eventual decrease as the supply of liana-free susceptible trees dwindles (Fig. 1B). The height of the curve depends on liana reproductive success per infested tree, expressed in the simplified model through the parameter β. The point at which this curve returns to zero reflects the abundance of liana-free trees *susceptible* to colonization, which may be less than the total abundance of liana-free trees in the real world, unlike the simplified model. Any potential density- or frequency-dependence in reproductive success would result in deviations from the simplified model that alter the shape of the curve from a simple parabola. We treat each of these features of the curve in detail in the following paragraphs.

FIGURE 1. (A) A simple box model for the prevalence of liana infestation in a tree community. Under the assumption that the total number of trees remains constant, the proportion of trees infested with lianas depends only on the rate at which liana-free trees are colonized by lianas (light-gray arrow), and the rates at which liana-infested trees die or lose lianas (dark-gray arrows). More generally (if total tree number is not fixed), it will also depend on the mortality and recruitment rates of liana-free trees (black arrows). These rates can all take various functional forms; the simple examples discussed in the text are shown here. (B) The flux of trees into the liana-infested pool (newly infested trees per time) is expected to be a unimodal function of the number of trees infested with lianas. It depends on both the availability of liana-infested individuals to serve as a source of colonization (increasing part of the curve), and the availability of liana-free individuals susceptible to colonization (decreasing part of the curve). (C) The flux of trees leaving the liana-infested pool encompasses cases in which lianas are lost from the tree (tree transitions from liana-infested to liana-free) and death of liana-infested trees. Both of these fluxes are expected to be continuously increasing functions of the number of liana-infested trees; in the simplest case treated here they are proportional. (D) The combination of unimodal fluxes in and proportional fluxes out of the liana-infested tree pool leads the net change in liana-infested trees to be a hump-shaped function of the number of liana-infested trees, and thus to an expected stable equilibrium abundance of trees infested with lianas (point), below which the net change is positive, and above which it is negative. For the specific functional forms given in (A) and constant total population size N, the equilibrium proportion infested is $\overline{T}/N = 1-(\upsilon+\alpha)/\beta$. All these curves and associated parameters will depend upon liana life history strategies (e.g., aggressive versus conservative, climbing mode), tree life history strategy (e.g., fast versus slow), and the environment. The peak of the colonization curve (B) may be increased by factors that increase liana fecundity, increase juvenile liana survival, increase trellises for lianas to climb to the canopy, and/or reduce tree defenses against lianas, depending on which factors limit liana infestation. The right-hand x-intercept of the colonization curve will be reduced by factors that reduce the proportion of liana-free trees susceptible to colonization. The steepness of the loss functions (C) and thus the magnitude of the outgoing flux will be decreased by factors that decrease the mortality of infested trees and decrease the rate of liana loss from infested trees, including less effective tree defenses to promote liana loss. Factors that increase the colonization flux or decrease the flux out of the infested pool will increase the equilibrium abundance of liana-infested trees.

The height of the colonization curve depends on what would be called *transmission* in disease ecology. Here, this reflects the combined influences of liana seed production and vegetative shoot production per host tree infested, and the odds that these offspring survive and succeed in finding and climbing a host tree. Seed production and vegetative shoot production per infested host depend closely on the liana load in the host tree—whether there are many liana leaves in favorable (sunlit) positions on the tree, or few in unfavorable (shaded) positions, for example. (The factors controlling liana load will be considered in the section Liana Load Within Individual Trees.) The colonization success per liana seed depends on liana regeneration requirements and the conditions in the forest understory—especially light levels—as well as on the availability of suitable trellises to the canopy along which the juvenile liana can grow. Different liana climbing strategies are limited to different substrates; for example, tendril climbers require small stems to grip, twining lianas cannot climb stems above a certain diameter; and adhesive climbers require a relatively rough surface and cannot move between branches or trees (Putz 1984b). The success per vegetative shoot running along the ground similarly depends on the availability of trellises, whereas the success of shoots deployed from one crown to another depends on sufficient proximity of other uninfested crowns (Putz 1984b), and thus can only fully be addressed with a spatially explicit model.

The point at which the colonization curve returns to zero depends on the proportion of liana-free trees that are susceptible to liana colonization. If all liana-free trees are susceptible to liana colonization, then the colonization rate reaches zero only when all trees are infested with lianas. However, some trees may be resistant to liana colonization. Canopy trees may be able to escape in size: If they grow a fat enough trunk and high enough lower branches sufficiently separated from other trees, then they cease to be amenable to colonization, at least by most types of lianas (Putz 1984a, Campbell and Newbery 1993). Trees may also be uncolonizable by virtue of possessing highly effective defenses against lianas, such as mutualist ants that clip liana tendrils (Janzen 1969, Tanaka and Itioka 2011). This uncolonizable host population parallels the resistant/recovered population in many disease models. In this case, the colonization curve will return to zero at a proportion infested equaling one minus the proportion of uncolonizable hosts.

There are several mechanisms that could introduce negative density-dependence to liana "transmission" beyond that due simply to the exhaustion of susceptible hosts, and thereby change the shape of the colonization curve. Much of liana transmission to new hosts is through vegetative propagation, which is short-distance (Schnitzer et al. 2012a). This local transmission results in clustering, such that trees close to liana-infested trees are more likely to be already infested, and less likely to be liana-free and susceptible to colonization, than is the average for the larger forest stand (Schnitzer et al. 2012b). Insofar as liana infestation reduces tree reproductive success (Visser et al. 2017), higher densities of lianas may also lead to higher local abundances of tree species uncolonizable by lianas. Both of these mechanisms reduce liana reproductive success in areas of high liana densities.

On the other hand, there are other mechanisms by which liana reproductive success may increase with the proportion of trees infested, leading to positive density-dependence. The presence of liana-infested trees may change the environment in ways that increase colonization success and thereby introduce positive density-dependence. Liana-infested trees may be more likely to lose branches or experience die-back, increasing understory light and consequently tree and liana sapling survival, thereby increasing trellis availability and the probability that a liana seed or vegetative shoot will make it to the canopy. Some of the same factors that increase the proportion of trees infested also tend to increase mean liana loads per infested host (treated in the section Liana Load Within Individual Trees), and this in turn may increase liana reproductive success per infested host. Further, we hypothesize that higher liana abundance is likely to be associated with a greater incidence of host coinfection (more than one liana on a single host), and that this in turn may favor more aggressive liana strategies due to a tragedy of the commons (addressed in the section Liana Load Within Individual Trees). Such more aggressive strategies involve higher pre-emption of host resources, higher liana loads, and higher liana reproductive output per infested host. If the resulting positive density-dependence, whatever its origins, is sufficiently extreme, it could generate alternative stable states of high and low liana prevalence.

Liana Loss

Countering the gains in newly infested trees due to colonization are losses when infested trees die or lose their lianas (Figs. 1A, 1C). Liana-infested trees can lose their lianas when liana-infested branches are dropped, when lianas are pulled out of the tree or severed by a neighboring branchfall or treefall, or when lianas die due to disease, senescence, or other causes. In the simplest case, the rate at which liana-infested trees lose lianas is assumed to be a constant unrelated to the number of trees infested, so that total fluxes are proportional to the number of trees infested (Fig. 1C). This rate (v in our simple model) depends on liana traits (e.g., climbing strategy), tree traits (e.g., tree architecture), and environmental factors (e.g., windspeed) (Putz 1984a). Hypothesized tree defenses to increase the liana loss rate include higher trunk and branch flexibility, self-pruning of branches and/or large leaves, and bark shedding, although there are few studies testing these ideas (Hegarty 1991).

Alternatively, liana loss rates (per tree) may increase or decrease with the proportion of trees infested. We expect mean liana loads on individual trees to increase with the proportion of trees infested. Higher liana loads on a given branch would be expected to increase branchfall rates and thus loss rates. On the other hand, insofar as higher liana loads are associated with more lianas on a given tree, the probability that a tree entirely loses all its lianas could decrease with increasing liana load. Thus, the total flux of trees from liana-infested to liana-free could increase more than proportionally or less than proportionally with the proportion of trees infested (faster or slower than linear increase in Fig. 1C), thereby altering the equilibrium proportion infested (Fig. 1D).

Host Demography

The literature on variation in the proportion of trees infested with lianas within and among sites has focused on differences in liana colonization and loss (e.g., Putz 1984a). However, host demographic rates also limit the proportion of trees infested with lianas. Most obviously, mortality of infested trees removes infested trees from the community, to be replaced by recruits that are initially liana-free. Thus, the higher the mortality rate of infested trees (α in our simple model), the lower the proportion that will be infested for any given liana colonization and liana loss rates (Visser et al. 2017). The mortality rate of infested trees depends on the environment, tree traits, and liana traits, and increases with liana load (Ingwell et al. 2010, Wright et al. 2015). Thus, insofar as mean liana loads are higher when the proportion of trees infested is higher, we might expect the total mortality of infested trees to rise more than proportionally with the proportion of trees infested with lianas (faster than the linear increase shown in Fig. 1C), thereby decreasing the equilibrium proportion infested (Fig. 1D).

Host demographic rates have additional effects on the proportion of trees infested that are not captured in our simple model. Higher baseline (liana-free) tree mortality rates mean shorter tree lifespans and less time for trees to accumulate lianas before they die, reducing the proportion of trees infested. Negative effects of lianas on tree growth (Clark and Clark 1990, Ingwell et al. 2010) will reduce transitions of liana-infested trees into the larger size classes, and thereby reduce the proportion of large trees that are liana-infested.

Synthesis

Liana colonization, liana loss, and host demography necessarily together determine the proportion of trees infested with lianas (Fig. 1A). A key challenge is to elucidate these rates, their variation with proportion infested (i.e., the shapes of the functions in Figs. 1B, 1C), and the controls on these rates—e.g., determining to what degree the colonization rate is limited by the availability of liana seeds and shoots, understory light, and/or understory vegetation that provides trellises to the canopy. This will then illuminate what factors are most important for determining variation in the proportion of trees infested with lianas among species and sites, as we show in Visser et al. (2018a). Regulation requires negative density-dependence, and an obvious source of negative density-dependence on the proportion of trees infested with lianas is the declining availability of liana-free susceptible host trees. Additional negative density-dependence may emerge if liana loss rates and/or infested tree mortality rates increase with proportion of trees infested, as they may if mean liana load increases in parallel (discussed in the next section). Importantly, there is potential for alternative stable states of high and low liana infestation levels if there is sufficient positive density-dependence in liana colonization pressure, and/or if the liana loss rate declines sufficiently with the proportion infested, as discussed previously.

Liana Load within Individual Trees

A tree's liana load—the mass of lianas within its crown and the amount of light they preempt—affects the contribution that lianas in that tree make to colonizing other trees, the probability the host loses its lianas, and host mortality, growth, and fecundity. Thus, understanding what determines liana load within a host tree is a critical component of understanding what determines liana abundance more generally. Just as the proportion of trees infested with lianas parallels the proportion of hosts infected in disease ecology, so liana load within a tree crown parallels pathogen load within the host. Liana loads can in part be understood in terms of evolution of virulence—that is, selection for lianas with the highest reproductive number, defined as new hosts infested per infested host (Anderson and May 1982). This selection takes place in the context of a fundamental tradeoff between the benefits of higher liana load and associated resource preemption against the cost to the lianas of negative effects this load imposes on its host (Ichihashi and Tateno 2011).

Below, we first consider factors that influence selection on what we call the *target liana load*—that is, the liana load a liana seeks to achieve. In discussing target liana load, we focus on *liana leaf area index (LAI)* within the host crown, defined here as the total leaf area of liana leaves divided by the crown area, although we recognize that there are other aspects of liana load as well. We address three categories of factors shaping target liana load: diminishing returns from additional liana leaves, negative feedbacks

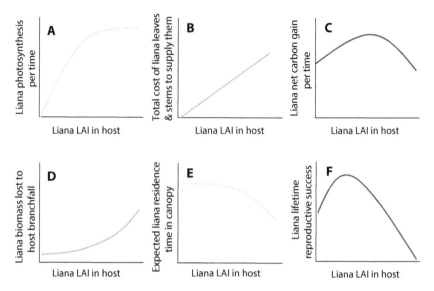

FIGURE 2. The hypothesized change in total benefits (light gray) and costs (dark gray) accruing to lianas within a particular host crown, and net impacts on liana success (black), as a function of the leaf area index (LAI) of lianas within that host crown. Liana LAI is the total liana leaf area on a given host divided by the total host crown area, and is used here as a measure of liana load. (A) Total photosynthesis per time is a saturating function of liana LAI because self-shading causes diminishing returns of additional liana leaves. (B) The total cost of liana leaves and the stems to supply them increases approximately proportionately with liana LAI. (C) Liana net carbon gain per time is a unimodal function of liana LAI, due to the combination of saturating returns and continually increasing costs. (D) Liana biomass losses to host branch fall increase with liana LAI, reflecting negative feedbacks on host tree branch retention. (E) Expected liana residence time in the canopy, and thus liana lifetime sun exposure (light access) and wind exposure (seed dispersal success), all decrease with liana LAI, reflecting negative feedbacks on host survival and growth. (G) Liana lifetime reproductive success is a unimodal function of liana LAI, with a peak at a lower LAI than that which maximizes liana net carbon gain per time (C). All of these curves will vary depending upon liana life history strategies (e.g., aggressive versus conservative, climbing mode), tree life history strategy (e.g., fast versus slow), and the environment. For example, photosynthesis at a given liana LAI (A) will be increased by a longer growing season, higher host light availability, and more aggressive liana leaf placement strategies, whereas the cost per liana LAI (B) will be increased by greater host tree height, and by greater per-height liana stem construction costs due to frost, drought, or a twining climbing strategy. Similarly, liana residence time (E) will be decreased by higher tree mortality rates and more negative effects of lianas on host tree survival.

from burdening host trees, and within-host competition among lianas. We then further consider factors that cause liana loads to be less than the target liana load.

Diminishing Returns of Higher Liana Leaf Area

Lianas, like most plants, tend to place their leaves disproportionately in the greatest-light environments they can reach. As a liana adds more leaves within a given host tree, accessible high-light positions are progressively exhausted, and additional leaves placed in the same crown are relegated to successively lower-light environments. Thus, benefits of additional leaves go down as the number of leaf layers increases, and total carbon gain of lianas on the host tree eventually saturates (Fig. 2A). At the same time, the marginal cost of leaves remains approximately constant, so that the total cost of leaves increases approximately proportionally with liana LAI (Fig. 2A). The combination of diminishing marginal returns and approximately constant marginal costs causes total liana net carbon gain per time to be a unimodal function of liana LAI (Fig. 2C). The diminishing returns of higher liana leaf area within a given crown are qualitatively the same for trees and for lianas. They have the potential to *regulate* liana LAI, just as they regulate tree LAI. We expect lianas to stop making leaves on a given host tree at or below some light level at which additional leaves are no longer a good investment and the total liana net carbon gain per time starts to decrease.

We hypothesize that the marginal costs of additional leaves are higher for climbing plants than for self-supporting trees, even though the total costs of supporting leaves are higher for trees. We thus further hypothesize that the liana LAI that maximizes net carbon gain of a resident liana on a host tree will be lower than the tree LAI that maximizes net carbon gain for the host. Our logic is that the marginal costs of leaves of canopy lianas include not only the costs of the leaves themselves, but also of the plumbing to the ground (xylem, phloem) required to supply these leaves with water and nutrients. In trees, we expect the size of the stem to be determined primarily by structural needs, and to be more than adequate for water and nutrient transport, so that the marginal cost of supplying an additional leaf with water are miniscule. In lianas, in contrast, the size of the stem is determined primarily by transport needs, and thus marginal costs for additional leaves are high, and become ever higher as the canopy becomes taller (and the stem becomes longer). The marginal cost of each additional liana leaf might decrease somewhat with increasing LAI if shade leaves are constructed more cheaply or require less water, but it is nonetheless considerably higher than the marginal cost for a tree. This difference is expected to lead lianas to have higher break-even light levels for leaf deployment than their host trees, with larger differences for taller hosts. Empirical observations are consistent with the prediction that lianas will have fewer leaf layers than trees. A review by Hegarty and Caballé (1989) found liana LAI of 1 to 3, whereas host tree LAI was 5 or more. This difference in LAI cannot be attributed to inherently lower shade tolerance in lianas: Even in a tropical forest where most canopy lianas have few or no leaves in the shade (Avalos et al. 2007), most liana species build and maintain shade leaves as juveniles in the understory (Gilbert et al. 2006).

Increasing Negative Feedbacks from Burdening the Host Tree

Additional negative density dependence of liana loads arises because higher liana loads impose increasing burdens on the host tree, and negative consequences for the host in turn feed back to resident lianas. First, the greater weight of more liana biomass in the tree increases the probability of branch-fall (Fig. 2D). Liana biomass that was on a branch that falls may be torn to the ground, imposing a cost on the respective liana or lianas. Or, in a best-case scenario for a liana, liana leaves that were on a branch that fell may be left hanging in the air in positions that are less optimal for light interception than those they occupied previously, positions that are likely to lead to dieback and reallocation of liana resources. In the worst scenario for a liana, the liana mass on the falling branch may take the entire liana with it or sever the stem linking the liana in the crown to the soil, removing the liana entirely from that host canopy, and potentially killing it. An increase in branchfall risk with liana load reduces liana residence time in the canopy, and thus expected associated future benefits (Fig. 2E).

More generally, greater liana load on the host tree will increase the mortality risk of the host and decrease its expected future growth, both of which have negative consequences for resident lianas. Most lianas place leaves where they pre-empt light that would otherwise be captured by the host tree, thereby reducing host carbon gain, reducing host growth and reproduction, and increasing host mortality (Visser et al. 2018b). The physical weight of higher liana loads also increases the structural burden on trees and the associated risks of treefall. Tree death in turn removes resident lianas from their advantageous positions in the canopy, whether immediately (treefall) or soon thereafter (standing dead tree) (Fig. 2E). Even if a resident liana survives the death of its host, it has lost or soon loses the scaffolding that provides it with access to light and suffers the consequences. Slowing tree growth has similar, although milder, negative consequences for resident lianas, as it reduces the future area of scaffolding that the resident lianas can benefit from in this host tree and increases the chance that light availability at the host crown will diminish as the host tree is overtopped by neighbors (Fig. 2F). Less favorable canopy positions also reduce the wind to which lianas are exposed, and thus reduce expected seed dispersal distances of the many lianas that are wind-dispersed (in tropical forests, lianas are considerably more likely to be wind-dispersed than are trees (Muller-Landau and Hardesty 2005)). When these negative feedbacks are taken into account, we expect the reproductive success of a liana on a given host tree will generally be maximized at a liana load that is lower than that which maximizes liana carbon gain per unit time (Ichihashi and Tateno 2011) (Figs. 2C, 2G). This is a classic case under which we expect evolution of reduced virulence in a parasite (Levin and Pimentel 1981, Anderson and May 1982).

It is important to recognize that there need not be a single optimal target liana load for all lianas at a site. Liana species co-occurring at the same site can vary widely in their host exploitation strategies and associated impacts on and feedbacks from host trees, paralleling variation in virulence among microbes co-occurring in the same host population. Such variation may in part arise in response to host heterogeneity, such as variation among co-occurring host tree species in response to liana infestation (Visser et al. 2018b), but can also emerge from game-theoretic dynamics even given homogenous host populations. Ichihashi and Tateno (2011) documented coordinated variation among four co-occurring temperate liana species in leaf light environments and effects on host growth. The most aggressive species placed more than half its leaves in very high-light environments (greater than 80% of full sun) and reduced host growth by 42%; the least aggressive placed 90% of its leaves in very-low-light environments (less than 20% of full sun) and had no effect on host growth; the two other species were intermediate in both respects. The liana species also varied in the number of host trees occupied by a single liana: The least aggressive species always occupied only a single tree, whereas the most aggressive species averaged 3.88 hosts per individual liana, and, of the two intermediate species, one averaged 1.59 and the other 2.56 hosts. A liana species that always occupied only a single tree would be expected to face especially strong negative feedbacks from burdening its host, and to have its interests relatively closely aligned with those of its host, to the point where it may even have an essentially commensal strategy, and an associated low target liana load (Ichihashi and Tateno 2011).

Within-Host Competition among Lianas

The above arguments regarding target liana loads implicitly assume that there is a single liana in any given host, but liana-infested trees often host more than one individual liana. Indeed, in one tropical site, 44% of infested trees hosted more than one *species* of liana (Visser et al. 2018b). Where there are multiple lianas in a single host, each liana bears the full cost of any restraint in its leaf deployment and proliferation, whereas the benefits of higher host survival and growth may be shared with other resident lianas. This can in principle set the stage for a tragedy of the commons favoring more aggressive host exploitation strategies and higher target liana loads (Nowak and May 1994, Mosquera and Adler 1998). Such dynamics could be mitigated if different lianas occupy different parts of a tree crown, and if feedbacks to branch growth and survival are largely localized. However, in many cases lianas intermingle in the crown. Further, the presence of one liana in a host crown often makes it easier for additional lianas to infest the tree, as the first liana stem can itself provide a route to access the canopy. Overall, it seems highly likely that the potential for within-host competition with other lianas substantially shapes selection on liana strategies, and that it is likely to favor more aggressive strategies.

Underachieving Target Liana Loads

Of course, realized liana loads may often be well below the theoretical target liana loads of the liana or lianas in a given crown. The clearest example of this is the presence of many trees with zero liana loads, which are extremely nonoptimal from the liana's perspective. The same forces that control the proportion of trees infested and lead many trees not to be infested, also contribute to "underinfestation". Most obviously, many trees with low liana loads may have been colonized relatively recently and be on their way to higher liana loads, with that process of liana spreading through the crown taking time.

In other cases, parts of trees may be inaccessible, limiting liana loads within those trees, just as there are some trees that are inaccessible to colonization. For example, if an upper tier of branches is separated from a lower tier by considerable vertical distance and accessible only via a large trunk, then lianas that cannot climb large trunks will be unable spread from the lower tier to the upper tier. Liana physiology may also limit where lianas place their leaves and thus the amount of light that is pre-empted; in particular, some species of lianas can deploy their leaves only close to trunks and major branches (e.g., ivy) (Putz and Holbrook 1991). Finally, mutualists may also play a role in making some areas of trees inaccessible, just as they make some trees inaccessible; for example, mutualist ants of epiphytes restrict liana spread in some emergent trees (Tanaka and Itioka 2011).

Synthesis

There are multiple sources of negative density-dependence that have the potential to regulate liana load within host crowns, by insuring that liana reproductive success declines with increasing liana load above a certain "target liana load" (Figs. 2C, 2G). This target liana load may vary among liana species, even within the same site, as there may be multiple optimal strategies. Multiple optimal strategies may emerge from variation in other liana traits influencing costs and benefits of different liana loads, from host heterogeneity, and simply from the game-theoretic nature of the underlying problem (Nowak and May 1994). In the end, however, such optimal strategies may or may not be particularly important in any given system, as liana loads may in practice often be well below target liana loads due to limitation by other factors, such as those that limit the proportion of trees infested (see the previous section, Proportion of Trees Infested with Lianas). All else being equal, a higher proportion of trees infested is expected to increase absolute colonization rates, and thereby increase realized liana loads relative to target liana loads, as lianas arrive in trees sooner and have more time to reach their target loads.

Proportion of Landscape in a Liana-Dominated, Low-Canopy State

Finally, at the landscape scale, we can think of areas as being in one of two states: either (1) tree-dominated forest, including tree-dominated canopy gaps, liana-free canopy trees, and liana-infested canopy trees, or (2) liana-dominated gaps, or more generally liana-dominated, low-canopy areas. In many types of forest, the entire area is tree-dominated, with no area in a liana-dominated, low-canopy state. In these landscapes, the liana-dominated low-canopy state must be unattainable and/or unviable (zero rate of transition into this state and/or infinite rate of transition out). In most other forests, liana-dominated gaps are rare, constituting a sort of temporarily arrested succession, into and out of which patches transition (Schnitzer et al. 2000). In the few landscapes in which liana-dominated, low-canopy states constitute a substantial fraction of total area, it appears that transitions between the states continue (Tymen et al. 2016), although there are areas in which the liana-dominated, low-canopy state appears to be highly persistent (Foster et al. 2008).

The frequency of the liana-dominated, low-canopy state at the landscape scale is governed by the rates of transition into and out of this state, and any frequency-dependence in these transitions (Fig. 3). In the simplest case, if the transition rates are density-independent, then there is a single and stable equilibrium proportion of the area expected to be in the liana-dominated, low-canopy state (solid and dashed lines in Figs. 3B–3D). If transitions in both directions are positively frequency-dependent, then alternative stable states are possible (dotted lines in Figs. 3E–3G). Such

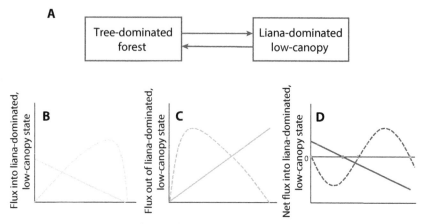

FIGURE 3. (A) The proportion of a landscape that is in a tree-dominated forest relative to the proportion in a liana-dominated, low-canopy state can be explained as a function of the rates of transition between the two, and how these vary with the relative proportions (if they do). Here tree-dominated forest encompasses tree-dominated low-canopy areas as well as areas with large trees, whether those trees are liana-infested or liana-free. (B) If the probability that a tree-dominated patch transitions to liana-dominated, low-canopy state is independent of the frequency of the two types of the landscape, then the flux of such transitions is proportional to the frequency of tree-dominated areas, and declines linearly with the proportion of liana-dominated areas (solid line). Alternatively, the transition rates may be frequency-dependent, reflecting dependence on the availability of seed sources as well, in which case the flux is a unimodal function of the frequency (dashed line) (C) Similarly, transitions out of the liana-dominated low-canopy state increase proportionally with the frequency of such patches if transition probabilities are frequency-independent (solid line), or may be a unimodal function if there is frequency-dependence (dashed line). (D) If neither of the transition probabilities is frequency-dependent (solid lines in (B) and (C)), then the net flux into the liana-dominated, low-canopy state will be a linear decreasing function of the frequency of this state (solid line in (D)), resulting in a stable equilibrium frequency (filled circle), which may be zero. If the transition probabilities are both frequency-dependent (dashed lines in (B) and (C)), then the net flux will be a more complex function of frequency (dashed line) and there may be alternative stable states of low and high frequencies of liana-dominated, low-canopy areas (open circles) as well as an unstable equilibrium x.

positive frequency-dependence is plausible, given that landscape-level abundance of a given vegetation type would be expected to feedback positively on seed availability and juvenile abundance, but has not been established.

A fundamental open question concerns the rates and controls on transitions between tree-dominated forest and liana-dominated low-canopy vegetation. For example, what is it that enables one or more trees to eventually "punch through" the surface of a liana-choked gap and grow to the canopy unencumbered by lianas? Are such escapes disproportionately by tree species with specific adaptations against lianas, such as mutualisms with ants that cut tendrils (Janzen 1969, Fiala et al. 1989)? Or do successful escape events generally follow some dieback of the dominant liana in that area of the gap, perhaps due to disease or senescence? It has been hypothesized that lianas are especially vulnerable to host-specialized natural enemies, perhaps because of their high frequency of vegetative propagation (Gentry 1991a). The dominance of kudzu and some other invasive lianas outside their native ranges—but not within them—is consistent with the idea that natural enemies of lianas likely play an important role in enabling escape from (and/or preventing transitions to) the liana-dominated, low-canopy state. It is also possible that disturbances or unusual environmental conditions are crucial for transitions in one direction or the other. Multiple authors have hypothesized that large patches of liana-dominated forest in South America originated after large-scale disturbance (Webb 1958, Balée and Campbell 1990).

Discussion

Lianas are critically important components of forest ecosystems, whose abundance has major consequences for forest carbon stores and cycling (van der Heijden et al. 2015), and for the relative performance of different tree species (Muller-Landau and Visser 2019). Yet ecologists remain very ignorant of liana strategies and the factors that control the relative abundance of lianas pursuing different strategies, which we argue is critical to understanding the overall abundance and impact of lianas in forests. There are central unanswered questions concerning the determinants of variation in liana abundance at the within-crown, within-forest, and landscape scales. These tie into unanswered questions concerning explanations for empirical variation in liana abundance with climate, forest age, and host tree life history strategy (Schnitzer 2005, DeWalt et al. 2015). The considerations discussed in this essay provide an organizing framework for contrasting hypotheses to explain such spatial variation in liana abundance (Tables 1,2).

Table 1. Consistency between Well-Established Patterns of Variation in Liana Abundance among Forests and Hypothesized Explanatory Mechanisms Motivated by Consideration of the Factors That Limit the Proportion of Trees Infested

PATTERNS CONSISTENT WITH MECHANISMS?	RELEVANT FIGURE	EMPIRICAL PATTERNS		
		TROPICAL FORESTS > TEMPERATE FORESTS (GENTRY 1991A, GENTRY 1991B, SCHNITZER 2005)	EARLY SUCCESSIONAL FORESTS > LATE SUCCESSIONAL FORESTS (DEWALT ET AL. 2000, BARRY ET AL. 2015)	DRY TROPICAL FORESTS > WET TROPICAL FORESTS (SCHNITZER 2005, DEWALT ET AL. 2010, DEWALT ET AL. 2015)
Higher densities of small understory trees increase the availability of trellises and thus liana colonization rates (Putz 1984b, Putz and Holbrook 1991, Balfour and Bond 1993).	1B	Yes. Tropical forests have more small understory trees.	Yes. Younger forests have more small understory trees.	Yes. Dry forests have more small understory trees.
Higher understory light levels increase liana juvenile survival and growth, increasing liana colonization rates (Hegarty 1991).	1B	No. Tropical forests have lower understory light availability.	No. Younger forests have lower understory light availability.	Yes. Dry forests have higher understory light availability.
Higher canopy tree mortality rates decrease the proportion of trees infested with lianas.	1C	No. Tropical forests have higher canopy tree mortality.	No. Younger forests have higher canopy tree mortality.	Yes. Dry forests have lower canopy tree mortality.

Hypothesized mechanisms

Table 2. Consistency between Well-Established Patterns of Variation in Liana Abundance among Forests and Hypothesized Explanatory Mechanisms Motivated by Consideration of the Factors That Limit Liana Load

Hypothesized mechanisms — Patterns consistent with mechanisms?	Relevant Figure	Empirical Patterns		
		Tropical forests > temperate forests (Gentry 1991a, Gentry 1991b, Schnitzer 2005)	Early successional forests > late successional forests (DeWalt et al. 2000, Barry et al. 2015)	Dry tropical forests > wet tropical forests (Schnitzer 2005, DeWalt et al. 2010, DeWalt et al. 2015)
Higher solar radiation increases expected carbon gain per leaf, favoring higher liana LAI.	2A	Yes. Tropical forests have higher solar radiation.	NA.	Yes. Dry forests have higher solar radiation.
Shorter canopies reduce the costs of supplying liana leaves with water and nutrients, favoring higher liana LAI.	2B	No. Temperate forests are on average shorter.	Yes. Early successional forests are shorter.	Yes. Dry forests are shorter.
Seasonal drought or frost increases the cost of supplying liana leaves with water and nutrients, favoring lower liana LAI.	2B	Yes. Temperate forests have seasonal frost.	NA	No. Dry forests have seasonal drought.
Host tree deciduousness enables phenological niche differentiation between lianas and trees, mitigating negative impacts of lianas on hosts, favoring higher liana LAI.	2D,E	No. Temperate forests have more deciduousness.	NA	Yes. Dry forests have more deciduousness.
Higher canopy tree turnover reduces the value of host survival to lianas and thus favors higher liana LAI.	2E	NA.	Yes. Early successional forests have higher turnover.	NA.

NA indicates that the hypothesized mechanism is not relevant to that comparison because the relevant factor does not vary strongly and systematically between the forest types being compared.

Globally, forests vary strongly in liana abundance overall and at all three of the scales distinguished here: the proportion of trees infested with lianas, the liana loads of infested trees, and the proportion of the landscape in a liana-dominated, low-canopy state. We suggest that variation in liana abundance can ultimately be explained by variation in the factors that regulate and limit lianas at one or more of these scales, with different factors potentially driving variation along different gradients. Rates of liana colonization, liana loss, and liana-infested tree mortality—which together determine the proportion of trees infested—all vary with the environment, liana species, and tree species (Visser et al. 2018a). Variation in these factors among forests naturally motivates hypotheses to explain variation in the proportion of trees infested with lianas and thus liana abundance, some of which were first advanced decades ago (Table 1). Similarly, the costs and benefits of different liana loads, as well as the degree to which target loads are realized, also vary with the environment, liana species, tree species, and their interaction. Variation in such factors along environmental gradients and between forest types naturally generates additional hypotheses to explain variation in liana loads and thus liana abundance (Table 2). Finally, expected rates of transition into and out of low-canopy, liana-dominated states, also vary strongly among forests and depend on the combination of liana and tree species traits. A better understanding of the controls on these transitions is critical to understanding variation in the frequency of such areas among forests.

Understanding what controls liana abundance is critical not only to understanding geographic variation in liana abundance and forest structure today, but also for predicting how liana abundance will respond to global change—another unanswered question. Liana abundance is increasing in Neotropical forests, and multiple hypotheses have been advanced to explain this pattern (Schnitzer and Bongers 2011). Increasing atmospheric carbon dioxide is hypothesized to favor lianas by increasing the survival and growth of juvenile lianas in the shaded understory (Korner 2009), which would be expected to increase liana colonization rates and thus the proportion of trees that are infested. Another hypothesis suggests climate change may drive increases in liana abundance as increases in evaporative demand from increasing temperatures outstrip increases in rainfall in many tropical areas, effectively making forests drier, consistent with higher liana abundances in drier tropical forests, a pattern that can itself be explained through a number of mechanisms (Schnitzer 2005) (Tables 1 and 2). Higher rates of disturbance, both natural and anthropogenic, also favor lianas as disturbances increase the abundance of early successional forests in which lianas thrive, and potentially increase transition rates to low-canopy, liana-dominated states. Finally, the introduction of nonnative lianas to areas lacking their natural enemies seems to threaten increased overall liana dominance in areas where these lianas become invasive (e.g., kudzu in the southeastern United States).

We hypothesize that a complete understanding of liana abundance and liana impacts on trees and forest ecosystem properties will require understanding what shapes the functional composition of liana communities. Liana species—even in the same forest—can vary widely in their traits (Asner and Martin 2012, Gallagher and Leishman 2012, Wyka et al. 2013) and host exploitation strategies, with important consequences for host tree growth and survival (Ichihashi and Tateno 2011). This parallels variation in virulence among co-occurring pathogens and parasites, with similar fundamental tradeoffs in which more aggressive resource acquisition by the liana comes at the cost of lower host growth and survival (Mosquera and Adler 1998). Changes in liana loads and liana abundance across sites are accompanied by shifts in the functional composition of liana communities, and specifically in the relative abundance of what might be considered more- versus less-virulent liana strategies. Different environments and tree communities select for different liana strategies and combinations of strategies, with liana communities themselves also shaping tree communities, contributing to a feedback loop. A game-theoretic framework is essential to understanding the liana strategies that emerge in any given forest. Explicit treatment of coinfection and host heterogeneity will also be necessary to understand liana functional composition in many forests. Such a research effort will also yield insights into liana diversity.

Here we have argued that tree–liana interactions can usefully be framed as host–parasite relationships, especially in the context of understanding the proportion of trees infested with lianas. Treating lianas as parasites ties into a large and relevant literature of models and empirical research in host–parasite interactions and disease ecology. However, although a host–parasite framework is broadly useful for understanding lianas, there are also competitive and even mutualistic aspects of the relationships of lianas with trees. Lianas compete with trees belowground for water and nutrients as classic competitors (Stewart and Schnitzer 2017). Many liana species also have free-standing juveniles that compete directly with tree saplings for regeneration opportunities in the understory and in canopy gaps (Schnitzer and Carson 2010, Stewart and Schnitzer 2017). Lianas can also have some positive effects on their hosts. Lianas may increase nutrient availability and soil quality beneath host crowns through the higher quality of their leaf litter, through mutualisms with nitrogen fixers, and by transporting nutrients from afar (Tang et al. 2012). The higher electrical conductivity of liana stems relative to trees also means that lianas can serve as lightning rods, diverting electrical current and thereby reducing the probability of tree death from a lightning strike (Gora et al. 2017). A full understanding of liana dynamics and abundance may require explicit consideration of these interactions.

Lianas have long fascinated biologists (Darwin 1865), and much has been written on liana physiology, diversity and distribution, and effects on trees and forest ecosystems (Putz and Mooney 1991, Isnard and Silk 2009, Schnitzer et al. 2015). Nonetheless, lianas remain relatively understudied compared to trees, due to their lesser silvicultural importance combined with the difficulty of measuring and modeling them. Simply censusing liana stem diameters requires far more complicated protocols than those for trees (Gerwing et al. 2006, Schnitzer et al. 2008). Further, measurements of stem rooting location and diameter are much less informative about lianas than they are about trees because biomass is less well-related to diameter for lianas (Gehring et al. 2004), and a single liana may extend great distances from its rooting point (Putz 1984b). Fortunately, new technologies make it easier to map lianas at the top of the canopy (Marvin et al. 2016), and to map the three-dimensional structure of the understory and the routes taken by liana stems within those canopies (Calders et al. 2015). Lianas also pose special challenges for modeling; we are not aware of a single published forest model or vegetation model that includes lianas (Verbeeck and Kearsley 2016), despite their known importance for tree recruitment, growth and survival, and forest dynamics. We argue here that many models from disease ecology can usefully be adapted for lianas (as we have shown in Visser et al. 2018a). The development of spatially explicit, individual-based models of forests that include lianas and liana-tree interactions would also aid in tackling questions of the controls on the abundance of lianas pursuing different regeneration and infestation strategies.

We tend to take it for granted that lianas are not more abundant, just as we take it for granted that the world is green. But just as the answer to the question "why is the world green?" is not self-evident, neither is the answer to the question of why lianas aren't more abundant. Natural history and considerations of basic biology provide us with insights into the processes that may play a role in governing liana abundance, as described in this essay. The key unanswered questions concern the relative roles of different factors within and across scales, and their importance for understanding variation in liana abundance today and in the future.

Acknowledgments

We thank Rae Winfree, Joe Wright, Ryan Chisholm, Bob Holt, Dave Tilman, and Marco Visser for thoughtful comments on earlier versions of this manuscript. This work also benefited from discussions with Simon Levin, Henry Horn, and members of the 2015–2016 Pacala lab.

References

Anderson, R. M., and R. M. May. 1982. Coevolution of hosts and parasites. Parasitology 85:411–426.

Asner, G. P., and R. E. Martin. 2012. Contrasting leaf chemical traits in tropical lianas and trees: implications for future forest composition. Ecology Letters 15:1001–1007.

Avalos, G., S. S. Mulkey, K. Kitajima, and S. J. Wright. 2007. Colonization strategies of two liana species in a tropical dry forest canopy. Biotropica 39:393–399.

Balée, W., and D. G. Campbell. 1990. Evidence for the successional status of liana forest (Xingu River basin, Amazonian Brazil). Biotropica 22:36–47.

Balfour, D. A., and W. J. Bond. 1993. Factors limiting climber distribution and abundance in a southern African forest. Journal of Ecology 81:93–100.

Barry, K. E., S. A. Schnitzer, M. van Breugel, and J. S. Hall. 2015. Rapid liana colonization along a secondary forest chronosequence. Biotropica 47:672–680.

Calders, K., G. Newnham, A. Burt, S. Murphy, P. Raumonen, M. Herold, et al. 2015. Non-destructive estimates of above-ground biomass using terrestrial laser scanning. Methods in Ecology and Evolution 6:198–208.

Campbell, E. J. F., and D. M. Newbery. 1993. Ecological relationships between lianas and trees in lowland rain-forest in Sabah, East Malaysia. Journal of Tropical Ecology 9:469–490.

Clark, D. B., and D. A. Clark. 1990. Distribution and effects on tree growth of lianas and woody hemiepiphytes in a Costa Rican tropical wet forest. Journal of Tropical Ecology 6:321–331.

Darwin, C. 1865. On the movements and habits of climbing plants. Journal of the Linnean Society 9:1–118.

DeWalt, S. J., S. A. Schnitzer, L. F. Alves, F. Bongers, R. J. Burnham, Z. Cai, et al. 2015. Biogeographical patterns of liana abundance and diversity. In: S. A. Schnitzer, F. Bongers, R. J. Burnham, and F. E. Putz, eds. Ecology of Lianas. Hoboken: Wiley, 131–146.

DeWalt, S. J., S. A. Schnitzer, J. Chave, F. Bongers, R. J. Burnham, Z. Q. Cai, et al. 2010. Annual rainfall and seasonality predict pan-tropical patterns of liana density and basal area. Biotropica 42:309–317.

DeWalt, S. J., S. A. Schnitzer, and J. S. Denslow. 2000. Density and diversity of lianas along a chronosequence in a central Panamanian lowland forest. Journal of Tropical Ecology 16:1–19.

Fiala, B., U. Maschwitz, T. Y. Pong, and A. J. Helbig. 1989. Studies of a south east Asian ant-plant association—protection of *Macaranga* trees by *Crematogaster borneensis*. Oecologia 79:463–470.

Foster, J. R., P. A. Townsend, and C. E. Zganjar. 2008. Spatial and temporal patterns of gap dominance by low-canopy lianas detected using EO-1 Hyperion and Landsat Thematic Mapper. Remote Sensing of Environment 112:2104–2117.

Gallagher, R. V., and M. R. Leishman. 2012. A global analysis of trait variation and evolution in climbing plants. Journal of Biogeography 39:1757–1771.

Gehring, C., S. Park, and M. Denich. 2004. Liana allometric biomass equations for Amazonian primary and secondary forest. Forest Ecology and Management 195:69–83.

Gentry, A. H. 1991a. Breeding and dispersal systems of lianas. In: F. E. Putz and H. A. Mooney, eds. The biology of vines. Cambridge: Cambridge University Press, 393–423.

Gentry, A. H. 1991b. The distribution and evolution of climbing plants. In: F. E. Putz and H. A. Mooney, eds. The biology of vines. Cambridge: Cambridge University Press, 3–49.

Gerwing, J. J., S. A. Schnitzer, R. J. Burnham, F. Bongers, J. Chave, S. J. DeWalt, et al. 2006. A standard protocol for liana censuses. Biotropica 38:256–261.

Gilbert, B., S. J. Wright, H. C. Muller-Landau, K. Kitajima, and A. Hernandez. 2006. Life history trade-offs in tropical trees and lianas. Ecology 87:1281–1288.

Gora, E. M., P. M. Bitzer, J. C. Burchfield, S. A. Schnitzer, and S. P. Yanoviak. 2017. Effects of lightning on trees: a predictive model based on in situ electrical resistivity. Ecology and Evolution 7:8523–8534.

Hairston, N. G., F. E. Smith, and L. Slobodkin. 1960. Community structure, population control and competition. American Naturalist 94:421–4215.

Hegarty, E. E. 1991. Vine-host interactions. In: F. E. Putz and H. A. Mooney, eds. The biology of vines. Cambridge: Cambridge University Press, 357–375.

Hegarty, E. E., and G. Caballé. 1989. Physiological ecology of mesic, temperate woody vines. In: F. E. Putz and H. A. Mooney, eds. Ecology of vines. Cambridge: Cambridge University Press, 313–335.

Ichihashi, R., and M. Tateno. 2011. Strategies to balance between light acquisition and the risk of falls of four temperate liana species: to overtop host canopies or not? Journal of Ecology 99:1071–1080.

Ingwell, L. L., S. J. Wright, K. K. Becklund, S. P. Hubbell, and S. A. Schnitzer. 2010. The impact of lianas on 10 years of tree growth and mortality on Barro Colorado Island, Panama. Journal of Ecology 98:879–887.

Isnard, S., and W. K. Silk. 2009. Moving with climbing plants from Charles Darwin's time into the 21st century. American Journal of Botany 96:1205–1221.

Janzen, D. H. 1969. Allelopathy by myrmecophytes: ant Azteca as an allelopathic agent of cecropia. Ecology 50:147–153.

Körner, C. 2009. Responses of humid tropical trees to rising CO_2. Annual Review of Ecology Evolution and Systematics 40:61–79.

Levin, S., D. Pimentel. 1981. Selection of intermediate rates of increase in parasite-host systems. American Naturalist 117:308–315.

Marvin, D. C., G. P. Asner, and S. S. Schnitzer. 2016. Liana canopy cover mapped throughout a tropical forest with high-fidelity imaging spectroscopy. Remote Sensing of Environment 176:98–106.

Mosquera, J., and F. R. Adler. 1998. Evolution of virulence: a unified framework for coinfection and superinfection. Journal of Theoretical Biology 195:293–313.

Muller-Landau, H. C., and B. D. Hardesty. 2005. Seed dispersal of woody plants in tropical forests: concepts, examples, and future directions. In: D.F.R.P. Burslem, M. A. Pinard, and S. Hartley, eds. Biotic interactions in the tropics. Cambridge: Cambridge University Press, 267–309.

Muller-Landau, H. C., and M. D. Visser. 2019. How do lianas and vines influence competitive differences and niche differences among tree species? Concepts and a case study in a tropical forest. Journal of Ecology 107:1469–1481.

Nowak, M. A., and R. M. May. 1994. Superinfection and the evolution of parasite virulence. Proceedings of the Royal Society of London B: Biological Sciences 55:81–89.

Pérez-Salicrup, D. R., V. L. Sork, and F. E. Putz. 2001. Lianas and trees in a liana forest of Amazonian Bolivia. Biotropica 33:34–47.

Putz, F. E. 1984a. How trees avoid and shed lianas. Biotropica 16:19–23.

Putz, F. E. 1984b. The natural history of lianas on Barro Colorado Island, Panama. Ecology 65:1713–1724.

Putz, F. E., and N. M. Holbrook. 1991. Biomechanical studies of vines. In: F. E. Putz and H. A. Mooney, eds. The biology of vines. Cambridge: Cambridge University Press, 73–97.

Putz, F. E., and H. A. Mooney, eds. 1991. The biology of vines. Cambridge: Cambridge University Press.

Schnitzer, S. A. 2005. A mechanistic explanation for global patterns of liana abundance and distribution. American Naturalist 166:262–276.

Schnitzer, S. A. 2018. Testing ecological theory with lianas. New Phytologist 220:366–380.

Schnitzer, S. A., and F. Bongers. 2011. Increasing liana abundance and biomass in tropical forests: emerging patterns and putative mechanisms. Ecology Letters 14:397–406.

Schnitzer, S. A., and W. P. Carson. 2010. Lianas suppress tree regeneration and diversity in treefall gaps. Ecology Letters 13:849–857.

Schnitzer, S. A., J. W. Dalling, and W. P. Carson. 2000. The impact of lianas on tree regeneration in tropical forest canopy gaps: Evidence for an alternative pathway of gap-phase regeneration. Journal of Ecology 88:655–666.

Schnitzer, S. A., S. A. Mangan, J. W. Dalling, C. A. Baldeck, S. P. Hubbell, A. Ledo, et al. 2012a. Liana abundance, diversity, and distribution on Barro Colorado Island, Panama. PLOS One 7:e52114.

Schnitzer, S. A., F. E. Putz, F. Bongers, and K. Kroening, eds. 2015. Ecology of lianas. Hoboken: Wiley.

Schnitzer, S. A., S. Rutishauser, and S. Aguilar. 2008. Supplemental protocol for liana censuses. Forest Ecology and Management 255:1044–1049.

Stewart, T. E., and S. A. Schnitzer. 2017. Blurred lines between competition and parasitism. Biotropica 49:433–438.

Tanaka, H. O., and T. Itioka. 2011. Ants inhabiting myrmecophytic ferns regulate the distribution of lianas on emergent trees in a Bornean tropical rainforest. Biology Letters 7:706–709.

Tang, Y., R. L. Kitching, and M. Cao. 2012. Lianas as structural parasites: A re-evaluation. Chinese Science Bulletin 57:307–312.

Terramura, A. H., W. G. Gold, and I. N. Forseth. 1991. Physiological ecology of mesic, temperate woody vines. In: F. E. Putz and H. A. Mooney, eds. The biology of vines. Cambridge: Cambridge University Press, 245–285.

Turchin, P. 1995. Population regulation: Old arguments and a new synthesis. In: N. Cappuccino and P. Price, eds. Population dynamics. Cambridge, Mass.: Academic Press, 19–40.

Tymen, B., M. Rejou-Mechain, J. W. Dalling, S. Fauset, T. R. Feldpausch, N. Norden, O. L. Phillips, B. L. Turner, J. Viers, and J. Chave. 2016. Evidence for arrested succession in a liana-infested Amazonian forest. Journal of Ecology 104:149–159.

van der Heijden, G.M.F., J. S. Powers, and S. A. Schnitzer. 2015. Lianas reduce carbon accumulation and storage in tropical forests. Proceedings of the National Academy of Sciences of the United States of America 112:13267–13271.

Verbeeck, H., E. Kearsley. 2016. The importance of including lianas in global vegetation models. Proceedings of the National Academy of Sciences of the United States of America 113:E4.

Visser, M. D., H. C. Muller-Landau, S. A. Schnitzer, H. de Kroon, E. Jongejans, and S. J. Wright. 2018a. A host-parasite model explains variation in liana infestation among co-occurring tree species. Journal of Ecology 106:2435–2445.

Visser, M. D., S. A. Schnitzer, H. C. Muller-Landau, E. Jongejans, H. de Kroon, L. S. Comita, S. P. Hubbell, and S. J. Wright. 2018. Tree species vary widely in their tolerance for liana infestation: A case study of differential host response to generalist parasites. Journal of Ecology 106:781–794.

Webb, L. J. 1958. Cyclones as an ecological factor in tropical lowland rain-forest, north Queensland. Australian Journal of Botany 6:220–228.

Wright, S. J., I.-F. Sun, M. Pickering, C. D. Fletcher, and Y.-Y. Chen. 2015. Long-term changes in liana loads and tree dynamics in a Malaysian forest. Ecology 96:2748–2757.

Wyka, T. P., J. Oleksyn, P. Karolewski, and S. A. Schnitzer. 2013. Phenotypic correlates of the lianescent growth form: a review. Annals of Botany 112:1667–1681.

The World Beneath Us

Making Soil Biodiversity and Ecosystem Functioning Central to Environmental Policy

Diana H. Wall and Ross A. Virginia

How and when will the diversity of species in soil and what they do for us become recognized and integrated into the discipline of ecology for use in global-scale climate change research and land use policy? This question is the basis for a key unsolved problem in ecology from our perspective, one that is globally relevant: the growing need for understanding how soil biodiversity and soil ecology contribute to the functioning of terrestrial ecosystems and the services they provide to society. There is increasing urgency for intensified ecological research and synthesis with a belowground focus because soils are at the center of multiple interacting global environmental issues, such as: land use change, loss of fertile land area, climate change, desertification, biodiversity loss (human, animal, and plant), disease suppression, and food production. Considerable evidence exists for example, that land use change and widespread soil degradation decreases species diversity and abundance in soils (Dirzo et al. 2014, Franco et al. 2016, Tsiafouli et al. 2015). Developing options to address these global scale challenges requires that we understand the role of soil organisms in ecosystems and acknowledge our dependence on soils and biodiversity. Here we present a few insights that we posit have contributed to dissuading ecologists from bringing soils, biodiversity, and the role of species to the forefront in ecological research and education.

Hypotheses on Biodiversity and Ecosystem Functioning

The hypotheses on the relationships between biodiversity and ecosystem functioning are difficult to describe and quantify at the species level (Balvanera et al. 2006, Kardol et al. 2016, Lawton 1994, Naeem 1998). Of the several hypotheses, the continuing dogma of species redundancy (Soliveres et al. 2016) dominates discussions of the role of individual species in soil functioning. It suggests that because of the massive, undescribed diversity in soils, ecosystem processes are insensitive to any species losses or shifts in species abundance. However, we argue that the focus on the unsolved problem of whether species in soil are redundant in an ecosystem process slows ecological recognition that there is a wealth of increasing and untapped knowledge available on individual species and their interactions within communities of soil biota. In fact, several studies have shown that species of soil fauna thought to be redundant actually have different effects on ecosystem functioning (Andriuzzi et al. 2016a, Faber and Verhoef 1991, Postma-Blaauw et al. 2005), consistent with the idiosyncratic hypothesis of biodiversity function (Lawton et al. 1994). For example, it is well known that soils harbor "keystone" species, whose losses or invasions affect the whole ecosystem (Crowther et al. 2013, Loss and Blair, 2011, Pelini et al. 2015, Schwarzmüller et al. 2015). Finally, it has been shown that biodiversity effects emerge from *interactions* between species in soil (Andriuzzi et al. 2016b, Heemsbergen et al. 2004, Wardle, 2006). Therefore, given that soil communities are increasingly under threat due to global changes, addressing when loss of single species or communities will diminish soil quality and ecosystem functioning is an increasingly urgent question for all terrestrial ecologists.

Soils Are Habitats

Soils are complex and extremely heterogenous habitats for communities. For example, for the USA and its territories, the United States Department of Agriculture has described 20,000 soil series (taxonomic units) that have considerable variation in soil age, soil physical and chemical characteristics, and vegetation cover. Soil scientists recognize that biota (plants, microbes, invertebrates, and vertebrates) are one of the five state variables that form soils (the others being parent material, relief, regional climate and time) along with the influences of human activities (e.g., acidic deposition) (Jenny 1980). Assemblages of soil organisms (microbes and invertebrates, or total soil taxa as observed with molecular technology) differ among soils because they are products of evolutionary and environmental (e.g., land use change) processes that form each soil.

There is evidence that soil characteristics and climate influence the biogeography of soil species at local, regional and continental scales (Freckman and Virginia 1997). We know, for example, that soil taxa do not necessarily follow plant diversity latitudinal gradients (Bardgett and van der Putten, 2014, Nielsen et al. 2014, Wu et al. 2011, van den Hoogen et al. 2019, Phillips et al. 2019). Research to identify the soil habitat factors that determine geographic ranges of soil biota at various taxonomic levels is ongoing (Kerfahi et al. 2016, Robeson et al. 2011, Wu et al. 2011), and with standardized sampling and molecular technologies should allow ecologists to generate new hypotheses on biogeographic distributions and evolutionary relationships. This information is key to understanding the adaptation and resilience of soil biota to key processes affecting soils.

For example, it is widely recognized that organisms are active participants in dynamic biogeochemical processes and in soil formation, both major ecosystem processes. In particular, fungi, plants, and large soil animals such as earthworms, termites, ants, dung beetles, and some mammals are all recognized as ecosystem engineers due to their regulation of soil structure and fertility. Fungi stabilize soil and redistribute nutrients (Rillig and Mummey, 2006, Wilson et al. 2009), and burrowing animals produce channels that facilitate water and air flow through soil (Spurgeon et al. 2013). Macroinvertebrates greatly enhance the decomposition of plant and animal organic matter (Gessner et al. 2010), and by affecting soil structure they influence plant growth and soil nutrient retention (Howison et al. 2016, van Groenigen et al. 2014); their work in concert with other smaller species of invertebrates and microbes leads to the formation of soil organic matter aggregates where soil carbon is stabilized and stored (Six et al. 2006). Gathering more information on ecological roles of the many species involved in soil formation in a range of climatic and geographic regions will be useful for determining whether species influence resilience of soils or are modifiers of an ecosystem service that is key for restoration of degraded soils. The extrapolation of visible aboveground impacts of soil biota to the subsurface formation and maintenance of soils, remains a critical challenge.

We Need to Know More about Species and Their Functional Roles

When we began our careers, a 'black box' was the designation for all life in soil and soil ecosystem functions. By simplifying the complexity of the many species and their interactions into soil foodwebs, significant progress has been made on determining the role of various functional groups in carbon and nutrient cycling. Generally, a representative species with known feeding preference and ecology is a basis for those organisms of similar morphology being placed in a functional or trophic group of detrital foodwebs (Barrett et al. 2008). However, all species within a functional group may not have the same function, or they may have more than one function, or their functions may change through time with development. Adding information on the identity and ecological roles of several species in each functional group will improve quantification on the relationships of soil organisms and their role in ecosystem function, such as soil carbon cycling. Our perspective is derived from more than 25 years of work in a low diversity field ecosystem, the McMurdo Dry Valleys of Antarctica. There, as we sampled across a vast landscape dominated by two- three nematode species, we categorized two species as being in the bacterial feeder functional group (consumers of soil microbes). As we collected data to characterize soil habitats and factors affecting distribution of these two species, it became apparent that species abundance of *Scottnema lindsayae* flourished in dry, saline, low organic carbon soils and was absent in the wetter, more carbon rich soils where *Plectus antarcticus* was dominant (Virginia and Wall 1999). This information on species identity informed the distributional maps for the dry valleys, informed estimates of species contribution to soil carbon cycling (Barrett et al. 2008), and studies on biodiversity shifts under climate changes where soil moisture and primary productivity were expected to respond (Andriuzzi et al. 2018).

In soils, however, as Kibblewhite et al. (2008) noted, the standard definitions of diversity and ecosystem function are so broad as to be poorly suited to characterizing soil processes. For example, measures such as soil respiration or decomposition rates over time encompass total biotic species and activity, whereas processes aboveground such as leaf respiration can be ascribed to a single species. This inability to easily ascribe the functioning of individual species in soils reinforces the concept that soil species are redundant in a functional group or for an ecosystem process (Shaw et al. 2018, Soliveres et al. 2016). Adding more information on the species of soil biodiversity and their functional relationships in ecosystem processes can lead to more precise testing of biodiversity–ecosystem functioning hypotheses and perhaps a new interdisciplinary paradigm of soil functioning will emerge.

Information on Individual Species, or Their Natural Histories and Hosts Is Available

The often-held perception that there is little information on species and their ecology in soil overlooks a vast body of knowledge that is available from an extensive number of disciplines that can inform ecosystem processes. Many species and their natural history (feeding habits, physiologies, life cycles, survival states, geographic ranges, hosts) have been studied for centuries, albeit in disciplinary silos focused on disease and food production (soil microbiology, biology, mycology, protozoology, human and animal parasitology, plant pathology, nematology, entomology). Researchers have examined species in soil to serve as models for developmental biology (e.g., the nematode *Caenorhabditis elegans*) and for studies of population and community dynamics in natural and managed systems; to identify predator–prey systems for use in biocontrol by understanding their life cycles as pests, pathogens, and parasites of humans, other animals, and plants; and as beneficial symbionts for forestry and agriculture where soil biota (rhizobia, mycorrhizae) have large effects on an ecosystem function (productivity).

A good example of the species specificity in some mesofauna in soils are plant parasitic nematodes whose species have highly evolved specific relationships with their hosts. Some species secrete cellulases and chitinases to break down and penetrate host root cell walls; there they have evolved differences in root feeding sites (meristem versus cortex), some species are sedentary, others migratory, or have parasitic ecologies acting as either endo- or ectoparasites. Their effects on plant metabolism (carbon allocation, water transfer, changes in root architecture, necrosis of roots or leaves) are well known to scientists and growers as dependent on the nematode species involved. Likewise, many of the microbes and invertebrates that prey on nematodes are known. Yet, despite this knowledge, there are few detailed US geographic maps of soils and their relation to key soil-borne nematode, bacterial, or fungal plant or human pathogens. It is clear that soil biota and many of their species are important in plant, animal, and human health. Ecologists have an opportunity to explore and cross the disciplinary boundaries and connect soil science, agriculture, microbial ecology, biogeochemistry, remote sensing, and geographic information systems such that the wealth of information on species is integrated to better understand the relationships of soil biodiversity and ecosystem functioning and their benefits to planet health. This approach would also allow detection of species movement, loss, or the establishment of invasive species.

Where Next?

Good functioning of soils is critical for life on earth now and will be a critical component of global policy decisions in the next several decades as natural resources and our food, water, landscapes, climate, and air are altered. Thus, we believe that the time is now for incorporating soils as a core subject within the discipline of ecology as has been better accomplished for sciences focused on freshwater, oceans, and the atmosphere. By relegating soils to being a fringe field of ecology, we will continue to inadequately address modifications to biodiversity and functioning for a major component of the earth that provides generation and renewal of soil fertility, clean air and water, and controls of disease, erosion, and biogeochemical cycling. Our perspective in polar research adds to this urgency as we see the polar land masses warmed, ice retreat, altered biodiversity, and other global changes that will accelerate biotic feedbacks affecting soil organic matter dynamics and greenhouse gas emissions. Treating soils as a major biotic habitat with feedbacks to multiple ecosystem functions will enable better options for land management at regional and global scales.

There has been rapid progress in soil science and soil ecology, particularly in soil biodiversity and ecosystem functioning, that is both noteworthy and encouraging. We use as a basis to gauge progress, priorities from Wall and Virginia (2000), that were modified from previous soil ecology priorities from the 1990s. The summary of these priorities from 2000 (in italics) and our updates on progress are as follows:

Development of new techniques, standardization of techniques for sampling and analysis and informatics to enhance the database on soil biodiversity. These advances have been rapid and have expanded the lists of biodiversity and functional roles in soils. Molecular techniques now allow unprecedented analysis of total soil biodiversity (prokaryotes and eukaryotes) from a single soil sample and comparison to other samples (approximately 600–800 soil samples) collected at the same time. Many of the species or operational taxonomic units are rare or new to science and the numbers of sequences per soil sample confirms the immense biodiversity in the world beneath us. These advances in technology will promote soil biodiversity analyses across landscapes, regions, and globally. Metagenomic approaches are also improving information on species and their functions and potential responses to global change (for example, analysis of functional genes across ecosystems and responses to climatic variables). Costs for the technology per sample are declining, which will encourage studies combining soil habitat (soil characterization) for each sample. Bioinformatics frameworks are also advancing and being tested. Needed for the future will be greater representation of the biodiversity in a soil sample because at present only a small fraction of the biodiversity in a soil sample is captured by techniques that often used less than a gram of soil. Whether soil species identifications based on morphological data will be combined with molecular characterizations to provide more robust databases is as yet unresolved. Because these advances primarily have been focused on microorganisms, a priority continues to be the combined study and analysis of both invertebrates and microbes (or prokaryotes and eukaryotes) from the same soil sample, to elucidate the relationship and influence of invertebrates on the microbial populations and diversity.

Development of interdisciplinary and international cross-site experiments and predictive models to quantify the relationship of soil biodiversity to eco-

system processes for use in global change policy. Progress in experiments and models is occurring, but it is still limited, and many cross-site experiments and models are done at the grass-roots level and are poorly funded. Priorities are to assure the incorporation of sites in developing nations and those ecosystems having faster rates of land use and other global changes. Inclusion of experts in soil-borne plant, human, and other animal diseases, as is happening with planetary or one health research, will be a basis for projecting future diversity change, disease potential, and other service changes to soils.

Development of syntheses of distributional patterns of soil biodiversity globally, to predict how soils should be maintained for sustainable use. There have been advances in this priority, such as formation of the Global Soil Biodiversity Initiative (GSBI) in 2011 as a grass-roots partnership of soil researchers from multiple disciplines working to enhance the use of soil biodiversity science and ecosystem services in global environmental policy and the sustainable management of terrestrial ecosystems. A 2016 product of this initiative has been maps of distributions of soil biodiversity and maps of global threats to soil biodiversity published in the Global Soil Biodiversity Atlas (Orgiazzi et al. 2016) that serve as the framework for hypotheses to be tested in the lab, field, and through modeling. Notably, the GSBI's influence contributed to favorable decisions relating to soil biodiversity from the United Nations Convention on Biological Diversity Fourteenth Conference of the Parties (COP 14) in Fall 2018. Future needs include long-term observational networks and synthesizing case studies of soil biodiversity and ecosystem functioning with other soil-related disciplines across management and spatial scales as a basis for sustainable management of terrestrial ecosystems.

These three priorities show that there has been sufficient knowledge for some time to include soils, soil biodiversity science, and ecosystem functioning as a central component of ecology in order to address more accurately multilinked global challenges (Wall and Six 2015). Research has continued on many of these priorities and the results are impressive although not complete. We reemphasize here what we proposed in 2000 as one of the most pressing research issues: We must increase training of ecologists to include working on species, their natural history, and ecological roles of soil-dwelling organisms. To do this will require a breadth of soil-related scientists integrated within ecology and engaged in bringing new technologies and ideas to the study of soil biodiversity and ecosystem functioning. Today's students are often disappointed by the minimal attention to soils and soil biodiversity presently included in ecology lectures but can be encouraged to fill that gap to prepare for future challenges.

Closing Thoughts

The black box of soil has opened and an exciting era for ecology awaits. It reveals soil and biodiversity characteristics that can be documented, functions that can be evaluated based on more precise species identifications, exciting new hypotheses on soil biodiversity and ecosystem functioning that can be more accurately tested at field scale and beyond, and presents a wealth of linkages and interdependence between biodiversity in soil and aboveground. Importantly, opening the black box has shown that we do not need to know all species in a soil to manage it, but more species-level information in soil food webs is key to better quantification of the most significant interacting functions. Opening the black box shows the multiple needs for research and syntheses on soil biodiversity databases, multifunctions of species within functional groups, cross-regional and continental distributional patterns, and inclusion of soil habitat characteristics with vegetation and wildlife management plans. This is a scientifically challenging and pivotal time to plan for the future. Looking to the next fifty years, we urge a new generation of ecologists to embrace soil and its related disciplines as a core component of our discipline of ecology to better inform and reverse terrestrial habitat degradation and preserve the many services provided by soil biodiversity.

Acknowledgments

We thank our numerous colleagues, especially our NSF McMurdo Dry Valley LTER colleagues, who influenced our thinking about soil biodiversity and ecosystems.

References

Andriuzzi, W. S., B. J. Adams, J. E. Barrett, R. A. Virginia, and D. H. Wall. 2018. Observed trends of soil fauna in the Antarctic Dry Valleys: Early signs of shifts predicted under climate change. Ecology 99:312–321. doi: https://doi.org/10.1002/ecy.2090.

Andriuzzi, W. S., P. T. Ngo, S. Geisen, A. M. Keith, K. Dumack, T. Bolger, et al. 2016a. Organic matter composition and the protist and nematode communities around anecic earthworm burrows. Biology and Fertility of Soils 52:91–100. doi:10.1007/s00374-015-1056-6.

Andriuzzi, W. S., O. Schmidt, L. Brussaard, J. H. Faber, and T. Bolger. 2016b. Earthworm functional traits and interspecific interactions affect plant nitrogen acquisition and primary production. Applied Soil Ecology 104:148–156. doi: https://doi.org/10.1016/j.apsoil.2015.09.006.

Balvanera, P., A. B. Pfisterer, N. Buchmann, J. S. He, T. Nakashizuka, D. Raffaelli, and B. Schmid. 2006. Quantifying the evidence for biodiversity effects on ecosystem functioning and services. Ecology Letters 9:1146–1156. doi: https://doi.org/10.1111/j.1461-0248 .2006.00963.x.

Bardgett, R. D., and W. H. van der Putten. 2014. Belowground biodiversity and ecosystem functioning. Nature 515:505–511. doi: https://doi.org/10.1038/nature13855.

Barrett, J. E., R. A. Virginia, D. H. Wall, and B. J. Adams. 2008. Decline in a dominant invertebrate species contributes to altered carbon cycling in a low-diversity soil ecosystem. Global Change Biology 14:1734–1744. doi: https://doi.org/10.1111/j.1365-2486 .2008.01611.x.

Crowther, T. W., D.W.G. Stanton, S. M. Thomas, A. D. A'Bear, J. Hiscox, T. H. Jones, J. Voříšková, P. Baldrian, and L. Boddy. 2013. Top-down control of soil fungal community composition by a globally distributed keystone consumer. Ecology 94:2518–2528. doi: https://doi.org/10.1890/13-0197.1.

Dirzo, R., H. S. Young, M. Galetti, G. Ceballos, N.J.S. Isaac, and B. Collen. 2014. Defaunation in the Anthropocene. Science 345:401–406. doi: https://doi.org/10.1126/science .1251817.

Faber, J. H., and H. A. Verhoef. 1991. Functional differences between closely-related soil arthropods with respect to decomposition processes in the presence or absence of pine tree roots. Soil Biology and Biochemistry 23:15–23.

Franco, A.L.C., M.L.C. Bartz, M. R. Cherubin, D. Baretta, C.E.P. Cerri, B. J. Feigl, D. H. Wall, C. A. Davies, and C. C. Cerri. 2016. Loss of soil (macro)fauna due to the expansion of Brazilian sugarcane acreage. Science of the Total Environment 563–564:160–168. doi: https://doi.org/10.1016/j.scitotenv.2016.04.116.

Freckman, D. W., and R. A. Virginia. 1997. Low-diversity antarctic soil nematode communities: distribution and response to disturbance. Ecology 78:363–369. doi: https://doi .org/10.1890/0012-9658(1997)078[0363:LDASNC]2.0.CO;2.

Gessner, M. O., C. M. Swan, C. K. Dang, B. G. McKie, R. D. Bardgett, D. H. Wall, and S. Hättenschwiler. 2010. Diversity meets decomposition. Trends in Ecology and Evolution 25:372–380. doi: https://doi.org/10.1016/j.tree.2010.01.010.

Heemsbergen, D. A., M. P. Berg, M. Loreau, J. van Hal, J. H. Faber, H. A. Verhoef. 2004. Biodiversity effects on soil processes explained by interspecific functional dissimilarity. Science 306:1019.

Howison, R. A., M. P. Berg, C. Smit, K. van Dijk, and H. Olff. 2016. The importance of coprophagous macrodetritivores for the maintenance of vegetation heterogeneity in an African savannah. Ecosystems 19:674–684. doi: https://doi.org/10.1007/s10021-016 -9960-7.

Jenny, H. 1980. The soil resource: Origin and behavior. Berlin: Springer Nature.

Kardol, P., H. L. Throop, J. Adkins, and M. A. de Graaff. 2016. A hierarchical framework for studying the role of biodiversity in soil food web processes and ecosystem services. Soil Biology and Biochemistry 102:33–36. doi: https://doi.org/10.1016/j.soilbio.2016.05 .002.

Kerfahi, D., B. M. Tripathi, D. L. Porazinska, J. Park, R. Go, and J. M. Adams. 2016. Do tropical rain forest soils have greater nematode diversity than High Arctic tundra? A metagenetic comparison of Malaysia and Svalbard. Global Ecology and Biogeography 25:716–728. doi: https://doi.org/10.1111/geb.12448.

Kibblewhite, M. G., K. Ritz, and M. J. Swift. 2008. Soil health in agricultural systems. Philosophical Transactions of the Royal Society B: Biological Sciences 363:685–701. doi: https://doi.org/10.1098/rstb.2007.2178.

Lawton, J. H. 1994. What do species do in ecosystems? Oikos 71:367–374.

Loss, S. R., and R. B. Blair. 2011. Reduced density and nest survival of ground-nesting songbirds relative to earthworm invasions in northern hardwood forests. Conservation Biology 25:983–992. doi: https://doi.org/10.1111/j.1523-1739.2011.01719.x.

Naeem, S. 1998. Species redundancy and ecosystem reliability. Conservation Biology 12:39–45. doi: https://doi.org/10.1046/j.1523-1739.1998.96379.x.

Nielsen, U. N., E. Ayres, D. H. Wall, G. Li, R. D. Bardgett, T. Wu, and J. R. Garey. 2014. Global-scale patterns of assemblage structure of soil nematodes in relation to climate and ecosystem properties. Global Ecology and Biogeography 23:968–978. doi: https://doi.org/10.1111/geb.12177.

Orgiazzi, A., R. D. Bardgett, E. Barrios, V. Behan-Pelletier, M. J. I. Briones, J.-L. Chotte, and G. B. De Deyn, 2016. Global soil biodiversity atlas. Brussels: European Commission. doi: https://doi.org/10.2788/2613.

Pelini, S. L., A. M. Maran, A. R. Chen, J. Kaseman, and T. W. Crowther. 2015. Higher trophic levels overwhelm climate change impacts on terrestrial ecosystem functioning. PLOS One 10:1–10. doi: https://doi.org/10.1371/journal.pone.0136344.

Phillips, H.R.P., C. A. Guerra, M.L.C. Bartz, M.J.I. Briones, G. Brown, T. W. Crowther, et al. 2019. Global distribution of earthworm diversity. Science. doi: 10.1126/science.aax4851.

Postma-Blaauw, M. B., F. T. de Vries, R.G.M. De Goede, J. Bloem, J. H. Faber, and L. Brussaard. 2005. Within-trophic group interactions of bacterivorous nematode species and their effects on the bacterial community and nitrogen mineralization. Oecologia 142:428–439. doi: https://doi.org/10.1007/s00442-004-1741-x.

Rillig, M. C., and D. L. Mummey. 2006. Mycorrhizas and soil structure. New Phytologist 171:41–53.

Robeson, M. S., A. J. King, K. R. Freeman, C. W. Birky, A. P. Martin, and S. K. Schmidt. 2011. Soil rotifer communities are extremely diverse globally but spatially autocorrelated locally. Proceedings of the National Academy of Sciences of the United States of America 108:4406–4410. doi: https://doi.org/10.1111/j.0954-6820.1944.tb15060.x.

Schwarzmüller, F., N. Eisenhauer, and U. Brose. 2015. "Trophic whales" as biotic buffers: Weak interactions stabilize ecosystems against nutrient enrichment. Journal of Animal Ecology 84:680–691. doi: https://doi.org/10.1111/1365-2656.12324.

Shaw, E. A., B. J. Adams, J. E. Barrett, W. B. Lyons, R. A. Virginia, and D. H. Wall. 2018. Stable C and N isotope ratios reveal soil food web structure and identify the nematode *Eudorylaimus antarcticus* as an omnivore–predator in Taylor Valley, Antarctica. Polar Biology 41:1013–1018. doi: https://doi.org/10.1007/s00300-017-2243-8.

Six, J., S. D. Frey, R. K. Thiet, and K. M. Batten. 2006. Bacterial and fungal contributions to carbon sequestration in agroecosystems. Soil Science Society of America Journal 70:555–569. doi: https://doi.org/10.2136/sssaj2004.0347.

Soliveres, S., F. van der Plas, P. Manning, D. Prati, M. M. Gossner, S. C. Renner, et al. 2016. Biodiversity at multiple trophic levels is needed for ecosystem multifunctionality. Nature 536:456–459. doi: https://doi.org/10.1038/nature19092.

Spurgeon, D. J., A. M. Keith, O. Schmidt, D. R. Lammertsma, and J. H. Faber. 2013. Land-use and land-management change: Relationships with earthworm and fungi communities and soil structural properties. BMC Ecology 13:1–13. doi: https://doi.org/10.1186/1472-6785-13-46.

Tsiafouli, M. A., E. Thébault, S. P. Sgardelis, P. C. de Ruiter, W. H. van der Putten, K. Birkhofer, et al. 2015. Intensive agriculture reduces soil biodiversity across Europe. Global Change Biology 21:973–985. doi: https://doi.org/10.1111/gcb.12752.

van den Hoogen, J., S. Geisen, D. Routh, H. Ferris, W. Traunspurger, D. A. Wardle, et al. 2019. Soil nematode abundance and functional group composition at a global scale. Nature. doi: 10.1038/s41586-019-1418-6.

van Groenigen, J. W., I. M. Lubbers, H. M. J. Vos, G. G. Brown, G. B. De Deyn, and K. J. van Groenigen. 2014. Earthworms increase plant production: a meta-analysis. Scientific Reports 4:1–7. doi: https://doi.org/10.1038/srep06365.

Virginia, R. A., and D. H. Wall. 1999. How soils structure communities in the Antarctic Dry Valleys. BioScience 49:973–983. doi: https://doi.org/10.2307/1313731.

Wall, D. H., and J. Six. 2015. Give soils their due. Science 347:695. doi: https://doi.org/10.1126/science.aaa8493.

Wall, D. H., and R. A. Virginia. 2000. The world beneath our feet: Soil biodiversity and ecosystem functioning. In: P. Raven and T. A. Williams, eds. Nature and human society: The quest for a sustainable world. Washington D.C.: National Academy Press, 225–241.

Wardle, D. A. 2006. The influence of biotic interactions on soil biodiversity. Ecology Letters 9:870–886. doi: https://doi.org/10.1111/j.1461-0248.2006.00931.x.

Wilson, G.W.T., C. W. Rice, M. C. Rillig, A. Springer, and D. C. Hartnett. 2009. Soil aggregation and carbon sequestration are tightly correlated with the abundance of arbuscular mycorrhizal fungi: Results from long-term field experiments. Ecology Letters 12:452–461. doi: https://doi.org/10.1111/j.1461-0248.2009.01303.x.

Wu, T., E. Ayres, R. D. Bardgett, D. H. Wall, and J. R. Garey. 2011. Molecular study of worldwide distribution and diversity of soil animals. Proceedings of the National Academy of Sciences 108:17720–17725. doi: https://doi.org/10.1073/pnas.1103824108.

PART V

ECOLOGY AND HEALTH

Ecology and Medicines

Andrew F. Read

This year, more than 2 million people in the richest countries in the world will be overwhelmed by evolution and die. At the heart of that carnage is an ecological process that is barely being studied. My choice for Unsolved Problem in Ecology: What regulates the population densities of drug resistant pathogens, parasites and cancer cells?

The Grand Challenge

One of the greatest triumphs of twentieth-century medicine was the discovery of drugs that could be used to treat infections and cancer (Greenwood 2008, Burch 2010, DeVita and DeVita-Raeburn 2015). But almost as soon as those wonder drugs were discovered, failures due to what we would now call drug resistance were observed. Every known cancer drug can fail for this reason, as can most antimicrobial drugs. Today, virtually all cancer deaths in rich countries are because therapeutically resistant populations of neoplastic cells come to so dominate a tumor that initially effective treatment no longer works (Aktipis et al. 2011). Likewise, drug-resistant strains of microbes increasingly challenge global health in settings as diverse as US hospitals, Mumbai slums, and animal food production systems. For cancers, resistance evolution plays out de novo in each patient. For infections, de novo evolution can be quite rare, but having emerged, drug resistant strains can rapidly spread globally. One estimate has it that by 2050, antimicrobial resistance will kill more people than cancer does today—if nothing is done (O'Neill 2014).

Clearly, there is much that can be done. Inventing new drugs is important, but a constant search for the nth-generation drug to treat resistance to the $(n-1)^{th}$-generation drug (Foo and Michor 2014) is not obviously sustainable, especially when resistance mechanisms get ever more generic and when it costs in excess of US\$1 billion to bring a new drug to market. Nondrug solutions have to be a top priority—particularly ways to attack infections and tumors with biologics, such as phage, vaccines, and immunotherapy (although many of those approaches will also drive antagonistic evolution in the target cell populations). And for infections, simple things such as hygiene and alterations to farm practices can help. But for all that, it is hard to imagine that medicine can continue to deliver health gains in the twenty-first century without heavy reliance on therapeutic chemotherapy. We need to figure out how to use current and next-generation drugs in a way that delivers therapy without also delivering drug resistance.

Human immunodeficiency virus (HIV) treatment shows that is possible (zur Wiesch et al. 2011). The right combination of drugs delivered with the right regularity and at the right doses prevents the evolution of resistance to antiviral drugs and makes HIV infection survivable. Apparently evolution-proof drug treatment therapies have also recently been developed for hepatitis C virus (Ke et al. 2015). These strategies work by preventing resistance arising in the first place. If for example there is 10^{-8} chance that a mutation conferring resistance to a drug will occur, the chance that an individual pathogen or tumor cell will simultaneously acquire resistance to n drugs with different modes of action is 10^{-8n}, a vanishingly small number. But patients are not always fully compliant with the right regimens (zur Wiesch et al. 2011, Ke et al. 2015), so periods of monotherapy sometimes result. And even where patients can be relied on, combination therapy is not always possible. There can be limited drug options, especially because contrasting modes of action are required. Cross-resistance readily evolves anyway in many cancers and infections, and in the case of infections, de novo resistance can be a small part of the problem. For example, multidrug-resistant tuberculosis is mostly caught from other people (Luciani et al. 2009). Moreover, for many bacterial infections there are often good data that patient health outcomes are not improved by combination therapy (reviewed by Woods and Read 2015). Where immunity is going to clear an infection anyway, asking a patient to swallow the extra cost and side effects of an additional drug to perhaps prevent the spread of resistance in a hospital is tricky stuff.

So combination therapy can provide a solution to the problem of drug resistance but, at least as it is currently formulated, not a universal one. We have to explore other ways to use drugs to treat patients while minimizing the resistance evolution. I contend that there will be a myriad of solutions, but to get at them, we need to add some serious ecology to current efforts in oncology, molecular genetics, pharmacology, and clinical microbiology.

The Ecological Challenge

Evolutionary rescue, the ability of a population under rapid decline to evolve traits to enable population recovery before extinction, is relatively well studied by evolutionary ecologists and geneticists not least in the con-

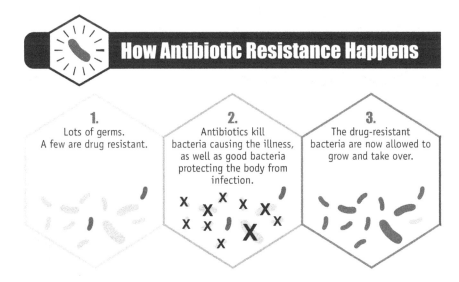

FIGURE 1. Public education schematic of the evolution of antibiotic resistance. Figure reprinted from and courtesy of the Centers for Disease Control and Prevention. http://www.cdc.gov/drugresistance/about.html.

text of climate change (Gonzalez et al. 2013). An unfortunate phrase in the context of human health, *evolutionary rescue* is nonetheless precisely what is happening when drug resistance emerges in infections and cancer. And at the heart of it all, is ecology (Uecker et al. 2014).

Figure 1 is the schematic used by the Centers for Disease Control and Prevention to explain to the public how antibiotic resistance evolves. The process is fundamentally the same for all the pathogens, parasites, and cancers on which we wage chemical warfare. In the absence of drug treatment, the population size of resistant cells is tiny. After aggressive drug treatment, the sensitive cells are removed and the resistant cells replicate, sometimes to life-threatening densities (and, in the case of infection, often to transmissible densities). Thus, therapy-sensitive populations prevent the replication of resistance. When treatment removes those populations, a massive expansion of the resistant cell population ensues. It is the resistant population that causes medical problems. This process can play out many times in a patient as treatment regimens are repeated or changed, but whatever the details, the key effect is the vast amplification of resistance. Before treatment, the sensitive population makes resistance so rare as to be of no concern (that is how we recognize a drug as being effective or useful in the first place). Afterwards, the ecology of the situation is so rearranged by chemotherapy that resistance has been amplified by many, many orders of magnitude.

Of course, the resistance arises in the first place by some sort of genetic event such as mutation or, in the case of bacterial infections, from horizontal gene transfer from nontarget organisms in the microbiome (Willmann and Peter 2017). Over the last few decades, science has generated vast catalogs of the genetic events that cause resistance in infections and tumors. But from the genetic details alone, we can infer rather little about what will happen once an individual pathogen or tumor cell has become resistant. Ecological forces determine its fate. We control those forces with our drugs. If we want to stop creating therapy-resistant cancers and pathogens, or deal with them once we have, we have to understand those ecological forces.

Drug resistance is thus a problem in applied ecology. When we want to manage endangered species, invasive species, or pest species, we need to first understand the ecological processes determining the dynamics of the populations of concern. Solutions come from that population science. In agriculture, where problems of resistance to insecticides and herbicides are legend, many solutions have come from studying the ecology of the target organisms (Radcliffe et al. 2009). The same must be true, surely, with drug-resistant tumor cells, pathogens, and parasites. Yet the ecology of resistance in patients is barely being studied. We often know in excruciating detail the genetic and cellular mechanics of drug action and resistance mechanisms. By contrast, our understanding of the ecology by which any resistance mechanism threatens patient health is rudimentary. I am not sure it is even yet at the level of Elton's *Animal Ecology*—published in 1927.

Competition

Let me illustrate that claim. The dominant ecological force that can account for the dynamics schematized in Figure 1 is competition. Most obviously, competition with sensitive progenitor cells or cells in the microbiome prevents resistance emerging once it has arisen. Competitive suppression explains why resistance is rare prior to treatment; competitive release accounts for the subsequent resistance explosion. Experiments bear out that interpretation. Figure 2 shows the ecological processes of competitive suppression and competitive release in play in experimental infections in my lab.

But, echoing earlier debates in ecology (Diamond 1975, Conner and Simberloff 1979), others do not think competition is important. For instance, Bruce Levin (quoted in Kupferschmidt 2016) says that because fitness costs of resistance are often not that high, competitive release is not a very strong force (see also Ankomah and Levin 2014). My view is

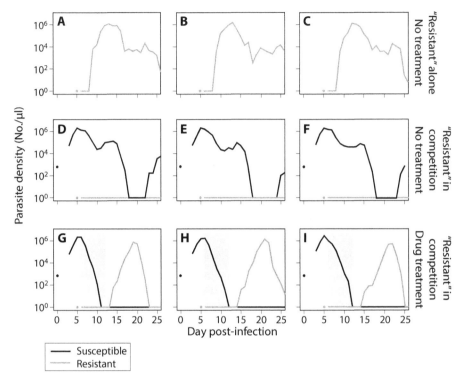

FIGURE 2. Competition in a mouse model of malaria. Kinetics of infections in nine mice infected with pyrimethamine-resistant (gray) and -sensitive parasites (black). Mice were infected with 25 resistant parasites (gray dots); some mice were infected five days earlier with approximately 1 million sensitive parasites (black dots). Mice G–I were treated with pyrimethamine for seven days (gray bars) to eliminate the sensitive parasites. Note that, formally, gray flat lines denote times at which densities were below polymerase chain reaction (PCR) detection. Otherwise proliferating populations of resistant parasites (mice A–C) are competitively suppressed by sensitive parasites (mice D–F). Drug treatment releases them from competitive suppression (mice G–I). Reprinted from Day, Huijben, and Read, Is selection relevant in the evolutionary emergence of drug resistance? *Trends in Microbiology* 23:126–133. Copyright © 2015, with permission from Elsevier.

that competition will still be important even if there are zero fitness costs to resistance, and that where experiments have been done (e.g., see Fig. 2), they clearly show competition. But without doubt, the importance of competition between drug-sensitive and drug-resistant cells in infections and tumors is an empirical question. A consensus cannot be reached until ecological experiments in a wide range of settings have been done. And incredibly, there isn't even consensus on how to do the relevant experiments. What are often called competition experiments in studies of antimicrobial resistance are often measures of relative growth rate in exponential growth phase, before any density dependence kicks in (Hughes and Andersson 2015).

Even if competition turns out to be a key ecological force, as I strongly believe it will, that is only a starting point. What is the mechanism of any competition? Resource, interference, or apparent (immunity-mediated) competition (Read and Taylor 2001)? My attempts to test for apparent competition in a mouse model of malaria, some of the only in vivo studies I am aware of, have proven frustratingly contradictory (Raberg et al. 2006, Barclay et al. 2008). Quite possibly in that particular host–pathogen system, several types of competitive interaction are going on at once, with their relative importance changing during the course of infection. More generally, I know of only one study looking at competition between resistant and sensitive lineages across resource gradients (Wale et al. 2017), the simplest and most fundamental ecological question. Are single nutrients limiting? Which ones? How does immunity modify that? Where is the density dependence coming from? For infections, when is competition with drug-sensitive progenitors that which is most important, and when is it competition with confecting strains or commensal members of the microbiome? Or is something else the reason we are not already neck deep in resistance?

The ecological processes controlling resistance in two arenas seem particularly important. With both tumors and bacterial biofilms, very strong competition must be going on; indeed, necrosis is common within tumors as cells die from lack of oxygen and glucose (Gillies et al. 2012). How do sensitive and resistant cells compete in those arenas? Is there competition on the growing edge of biofilms and tumors or are those regions so resource rich that there is no density dependence? Therapeutic drug concentrations can be very high on the outside of biofilms and tumors; how does that modify any competition? If we understood the ecology in those settings, and how pathogens and cells adapt to them, we could make more informed decisions about dosing regimens, choice of drugs and drug combinations, and of nonchemotherapeutic solutions (Day et al. 2015, Gatenby and Brown, 2017, Hansen et al. 2017, Hochberg 2018).

More generally, competition is quite possibly the only major natural force preventing the evolution of drug resistance. I have trouble imaging what else stops resistance emerging in the absence of treatment. Complex adaptive valleys that cannot be crossed in the absence of drugs? That is competition. Drift? That is demographics. Waiting time for mutations? They seem to come along pretty fast when we use drugs. Whatever: some natural forces stop resistance spreading—otherwise we would not have anything we call a drug. What is that natural force? Can we intensify it? Can we harness it (Wale et al. 2017)?

Coda

Every day, oncologists battle to keep their patients alive. When they lose that battle (as they will almost 600,000 times this year in the United States alone), they lose it because chemotherapy has profoundly remodeled the ecology of a tumor. We know next to nothing about that ecology. It is of course easier to sequence cells than to measure growth rates while manipulating resource gradients, controlling immunity, hormones, cell–cell interactions, and chemotherapy. But not so long ago, the idea of getting a complete DNA sequence for a single tumor cell was fantasy. I hope that in the not-too-distant future, we will be able to peer into a tumor or a biofilm, observe the relevant natural history and do decisive ecological experiments. I bet it will be hugely interesting. And save lives.

Acknowledgments

My thinking on these issues has been shaped over the years by many people. I particularly thank J. de Roode, who first suggested we use drugs to manipulate within-host competition; the many members of my group who subsequently got sucked in to doing it; T. Day, E. Hansen, D. Kennedy, and R. Woods, who have done so much to help clarify my recent thinking; and I. Hastings, B. Levin, and N. White, who continue to persuade me that ecology is not obvious.

References

Aktipis, C. A., V. S. Y. Kwan, K. A. Johnson, S. L. Neuberg, and C. C. Maley. 2011. Overlooking evolution: A systematic analysis of cancer relapse and therapeutic resistance research. PLOS One 6:9. doi: https://doi.org/10.1371/journal.pone.0026100.

Ankomah, P., and B. R. Levin. 2014. Exploring the collaboration between antibiotics and the immune response in the treatment of acute, self-limiting infections. Proceedings of the National Academy of Sciences of the United States of America 111(23):8331–8. doi: https://doi.org/10.1073/pnas.1400352111.

Barclay, V. C., L. Raberg, B.H.K. Chan, S. Brown, D. Gray, and A. F. Read. 2008. CD4+T cells do not mediate within-host competition between genetically diverse malaria parasites. Proceedings of the Royal Society B: Biological Sciences 275:1171–1179. doi: https://doi.org/10.1098/rspb.2007.1713.

Burch, D. 2010. Taking the medicine: A short history of medicine's beautiful idea and our difficulty swallowing it. London: Vintage.

Conner, E., and D. Simberloff. 1979. The assembly of species communities: Chance or competition? Ecology 60:1132–1140.

Day, T., V. Huijben, and A. F. Read. 2015. Is selection relevant in the evolutionary emergence of drug resistance? Trends in Microbiology 23(3):126–133. doi: https://doi.org/10.1016/j.tim.2015.01.005.

DeVita, V., and E. DeVita-Raeburn. 2015. The death of cancer. New York: Sarah Crichton Books.

Diamond, J. 1975. Assembly of species communities. In: M. Cody and J. Diamond, eds. Ecology and evolution of communities. Cambridge, Mass.: Harvard University Press, 342–444.

Elton, C. 1927. Animal ecology. Oxford: Oxford University Press.

Foo, J., and F. Michor. 2014. Evolution of acquired resistance to anti-cancer therapy. Journal of Theoretical Biology 355:10–20. doi: https://doi.org/10.1016/j.jtbi.2014.02.025.

Gatenby, R. A., and J. S. Brown. 2017. The evolution and ecology of resistance in cancer therapy. Cold Spring Harbor Perspectives in Medicine. 2017. doi: 10.1101/cshperspect.a033415.

Gillies, R. J., D. Verduzco, and R. A. Gatenby. 2012. Evolutionary dynamics of carcinogenesis and why targeted therapy does not work. Nature Reviews Cancer 12:487–493. doi: https://doi.org/10.1038/nrc3298.

Gonzalez, A., O. Ronce, R. Ferriere, and M. E. Hochberg. 2013. Evolutionary rescue: An emerging focus at the intersection between ecology and evolution. Philosophical Transactions of the Royal Society B: Biological Sciences 368:20120404. doi: https://doi.org/10.1098/rstb.2012.0404.

Greenwood, D. 2008. Antimicrobial drugs: Chronicle of a twentieth century medical triumph. Oxford: Oxford University Press.

Hansen, E. A., R. J. Woods, and A. F. Read. 2017. How to use a chemotherapeutic agent when resistance to it threatens the patient. PLoS Biology 15:e2001110. https://doi.org/10.1371/journal.pbio.2001110.

Hochberg, M. E. 2018. An ecosystem framework for understanding and treating disease. Evolution, Medicine and Public Health. 2018:270–286. https://doi.org/10.1093/emph/eoy032.

Hughes, D., and D. I. Andersson. 2015. Evolutionary consequences of drug resistance: Shared principles across diverse targets and organisms. Nature Reviews Genetics 16:459–471. doi: https://doi.org/10.1038/nrg3922.

Ke, R. A., C. Loverdo, H. F. Qi, R. Sun, and J. O. Lloyd-Smith. 2015. Rational design and adaptive management of combination therapies for Hepatitis C virus infection. PLOS Computational Biology 11:20. doi: https://doi.org/10.1371/journal.pcbi.1004040.

Kupferschmidt, K. 2016. Resistance fighters. Science 352:758–761.

Luciani, F., S. A. Sisson, H. L. Jiang, A. R. Francis, and M. M. Tanaka. 2009. The epidemiological fitness cost of drug resistance in *Mycobacterium tuberculosis*. Proceedings of the National Academy of Sciences of the United States of America 106:14711–14715. doi: https://doi.org/10.1073/pnas.0902437106.

O'Neill, J. 2014. Antimicrobial resistance: Tackling a crisis for the health and wealth of nations. https://amr-review.org/sites/default/files/AMR%20Review%20Paper%20-%20Tackling%20a%20crisis%20for%20the%20health%20and%20wealth%20of%20nations_1.pdf.

Raberg, L., J. C. de Roode, A. S. Bell, P. Stamou, D. Gray, and A. F. Read. 2006. The role of immune-mediated apparent competition in genetically diverse malaria infections. American Naturalist 168:41–53.

Radcliffe, E. B., W. D. Hutchison, and R. Cancelado, eds. 2009. Integrated pest management: Concepts, tactics, strategies and case studies: Cambridge: Cambridge University Press.

Read, A. F., and L. H. Taylor. 2001. The ecology of genetically diverse infections. Science. 292:1099–1102.

Uecker, H., S. P. Otto, and J. Hermisson. 2014. Evolutionary rescue in structured populations. American Naturalist 183:E17–E35. doi: https://doi.org/10.1086/673914.

Wale, N., D. G. Sim, M. J. Jones, R. Salathe, T. Day, and A. F. Read. 2017. Resource limitation prevents the emergence of drug resistance by intensifying within host competition. Proceedings of the National Academy of Science USA 114:13774–13779.

Wale, N., D. G. Sim, and A. F. Read. 2017. A nutrient mediates intraspecific competition between rodent malaria parasites in vivo. Proceedings of the Royal Society of London Series B 284:20171067.

Willmann, M., and S. Peter. 2017. Translational metagenomics and the human resistome: confronting the menace of the new millennium. Journal of Molecular Medicine 95:41–51. doi: https://doi.org/10.1007/s00109-016-1478-0.

Woods, R., and A. F. Read. 2015. Clinical management of resistance evolution in a bacterial infection: a case study. Evolution, Medicine and Public Health 2015:281–288.

zur Wiesch, P. A., R. Kouyos, J. Engelstädter, R. R. Regoes, and S. Bonhoeffer. 2011. Population biological principles of drug-resistance evolution in infectious diseases. Lancet Infectious Diseases 11:236–247.

Six Wedges to Curing Disease

Michael E. Hochberg

One of the great challenges in ecology and evolutionary biology is to explain disease, whether caused by infectious agents such as parasites and pathogens, or by the deterioration or transformation of cellular behavior and function, a prime example of the latter being cancer. Decades of observation and research suggest that successfully treating disease requires insights into how the environment mediates the interactions between individuals and biological etiological agents (BEAs) of disease such as parasites, pathogens and cancer cells. A major finding is that single factor, targeted therapies are not only likely to fail in controlling or eradicating many BEAs, but are also likely to select for resistance, reducing options for subsequent treatment attempts, and in cases of infectious BEAs, rendering therapeutic agents (e.g., antibiotics) obsolete.

I argue that meeting the growing challenge of treating disease in agriculture and animal husbandry, in protected and domesticated species, wildlife, and in the human population will require a fundamental understanding of ecological interactions at sites of infection or disease. I discuss different ways in which components of such *disease ecosystems* mediate BEA and therapeutic dynamics and resistance evolution and derive a very simple mathematical criterion for therapeutic success. I then touch on how fundamental insights as revealed by the processes of evolutionary rescue and competitive release can help understand why therapies succeed or fail. Finally, I present six "wedges" that can each contribute alone, or as part of multipronged approaches to successfully treating disease.

The Magic Bullet

Despite remarkable progress in prevention and treatment, the impacts of diseases in agriculture and animal husbandry, and on protected and domesticated species, wildlife, and human health are likely to intensify into the twenty-first century. Humans in particular are increasingly in contact with each other and with wildlife, treated with drug regimens that select for resistance, and adopt lifestyles or are exposed to environments that render them more susceptible to infectious diseases and more likely to get cellular diseases such as cancers.

For many clinicians the Holy Grail is to discover the Ehrlichian "magic bullet"—a drug that will target the BEAs and cure disease. Intuitively, the most effective way to reduce BEAs is to "hit hard and fast." That is, more drug means more kill within the tolerance limits of the patient. Rapid administration of the drug means forestalling further BEA growth and associated symptoms, but also reducing the probability of the appearance of resistant mutant strains. Despite decades of research on pest and disease control (much of it with an ecological basis), this approach and the many alternatives in following pages remain highly controversial (e.g., Read et al. 2011).

The magic bullet may be shortsighted for another reason. What will cure most patients may not be optimal for the general population in which resistant variants may emerge and spread.[1] This problem is in many ways similar to the objectives of classical pest control (Greene and Reid 2013), where short-term success is to attain economic targets for the treated area (e.g., an agricultural field), subject to meeting the global objective of slowing the spatial spread of resistance (Vacher et al. 2003). Although human cancers are not transmitted between individuals, the concept from pest control of spatial resistance management could apply to treating metastasis (Fu et al. 2015), and certain authors have argued that certain targeted therapies could be counterproductive to treatment success (Gillies et al. 2012).

Curing Disease Is an Ecological Problem

Some of the diseases that present the greatest threats to human health are caused by microparasites, including viruses, bacteria, and protozoa, and macroparasites such as helminth worms. Microparasites in particular are characterized by very large populations and therefore considerable potential for rapid evolutionary responses. When a drug is applied to a large, diverse population, the response will be determined in part by the absolute (growth) and relative (selection) impacts on sensitive and resistant subpopulations. Given that resistance depends on genetic expression, including epigenetics, this means that environment, and more generally ecology, needs to be incorporated into our understanding of why chemotherapies (against microparasites, macroparasites, or cancers) succeed or fail.

Disease control within the organism is an ecological problem but is rarely seen as such. Rather, research and application are overwhelmingly

Table 1. Some Basic Challenges to Treating Patients and Disease Management over the Broader Population for Several Important BEAs Affecting the Human Population

BEA	INDIVIDUAL PATIENTS	BROADER POPULATION
Bacterial pathogens	Large, diverse populations; hypermutator strains; protective biofilms; multidrug resistance	Horizontal gene transfer of virulence factors and resistance from other bacterial species
Plasmodium	Evades host immune system; resistant strains transmitted to host; distinguishing Plasmodium species to determine specific treatment	Difficulty in controlling the vector
HIV	High within-host population turnover and generation of antigen diversity; uses its enemy (white blood cells) to replicate	Prevention through safe-sex and no needle sharing
Cancer	Large, diverse populations; genomic instability; dormancy; refractory cancer stem cells; limited drug amounts due to toxicity; evasion of immune system	

focused on the direct interaction between the therapy (typically a drug) and the BEA. Treatment success means that the drug contributed to clearing the BEA, whereas failure would suggest that the drug choice, dose, and/or schedule was in error, and/or that resistant strains were present. The major omission from this perspective is environment. Indeed, the very same interactions found in terrestrial and aquatic systems are found *within* diseased hosts: intra- and interspecific competition, resource limitation, mutualism, facilitation, and predation.

Here I argue for a framework integrating abiotic and biotic interactions involving BEAs and drugs within *disease ecosystems*. I briefly discuss the basis for this concept. I then present several mechanisms associated with therapeutic failure, all of which involve BEA clonal escape. A criterion for therapeutic success integrating clonal escape is then formalized in a very simple mathematical expression, and I discuss parallels with the concepts of evolutionary rescue and competitive release. Finally, I present six complementary strategies or "wedges" towards improved control, their ecological basis, and why they should be considered in future theoretical developments.

My approach is largely review, and I make reference where appropriate to individual treatment strategies coming from bacterial infections, *Plasmodium*, human immunodeficiency virus (HIV), and cancers. Table 1 presents a list of some of the basic factors associated with these BEAs that make them particularly challenging to treat in individuals and to control at a population level. Comparing and contrasting insights from these different literatures could lead to a greater fundamental understanding of disease ecosystems, how drugs affect them, and translation to clinics.

The Disease Ecosystem

Despite many differences between parasites and tumor cells, one broad similarity is that both live in *disease ecosystems*. The disease ecosystem encompasses birth, growth, survival, and interindividual and interpopulational interactions such as predation (immune system), cooperation (cell–cell signals, tissue architecture, cell function), competition (both direct (as in the case of many parasites) and indirect for limiting resources (host cells, glucose, oxygen)), resource replenishment (angiogenesis), detritivory (phagocytosis), and external intervention in the form of therapy. The latter is tempered by the idiosyncratic nature of information gathering and translation into a rational treatment. Related perspectives have recently been discussed for parasites (Rynkiewicz et al. 2015), pathogenic bacteria (e.g., Conrad et al. 2013), and cancer (Basanta and Anderson 2013, Merlo et al. 2006).

Figure 1 shows an oversimplified depiction of one type of disease ecosystem: the cancer microenvironment. Tumor growth depends importantly on resource flows through the three-dimensional mass of uncontrollably dividing, motile cells (Alfarouk et al. 2013). Cells that are only a few millimeters' distance from the nearest capillary will experience hypoxic stress and lactic acid accumulation, due to deficits in oxygen and glucose. Cells tend to become increasingly hypoxic and anoxic towards the center of a tumor and under prolonged stress they form a necrotic core (Smallbone et al. 2007, Gillies and Gatenby 2007). Stressed cells may send signals to the local vascular system to extend capillaries into the tumor (angiogenesis), which, if successful, promotes further tumor growth (Folkman 2006). From a therapeutic point of view, the challenge is to arrest angiogenesis while promoting conditions for drug diffusion into the tumor (which *increases* with angiogenesis). Those cells experiencing insufficient doses of the drug or in a dormant state (Fig. 1, yellow cell) may be the source of future relapse, as are cells receiving the full drug dose, but harboring chemoresistance mutations (Fig. 1, blue cells; also, see later). Little is known about how resource dynamics and cell-to-cell signaling involving BEAs and healthy cells determine the emergence and fates of refuge cells and resistant cells. Finally, the predators in the disease ecosystem are different components of the innate and adaptive immune responses that are most functional at the incipient stages of

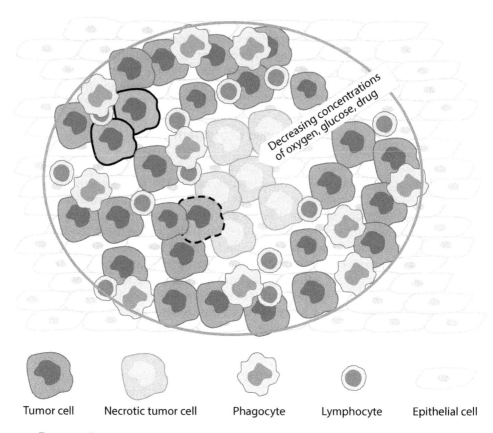

Decreasing concentrations of oxygen, glucose, drug

| Tumor cell | Necrotic tumor cell | Phagocyte | Lymphocyte | Epithelial cell |

FIGURE 1. Disease ecosystem for tumor growth. In this oversimplified representation of the tumor microenvironment, tumor cells can grow as long as they have sufficient access to oxygen and glucose and sufficient space. Tumor cells compete among each other and with surrounding healthy epithelial cells. The intensity of competition increases with cell–cell proximity and distance from the nearest blood capillary (not shown). In the absence of sufficient resources (away from blood supply), cells become necrotic. Tumor cells may be predated by components of the immune response (phagocytes and lymphocytes). Fibroblasts (not shown) modify the disease ecosystem through the production of structural tissue (extracellular matrix) and growth factors, facilitating angiogenesis (pericyte and endothelial cell production), inflammation, immune evasion, tumor growth, and metastasis. Tumor cells may escape the immune system (not shown) or therapies—the latter through, for example, genetic resistance (solid black outlines) or limited drug contact (dashed black outline).

tumorigenesis, becoming less effective or failing in conjunction with tumor escape (Merlo et al. 2006).

Largely analogous reasoning holds for nonself BEAs (e.g., HIV, bacterial pathogens, *Plasmodium*), although each has its own spatial, behavioral, and functional specificities. Moreover, the caricature in Figure 1 is centered on local interactions between BEAs, host cells and tissues, and drugs. A more realistic representation would include regional and global host interactions (e.g., immune, hormonal, other disease foci, effects on host health and behavior), and in cases of infectious disease, interactions with the external environment. Understanding the differences and commonalities between different BEA ecosystems will form a part of developing theories and predictions for how diseases can be cured.

BEA Escape

Beyond misdiagnosis or misappropriations in the choice of (or patient intolerance to) drugs, the main generic cause for treatment failure is BEA escape. BEAs cause disease, morbidity, and mortality in part because they either escape natural host controls or amplify them (e.g., cytokine storms). BEA escape from drugs bears limited resemblance to parasite escape from immune systems or pests escaping natural enemies in biological control. This is in part because drugs are nonadaptive, whereas the body has a wide array of checkpoints for detecting and destroying diseased cells, and the immune system for detecting nonself antigens and the subsequent disabling of parasites.

BEA escape from chemotherapies can occur in one or more of several ways. First, BEAs may produce quiescent or dormant states. Examples include quiescent cancer stem cells (Merlo et al. 2006) and bacterial persister cells (Balaban 2004). Second, BEAs may be tolerant, whereby mechanisms such as reduced receptor sensitivity or density limit drug impact, as cases of drug-tolerant cells in certain cancers (Sharma et al. 2010). Third, a BEA may directly resist a drug through a molecular change or a change in receptors and/or drug transport. For instance, bacteria may indirectly resist antibiotics through the upregulation of efflux pumps (Webber 2002), or directly resist them via drug modification or inactivation (Blair et al. 2015). Fourth, some individual agents may be less attainable by the drug than others, either due to spatial refuges or active escape through plastic behaviors. An example is limited drug diffusion through tumors leading to spatial refuges and resistance (Fu et al. 2015, Trédan 2007). Fifth, agents may naturally have high mutation rates, or be induced or selected to have increased mutation rates, such as in hypermutator bacteria (Oliver et al. 2004), HIV (Cuevas et al. 2015), and certain cancers (Tomlinson et al. 1996). Tumors in particular are thought to contain numerous resistant variants by the time neoplasms are typically discovered (Diaz et al. 2012). The preceding five mechanisms are largely inferred from *in vitro* experiments; their detailed roles in treatment failures are largely unknown (e.g., Mwangi et al. 2007).

A Simple Criterion

The primary objective of chemotherapy is to affect BEAs so as to durably allay or eliminate disease impacts on health. Outcomes are challenging to predict because the system is potentially very complex, involving how the BEA interacts with itself, with other BEAs (e.g., in coinfections) and with healthy cells, but also with other ecosystem components such as resource availability (space, nutrients), habitat quality, the microbiome, and the immune system. The addition of one or more drugs potentially influences these interacting compartments either directly or indirectly, creating possible feedbacks, and in so doing will make predictions based on individual components in isolation, overly simplistic.

The complexity summarized above indicates that simple criteria, such as reducing the BEA net population growth rate to below zero, are probably not useful for understanding chemotherapeutic success and failure. Rather, a model incorporating the key phenomena of BEA escape may provide significant insight into the basic control problem. The model can be modified to incorporate different features of the disease ecosystem.

We assume n distinct asexual, haploid classes of BEA where, for phenotype i, the current population size, maximum growth rate, density-dependent limitation (e.g., through competition and/or predation), and reductions in birth (or increases in mortality) from chemotherapy are respectively $N_{i,t}$, λ_i, $f_{i,t}$ and φ_i. Note that φ_i encapsulates both the sensitivity of a strain to the drug (given inherent levels of tolerance or resistance), and the eventuality that the strain is also protected due to spatial refuges.

Assuming a large initial population $N_{i,0}$ of the most frequent clone, such that population changes are dominated by deterministic rates, the population at a short time interval Δ into the future is

$$N_{i,t+\Delta} = \lambda_i \, \varphi_i \, f_{i,t}\{N_{i,t}\}N_{i,t}.$$

Further, assuming that density dependence (intrapopulation competition) in the BEA becomes negligible soon after the chemotherapy begins, and resistant or protected variants are initially rare, then after a treatment period $x\Delta$, clone i is approximated by

$$N_{i,t+x\Delta} = (\lambda_i \, \varphi_i)^x \, N_{i,0}.$$

Clearly if $\phi_i < 1/\lambda_i$ for all i, then some level of reduction in density is ensured. The criterion for achieving population control below a target density, T, predicted to result in success is therefore

$$W = \Sigma_i (\lambda_i \, \varphi_i)^x \, N_{i,0} < T. \tag{1}$$

Importantly, the objective $W < T$ may be initially met, simply because refuge populations are small and do not appreciably grow during the initial phases of the therapy. Should these populations not be sufficiently held in check thereafter, then eventually they will reemerge, and potentially falsify the success criterion 1. This may happen should subpopulations escape drug control and/or if the immune system is not sufficiently effective.

Notice the following. First, because this is an exponential population process, the required φ to achieve control is highly sensitive to the multiplier λ. This reflects the compounding effect of insufficiently checked exponential growth. Second, meeting criterion 1 requires that *all* genotypes are sufficiently controlled; this sufficiency will depend on the densities, biological characteristics of each clone present, and disease ecosystem interactions. Third, if a single resistant mutant is present at the beginning of the therapy, then, should it emerge, it takes a minimum of $\ln(T)/\ln(\lambda_m)$ time units for criterion 1 to be rejected. And fourth, the immune system is an important factor that potentially intervenes to reduce λ, whereby therapy accompanied by immune responses are likely to be more effective than either alone in reducing nonself parasites. In contrast, by the time a cancer is discovered (e.g., poses a health threat), the tumor has likely escaped natural immune control and would require specific immunotherapies to contribute to BEA suppression beyond the effects of a chemotherapeutic agent (or radiation therapy).

This model is clearly oversimplified and the inclusion of further realistic positive or negative density-dependent interactions such as competition between tumor cells and surrounding healthy cells could influence the therapeutic outcome. For example, Sottoriva et al. (Sottoriva et al. 2015) considered the emergence and growth of competing clones in colorectal cancers. They showed that the initial appearance of a set of driver mutations resulted in the dominance of these clones in tumor growth, even when more fit strains subsequently emerged. This is because early emerging clones gain a priority effect (a "head start"), and as the population carrying capacity is approached, clonal per capita growth rate and per capita beneficial mutation rate both decrease, resulting in slower clonal turnover.

Rescue and Release

Before proposing a framework towards improved treatment outcomes that incorporates the dynamics of disease ecosystems, I describe two complementary phenomena that are useful in understanding treatment failures.

In *evolutionary rescue*, the effects of a drug drive the sensitive population towards extinction, but resistant variants that are already present or emerge during the treatment "rescue" the population (Gonzalez et al. 2013). The evolution of BEA resistance under high drug doses is a textbook example of evolutionary rescue, although not all cases of resistance evolution need involve evolutionary rescue. Specifically, drugs may only partially impact the sensitive subpopulation due to e.g., BEA plasticity, spatial refuges, or underdosing. It is also possible that drug resistance evolves in situations where extinction would not have occurred in the first place (e.g., Gullberg 2011, Andersson and Hughes 2014). In the context of improving chemotherapies, if the initial BEA population is small and/or genetic variation low, then it may be best to treat with high doses until the BEA is cleared. This assumes that the therapy sufficiently decreases the rate of evolution (i.e., lowers the number of births and it is not mutagenic) and does not induce refuge behavior such as dormancy observed in some cancers (Sharma et al. 2010).

In *competitive release*, populations compete for space or limiting resources, and the reduction of one or more competitors contributes to the growth of other, less affected (e.g. resistant) populations (Greene and Reid 2013). Day et al. (2015) differentiated competitive release from competitive suppression, the latter occurring when one competitor suppresses another through a shared density-dependent response, such as an immune reaction or environmental degradation (Huijben et al. 2013). With a few exceptions, competitive release is not well understood in the context of disease ecosystems. For example, Wargo et al. (2007) showed an effect of treatment duration on the magnitude of competitive release in *Plasmodium*. Pena-Miller et al. (2013) used modeling and experiments of antibiotic combination therapies to show that failure by the first antibiotic strongly favors the growth rate and emergence of resistant clones, which, in turn, evade control by a second drug. One of the dangers in competitive release (and indeed evolutionary rescue) is that the expanding resistant population may evolve compensatory traits (MacLean et al. 2010, Schulz et al. 2010) or acquire other fitness-enhancing traits, such that proliferation potential is comparable or even greater than that of the original population, as has been hypothesized in certain cancer relapses (Ding et al. 2012).

Ecological and Evolutionary Wedges to Vanquish Disease

As described above, the conventional wisdom in treating disease is to hit hard and fast. The main constraints on dose and duration are toxicity and treatment costs. The risk of this approach is treatment failure and, in the case of infectious BEAs, the spread of resistant strains to other individuals.

Fundamental research aimed at understanding treatment outcomes needs to incorporate the dose debate into a more integrated framework based on the disease ecosystem. I describe six non–mutually exclusive variables and strategies (*wedges*) that could contribute to improved treatment outcomes (Table 2).

1. *Dosing.* Growth inhibition and kill assays are mainstays of pharmacodynamics. However, drugs are not static entities once introduced into the body: They are heterogeneously distributed, and may be modified or inactivated, and excreted (pharmacokinetics). For example, Foo et al. (2012) studied combined evolutionary and pharmacokinetic models of the emergence of cell resistance to erlotinib in patients with epidermal growth factor receptor (EGFR)–mutant lung cancers. They found that pulsed dosing could control tumors with resistant clones, but that sufficiently long treatment holidays risked enabling the emergence of *de novo* resistant clones. Ankomah and Levin (2014) employed models of pharmacodynamics, focusing on the dose and duration of antimicrobial therapies in systems with innate and adaptive immune responses. They found that high doses used until the immune response finished clearing the infection were superior to less intense strategies (but see Day and Read 2016 for contrasting findings). Akhmetzhanov and Hochberg (2015) used computational models of tumor growth, based on empirical parameter estimates to derive minimal levels of continuous chemotherapy that would extend patient life expectancy over that achieved with high-dose therapies. They found that optimal dosing should no more than counteract the growth of the BEA clone with the highest fitness, otherwise fully resistant clones would likely emerge, resulting in possible treatment failure.

Although there is considerable empirical work on how drug dosing potentially affects treatment outcome (for a literature survey, see Day and Read 2016), it does not provide a clear signal as to whether "hit hard" approaches are superior to alternative dosing schedules (Kouyos et al. 2014). For example, Negri et al. (2000) demonstrated how strains of *Escherichia coli* with different levels of resistance to the antibiotic cefotaxime persisted at different drug doses. They identified the *selective window*: doses that eliminate sensitive strains and favor relatively resistant ones. Work on *Plasmodium* has shown sensitivity to treatment dose (Huijben et al. 2013) and Wargo et al. (2007) demonstrated an effect of treatment

Table 2. Opportunities and Potential Risks in Treating Disease Associated with Each of the Six Wedges

WEDGE	OPPORTUNITIES	POTENTIAL RISKS (COMMENTS)
1. Dosing	Hit hard, fast and sufficiently long if probabilities of sensitive BEA elimination are high; resistance low; immune system completes clearance	Toxicity and selection for resistance (modulate dose and schedule or manage using one or more of wedges 2–6)
2. Combinations	Breakthrough refuges (phenotypic, dormancy, spatial, resistance); not only reduces BEAs compared to any drug used separately, but also may control drug resistance; combinations can be pre-evaluated for their synergy and used at lower doses than either drug separately	The development of multiple resistances (Choose agents, doses and schedules whereby one agent is the principal treatment and the other acts as a supplement or adjuvant: wedge 4)
3. Increasing resistance costs	Targets the cause of many chemotherapeutic failures by reducing or eliminating resistant phenotypes before a therapy commences	Pretherapy clinical side effects or causing significantly delays in administration of the drug (alternate with chemotherapies: wedges 2, 6)
4. Dynamic agents	Particularly useful as targeted therapies, to grow and persist in disease ecosystem, and counter evolved BEA resistance	Host range expansion to commensal microorganisms (bacteriophage); removal by the immune system; more time required for success (Combine with more traditional drugs (wedge 2) or natural components of disease ecosystem (wedge 5)
5. Harnessing immune responses and quenching resources	Harness immune system, competition (public good quenching, nutrient limitation), and altering environmental stresses (reduced inflammation)	Autoimmune reactions (also risks in wedge 4) (Most promising if used in combination with other therapies; wedge 2)
6. Adapting	Potential when the BEA can be directly or indirectly monitored and treatments be modified in type or dose, or adapted in parallel (e.g., phage training)	Inability to manage or eradicate infection

duration on the emergence of resistance. A corollary to dosing (intensity, schedule, and duration) is whether the therapy should actually kill BEAs (cytotoxicity) or rather render them unable to proliferate (cytostaticity). Theory shows that these alternatives can have contrasting effects on dynamics (e.g., Lorz et al. 2013), suggesting that interactions with other components in the disease ecosystem, such as resource competition or predation by the immune system, may result in different therapeutic outcomes depending on the extent to which therapy kills cells or arrests cells.

2. *Combination therapies*. Employing two or more therapeutic agents to treat a disease has shown promise for certain cancers (Vanneman and Dranoff 2012), bacterial diseases (Tamma et al. 2012) and HIV (Bartlett et al. 2006) (see, also examples in zur Wiesch et al. 2011). The basis is twofold. First, combination therapies can be devised to foil one or more escape responses, including spatial refuges, effect pathways, and resistance genes (e.g., Fitzgerald et al. 2006). The underlying idea is that total population coverage by two or more therapies is greater than any one alone. For example, Komarova and Wodarz (2005) parameterized mathematical models to show how the number of drugs and probabilities of resistance to each inform how many drugs are necessary to successfully treat cancers. Examples of combined approaches also abound in the antimicrobial literature; for example, the use of multiple phage types or combinations of phages and antibiotics against certain pathogenic bacteria (Chan et al. 2013, Torres-Barceló and Hochberg 2016, but see Betts et al. 2016). Second, the order, schedule, and dose of each anti-BEA can be adjusted for maximal effect (Torres-Barceló et al. 2014). Maximal effect is a complex quantity that integrates the impact on BEA numbers, the probability of resistance, and allaying disease severity, while avoiding toxicity issues (especially for cancer therapies).

Scheduling can be an important parameter in achieving maximal effect, for instance when therapeutic agents interfere with one another if administered together. For example, Torres-Barcelo et al. (2014) showed how the delay between additions of a bacteriophage and an antibiotic is key in maximizing impact and minimizing resistance in populations of *Pseudomonas aeruginosa*. This is due in part to the density-dependent nature of phage multiplication and impact on bacterial hosts. Roemhild et al. (2015) demonstrated how the order of sublethal doses of antibiotics could have substantial effects on bacterial populations and resistance (see also Pena-Miller et al. 2013). Combination therapies may also exploit characteristics of the disease ecosystem and of BEAs themselves. Examples include the sequential use of chemotherapy and immunotherapy (Antonia et al. 2006), and poly(ADP-ribose) polymerase (PARP) inhibitors capitalizing on synthetic lethality in certain breast and ovarian cancers (Rottenberg et al. 2008).

3. *Increasing the costs of resistance.* One underexplored approach to improve treatment outcomes is to increase the costs of resistance. The idea is to use specific interventions so that sensitive BEAs competitively control or even eliminate resistant populations. Enriquez-Navas and coworkers (2015) have recently argued that for some cancers, chemotherapies could be alternated with "fake drugs" that serve only to tip the competitive balance in favor of sensitive cells. This idea has considerable appeal, but the argument could be made that rapid treatment with a fake drug should be initiated *before* the real chemotherapy is even applied. The reasoning is as follows. The resistant cell population is expected to be at its lowest numbers before treatment. This is due both to resistant cells growing in the treatment environment (their absolute fitness is greater than one), and to their growth being enhanced by reduced competition with decreased numbers of sensitive cells (i.e., competitive release). The sensitive cell population will be at its maximum numbers and maximum competitive impact on resistant cells before treatment. Increasing the cost of resistance through a fake drug will further tip the balance in favor of sensitive BEAs. This paves the way for the administration of a bolus of the real drug to clear the infection or the cancer, or if not completely successful, to follow the real drug with cycling fake drug treatments, as in Enriquez-Navas et al. 2015.

4. *Dynamic agents.* Drug specificity is one of the mainstays of Ehrlichian magic bullets. However, it is also a shortcoming because, as discussed above, targeting one or a small number of vulnerable traits can select for resistance (Gillies et al. 2012). An alternative is to capitalize on the diversity of "living" agents and their potential to overcome the evolution of resistance *in situ,* or in treating previously evolved multidrug-resistant pathogens. Examples include lytic phages against pathogenic bacteria (Pirnay et al. 2011, Viertel et al. 2014) and oncolytic viruses (Russell et al. 2012). For example, Wodarz and Komarova (2009) developed and analyzed mathematical models to show that tumor control by oncolytic viruses depended importantly on virus growth and diffusion to refuge cancer cell populations. Others have argued for a "Trojan horse" strategy, whereby avirulent BEA variants could be used to control more virulent strains (Brown et al. 2009) and greatly reduce the risks of resistance evolution. An interesting related development is to employ bacteriophages as adjuvants to select for increased sensitivity to antibiotics (Chan et al. 2016).

A prime advantage of dynamic agents, and bacteriophages in particular, is self-amplification and adaptation to resistant hosts *in situ.* Because the diversity of phages is virtually limitless, they can be combined into cocktails to maximize overall strain coverage and effect on the target bacterium (Chan et al. 2013). Bacteriophages can either treat established infections or be used prophylactically ("lying in wait") so as to prevent pathogens from colonizing incisions or wounds (Kutter et al. 2010). When bacterial densities are low (such that phage self-amplification is limited) or difficult to attain, the phage inoculum can be increased, functioning more like a conventional antimicrobial drug.

5. *Tweaking different interactions in the disease ecosystem.* As introduced above, one of the cornerstones of ecology—that individuals interact with each other and the environment—is all but neglected in chemotherapies. There are many possible interventions into the disease ecosystem that could increase impacts on BEAs. For instance, immune systems ("predators") can be harnessed to combat malignant cells (e.g., Childs and Carlsten 2015). Empirical work shows that one way to accomplish this is through dynamic agents (Eriksson et al. 2009), whereas other immune stimulations prevent BEAs from becoming established in the first place (Barr et al. 2013). Other studies include those that show how healthy cells may help suppress neighboring tumors (Maley et al. 2004), or that the evolutionary rates and adaptability of BEAs may be slowed (Kostadinov et al. 2013), or how factors as diverse as resource bases, virulence factors, host tolerance, microbiomes, and cell–cell cooperation may be targeted to improve therapeutic outcomes (Jansen et al. 2015, Fig. 1; Ross-Gillespie et al. 2015; Vale et al. 2016).

6. *Adapt to the situation.* The conventional approach of treating disease is to employ information based on symptoms, eventual identification of the BEA, and patient characteristics to devise a one-off strategy with the objective of improvement or cure. A different approach is to use early-warning signs, and initial and ongoing information to predict disease dynamics and adapt treatments so as to manage or eradicate the BEA. For example, *adaptive therapy* has the dual objectives of an acceptable cancer burden and minimal resistance evolution (recently reviewed in Nichol et al. 2015). In the context of phage therapy, Pirnay et al. (2011) distinguished general or *prêt-à-porter* strategies and more-personalized *sur mesure* approaches. One example of the latter is adaptive *phage training*, whereby lytic phages are evolved before or during a therapy in the laboratory, so as to improve their impact on particular bacterial strains, extend the spectrum of hosts attacked, or anticipate bacterial resistance (Betts et al. 2013, 2014) (see also Betts et al. 2014 for a related discussion of *steering evolution*). Once training is complete, the trained isolates are then reintroduced into the disease ecosystem.

More generally, assessing possible "Plan B" strategies *before* a treatment actually begins is not only sensible, but may also influence the choice of Plan A. Forecasts of ecology and evolution in the disease ecosystem should guide such choices. For example, the use of a risky Plan A with no Plan B should Plan A fail (e.g., the treatment of a multidrug-resistant bacterium with a last resort antibiotic), may be inferior to the employment of a less-aggressive, slower-acting Plan A (the use of a phage cocktail) that has an ecologically and evolutionarily sensible Plan B (adaptively trained bacteriophage) should Plan A be partially successful or fail.

The Greater Community

The approaches presented above focus on the disease ecosystem within the individual patient. For infectious diseases, one must also consider the larger population for at least two reasons. First, in the short term, failure to sufficiently control an infection may result in the transmission of sensitive, or even worse, resistant strains. Second, in the longer term, the continued use of a single antimicrobial over a sufficiently large population will invariably select for the emergence and spread of resistance. Heesterbeek et al. (2015) presented a framework that reveals the complexity of managing infectious disease systems, whereby models (based on predictive epidemiological parameters, such as the basic reproductive rate), data, and policy are integrated into a plan of action, and improved iteratively. These models have the objective of control (while preventing or delaying resistance) or, if possible, eradication. Application of one or more of the wedges proposed here should form part of disease control and eradication policies as, by reducing the probability of resistance emergence, they may prevent or attenuate the spread of epidemics in the wider population. Nevertheless, more research is needed to understand how attempting to cure individual patients is or is not optimal for the population, and inversely, how optimal population-level programs may result in some patients not being cured, whereas they would have been in a patient-centered approached. The six wedges presented here, when used singly or in combination may provide a way forward to optimize both individual outcomes and population-level outcomes.

Conclusions

The complexity of the disease ecosystem provides numerous opportunities for the employment of novel therapeutic approaches. Real disease ecosystems and the mechanisms of BEA escape are evidently much more complex than presented here (see e.g., Table 2 in Gonzalez-Angulo et al. 2007), meaning that compromises are necessary between sufficient realism and conceptual and computational tractability. This will be particularly challenging in situations where rapid decisions are needed and for which accurate, rich information about the BEA is difficult to obtain or difficult to process (Woods and Read 2015). Nevertheless, the six wedges proposed here, based on a fundamental understanding of ecology and evolution in the disease ecosystem, provide a robust armory, which can be used singly or in combination to cure disease.

Acknowledgments

Thanks to Andy Dobson and Troy Day for comments on the manuscript and to Sam Brown, Robert Gatenby and Paul Turner for discussions. The author acknowledges the McDonnell Foundation (Studying Complex Systems Research Award No. 220020294) and the Institut National du Cancer (2014-1-PL-BIO-12-IGR-1) for funding.

Notes

The present paper was completed in September 2016. It is the basis of the recent study published in *Evolution, Medicine and Public Health* (Hochberg 2018).

1. In the context of infectious disease, transmission and spread are to other hosts, whereas, in cancer (with the exception of transmissible cancers in dogs, Tasmanian devils, and certain bivalve species) spread is either through local or distant metastasis.

References

Akhmetzhanov, A. R., and M. E. Hochberg. 2015. Dynamics of preventive vs post-diagnostic cancer control using low-impact measures. Elife 4:e06266. doi: https://doi.org/10.7554/eLife.06266.

Alfarouk, K. O., M. E. Ibrahim, R. A. Gatenby, and J. S. Brown. 2013. Riparian ecosystems in human cancers. Evolutionary Applications 6:46–53.

Andersson, D. I., and D. Hughes. 2014. Microbiological effects of sublethal levels of antibiotics. Nature Reviews Microbiology 12:465–478.

Ankomah, P., and B. R. Levin. 2014. Exploring the collaboration between antibiotics and the immune response in the treatment of acute, self-limiting infections. Proceedings of the National Academy of Sciences of the United States of America 111:8331–8338.

Antonia, S. J., N. Mirza, I. Fricke, A. Chiappori, P. Thompson, N. Williams, et al. 2006. Combination of p53 cancer vaccine with chemotherapy in patients with extensive stage small cell lung cancer. Clinical Cancer Research 12:878–887.

Balaban, N. Q. 2004. Bacterial persistence as a phenotypic switch. Science 305:1622–1625.

Barr, J. J., M. Youle, and F. Rohwer. 2013. Innate and acquired bacteriophage-mediated immunity. Bacteriophage 3:e25857.

Bartlett, J. A.., M. J. Fath, R. Demasi, A. Hermes, J. Quinn, E. Mondou, et al. 2006. An updated systematic overview of triple combination therapy in antiretroviral-naive HIV-infected adults. AIDS 20:2051–2064.

Basanta, D., and A.R.A. Anderson 2013. Exploiting ecological principles to better understand cancer progression and treatment. Interface Focus 3. https://royalsocietypublishing.org/doi/full/10.1098/rsfs.2013.0020.

Betts, A., D. R. Gifford, R. C. MacLean, and K. C. King. 2016. Parasite diversity drives rapid host dynamics and evolution of resistance in a bacteria-phage system. Evolution 70:969–978. doi: https://doi.org/10.1111/evo.12909.

Betts, A., M. Vasse, and O.M.E. Kaltz. 2013. Back to the future: Evolving bacteriophages to increase their effectiveness against the pathogen *Pseudomonas aeruginosa* PAO1. Evolutionary Applications 6:1054–1063.

Betts, A., O. Kaltz, and M. E. Hochberg. 2014. Contrasted coevolutionary dynamics between a bacterial pathogen and its bacteriophages. Proceedings of the National Academy of Sciences of the United States of America 111:11109–11114.

Blair, J.M.A., M. A. Webber, A. J. Baylay, D. O. Ogbolu, and L.J.V. Piddock. 2015. Molecular mechanisms of antibiotic resistance. Nature Reviews Microbiology 13:42–51.

Brown, S. P., S. A. West, S. P. Diggle, and A. S. Griffin. 2009. Social evolution in microorganisms and a Trojan horse approach to medical intervention strategies. Philosophical Transactions of the Royal Society London B: Biological Sciences 364:3157–3168.

Chan, B. K., M. Sistrom, J. E. Wertz, K. E. Kortright, D. Narayan, and P. E. Turner. 2016. Phage selection restores antibiotic sensitivity in MDR Pseudomonas aeruginosa. Scientific Reports 6:26717.

Chan, B. K., S. T. Abedon, and C. Loc-Carrillo. 2013. Phage cocktails and the future of phage therapy. Future Microbiology 8:769–783.

Childs, R. W., and M. Carlsten. 2015. Therapeutic approaches to enhance natural killer cell cytotoxicity against cancer: the force awakens. Nature Reviews Drug Discovery 14:487–498.

Conrad, D., M. Haynes, P. Salamon, P. B. Rainey, M. Youle, and F. Rohwer. Cystic fibrosis therapy: A community ecology perspective. American Journal of Respiratory Cell and Molecular Biology 8:50–156.

Cuevas, J. M., R. Geller, R. Garijo, J. López-Aldeguer, and R. Sanjuán. 2015. Extremely high mutation rate of HIV-1 in vivo. PLOS Biology 13:e1002251.

Day, T., S. Huijben, and A. F. Read. 2015. Is selection relevant in the evolutionary emergence of drug resistance? Trends in Microbiology 23:126–133.

Day, T., and A. F. Read. 2016. Does high-dose antimicrobial chemotherapy prevent the evolution of resistance? PLOS Computational Biology 12:e1004689.

Diaz, L. A., R. T. Williams, J. Wu, I. Kinde, J. R. Hecht, J. Berlin, et al. 2012. The molecular evolution of acquired resistance to targeted EGFR blockade in colorectal cancers. Nature 486:537–540.

Ding, L., T. J. Ley, D. E. Larson, C. A. Miller, D. C. Koboldt, J. S. Welch, et al. 2012. Clonal evolution in relapsed acute myeloid leukaemia revealed by whole-genome sequencing. Nature 481:506–510.

Enriquez-Navas, P. M., J. W. Wojtkowiak, and R. A. Gatenby. Application of evolutionary principles to cancer therapy. Cancer Research 75:4675–4680.

Eriksson, F., P. Tsagozis, K. Lundberg, R. Parsa, S. M. Mangsbo, M.A.A. Persson, et al. 2009. Tumor-specific bacteriophages induce tumor destruction through activation of tumor-associated macrophages. Journal of Immunology 182:3105–3111.

Fitzgerald, J. B., B. Schoeberl, U. B. Nielsen, and P. K. Sorger. 2006. Systems biology and combination therapy in the quest for clinical efficacy. Nature Chemical Biology 2:458–466.

Folkman, J. Angiogenesis. 2006. Annual Review of Medicine 57:1–18.

Foo, J., J. Chmielecki, W. Pao, and F. Michor. 2012. Effects of pharmacokinetic processes and varied dosing schedules on the dynamics of acquired resistance to erlotinib in EGFR-mutant lung cancer. Journal of Thoracic Oncology 7:1583–1593.

Fu, F., M. A. Nowak, and S. Bonhoeffer. 2015. Spatial heterogeneity in drug concentrations can facilitate the emergence of resistance to cancer therapy. PLOS Computational Biology 11:e1004142.

Gillies, R. J., and R. A. Gatenby. 2007. Hypoxia and adaptive landscapes in the evolution of carcinogenesis. Cancer and Metastasis Reviews 26:311–317.

Gillies, R. J., D. Verduzco, and R. A. Gatenby. 2012. Evolutionary dynamics of carcinogenesis and why targeted therapy does not work. Nature Reviews Cancer 12:487–493.

Gonzalez, A., O. Ronce, R. Ferriere, and M. E. Hochberg. 2013. Evolutionary rescue: an emerging focus at the intersection between ecology and evolution. Philosophical Transactions of the Royal Society B: Biological Sciences 368:20120404.

Gonzalez-Angulo, A. M., F. Morales-Vasquez, and G. N. Hortobagyi. 2007. Overview of resistance to systemic therapy in patients with breast cancer. Advances in Experimental Medicine and Biology 608:1–22.

Greene, S. E., and A. Reid. 2013. Moving targets: Fighting the evolution of resistance in infections, pests, and cancer. Microbe 8:279–285.

Gullberg, E., S. Cao, O. G. Berg, C. Ilbäck, L. Sandegren, D. Hughes, et al. 2011. Selection of resistant bacteria at very low antibiotic concentrations. PLOS Pathogens 7:e1002158.

Heesterbeek, H., R. M. Anderson, V. Andreasen, S. Bansal, D. De Angelis, C. Dye, et al. 2015. Modeling infectious disease dynamics in the complex landscape of global health. Science 347:aaa4339.

Hochberg, M. E. 2018. An ecosystem framework for understanding and treating disease. Evolution, Medicine, and Public Health 2018:270-286.

Huijben, S., A. S. Bell, D. G. Sim, D. Tomasello, N. Mideo, T. Day, et al. 2013. Aggressive chemotherapy and the selection of drug resistant pathogens. PLOS Pathogens 9:e1003578.

Jansen, G., R. Gatenby, and C. A. Aktipis. 2015. Opinion: Control vs. eradication. Applying infectious disease treatment strategies to cancer. Proceedings of the National Academy of Sciences of the United States of America 112:937–938.

Komarova, N. L., and D. Wodarz. 2005. Drug resistance in cancer: Principles of emergence and prevention. Proceedings of the National Academy of Sciences of the United States of America 102:9714–9719.

Kostadinov, R. L., M. K. Kuhner, X. Li, C. A. Sanchez, P. C. Galipeau, T. G. Paulson, et al. 2013. NSAIDs modulate clonal evolution in Barrett's esophagus. PLOS Genetics 9:e1003553.

Kouyos, R. D., C.J.E. Metcalf, R. Birger, E. Y. Klein, P.A.Z. Wiesch, P. Ankomah, et al. 2014. The path of least resistance: Aggressive or moderate treatment. Proceedings of the Royal Society B: Biological Sciences 281:20140566.

Kutter, E., D. De Vos, G. Gvasalia, Z. Alavidze, L. Gogokhia, S. Kuhl, et al. Phage therapy in clinical practice: Treatment of human infections. Current Pharmaceutical Biotechnology 11:69–86.

Lorz, A., B. Perthame, T. Lorenzi, M. E. Hochberg, and J. Clairambault. 2013. Populational adaptive evolution, chemotherapeutic resistance and multiple anti-cancer therapies. ESAIM: Mathematical Modelling and Numerical Analysis 47:377–399.

MacLean, R. C., A. R. Hall, G. G. Perron, and A. Buckling. 2010. The population genetics of antibiotic resistance: integrating molecular mechanisms and treatment contexts. Nature Reviews Genetics 11:405–414.

Maley, C. C., B. J. Reid, and S. Forrest. 2004. Cancer prevention strategies that address the evolutionary dynamics of neoplastic cells: Simulating benign cell boosters and selection for chemosensitivity. Cancer Epidemiology, Biomarkers & Prevention 13:1375–1384.

Merlo, L.M.F., J. W. Pepper, B. J. Reid, and C. C. Maley. 2006. Cancer as an evolutionary and ecological process. Nature Reviews Cancer 6:924–935.

Mwangi, M. M., S. W. Wu, Y. Zhou, K. Sieradzki, H. de Lencastre, P. Richardson, et al. 2007. Tracking the *in vivo* evolution of multidrug resistance in *Staphylococcus aureus* by whole-genome sequencing. Proceedings of the National Academy of Sciences of the United States of America 104:9451–9456.

Negri, M.-C., M. Lipsitch, J. Blazquez, B. R. Levin, and F. Baquero. 2000. Concentration-dependent selection of small phenotypic differences in TEM β-lactamase-mediated antibiotic resistance. Antimicrobial Agents and Chemotherapy 44:2485–2491.

Nichol, D., P. Jeavons, A. G. Fletcher, R. A. Bonomo, P. K. Maini, J. L. Paul, et al. 2015. Steering evolution with sequential therapy to prevent the emergence of bacterial antibiotic resistance. PLOS Computational Biology 11:e1004493.

Oliver, A., B. R. Levin, C. Juan, F. Baquero, and J. Blazquez. 2004. Hypermutation and the preexistence of antibiotic-resistant *Pseudomonas aeruginosa* mutants: Implications for susceptibility testing and treatment of chronic infections. Antimicrobial Agents and Chemotherapy 48:4226–4233.

Pena-Miller, R., D. Laehnemann, G. Jansen, A. Fuentes-Hernandez, P. Rosenstiel, H. Schulenburg, et al. 2013. When the most potent combination of antibiotics selects for the greatest bacterial load: the smile-frown transition. PLOS Biology 11:e1001540.

Pirnay, J. P., D. De Vos, G. Verbeken, M. Merabishvili, N. Chanishvili, M. Vaneechoutte, et al. 2011. The phage therapy paradigm: prêt-à-porter or sur-mesure? Pharmaceutical Research 28:934–937.

Read, A. F., T. Day, S. Huijben. 2011. The evolution of drug resistance and the curious orthodoxy of aggressive chemotherapy. Proceedings of the National Academy of Sciences of the United States of America 108:10871–10877.

Roemhild, R., C. Barbosa, R. E. Beardmore, G. Jansen, H. Schulenburg. 2015. Temporal variation in antibiotic environments slows down resistance evolution in pathogenic *Pseudomonas aeruginosa*. Evolutionary Applications 8:945–955.

Ross-Gillespie, A., Z. Dumas, and R. Kümmerli. 2015. Evolutionary dynamics of interlinked public goods traits: an experimental study of siderophore production in *Pseudomonas aeruginosa*. Journal of Evolutionary Biology 28:29–39.

Rottenberg, S, J. E. Jaspers, A. Kersbergen, E. van der Burg, A.O.H. Nygren, S.A.L. Zander, et al. 2008. High sensitivity of BRCA1-deficient mammary tumors to the PARP inhibitor AZD2281 alone and in combination with platinum drugs. Proceedings of the National Academy of Sciences of the United States of America 105:17079–17084.

Russell, S. J., K.-W. Peng, and J. C. Bell. 2012. Oncolytic virotherapy. Nature Biotechnology 30:658–670.

Rynkiewicz, E. C., A. B. Pedersen, and A. Fenton. 2015. An ecosystem approach to understanding and managing within-host parasite community dynamics. Trends in Parasitology 31:1–10.

Schulz zur Wiesch, P., J. Engelstadter, and S. Bonhoeffer. 2010. Compensation of fitness costs and reversibility of antibiotic resistance mutations. Antimicrobial Agents and Chemotherapy 54:2085–2095.

Sharma, S.V., D. Y. Lee, B. Li, M. P. Quinlan, F. Takahashi, S. Maheswaran, et al. 2010. A chromatin-mediated reversible drug-tolerant state in cancer cell subpopulations. Cell 141:69–80.

Smallbone, K., R. A. Gatenby, R. J. Gillies, P. K. Maini, and D. J. Gavaghan. 2007. Metabolic changes during carcinogenesis: potential impact on invasiveness. Journal of Theoretical Biology 244:703–713.

Sottoriva, A., H. Kang, Z. Ma, T. A. Graham, M. P. Salomon, J. Zhao, et al. 2015. A Big Bang model of human colorectal tumor growth. Nature Genetics 47:209–216.

Tamma, P. D., S. E. Cosgrove, and L. L. Maragakis. 2012. Combination therapy for treatment of infections with gram-negative bacteria. Clinical Microbiology Reviews 25:450–470.

Tomlinson, I. P., M. R. Novelli, and W. F. Bodmer. 1996. The mutation rate and cancer. Proceedings of the National Academy of Sciences of the United States of America 93:14800–14803.

Torres-Barceló, C., F. I. Arias-Sánchez, M. Vasse, J. Ramsayer, O. Kaltz, and M. E. Hochberg. 2014. A window of opportunity to control the bacterial pathogen *Pseudomonas aeruginosa* combining antibiotics and phages. PLOS One 9:e106628.

Torres-Barceló, C., and M. E. Hochberg. 2016. Evolutionary rationale for phages as complements of antibiotics. Trends in Microbiology 24:1–8.

Trédan, O., C. M. Galmarini, K. Patel, and I. F. Tannock. 2007. Drug resistance and the solid tumor microenvironment. Journal of the National Cancer Institute 99:1441–1454.

Vacher, C., D. Bourguet, F. Rousset, C. Chevillon, and M. E. Hochberg. 2003. Modelling the spatial configuration of refuges for a sustainable control of pests: a case study of *Bt* cotton. Journal of Evolutionary Biology 16:378–387.

Vale, P. F., L. McNally, A. Doeschl-Wilson, K. C. King, R. Popat, M. R. Domingo-Sananes, et al. 2016. Beyond killing: Can we find new ways to manage infection? Evolution, Medicine, and Public Health 2016:148–157.

Vanneman, M., and G. Dranoff. 2012. Combining immunotherapy and targeted therapies in cancer treatment. Nature Reviews Cancer 12: 237–251.

Viertel, T. M., K. Ritter, and H.-P. Horz. 2014. Viruses versus bacteria-novel approaches to phage therapy as a tool against multidrug-resistant pathogens. Journal of Antimicrobial Chemotherapy 69:2326–2336.

Wargo, A. R., S. Huijben, J. C. de Roode, J. Shepherd, and A. F. Read. 2007. Competitive release and facilitation of drug-resistant parasites after therapeutic chemotherapy in a rodent malaria model. Proceedings of the National Academy of Sciences of the United States of America 104:19914–19919.

Webber, M. A. 2002. The importance of efflux pumps in bacterial antibiotic resistance. Journal of Antimicrobial Chemotherapy 51:9–11.

Wodarz, D., and N. Komarova. 2009. Towards predictive computational models of oncolytic virus therapy: Basis for experimental validation and model selection. PLOS One 4:e4271.

Woods, R. J., and A.F. Read. 2015. Clinical management of resistance evolution in a bacterial infection. Evolution, Medicine, and Public Health 2015:281–288.

zur Wiesch, P. A., R. Kouyos, J. Engelstädter, R. R. Regoes, and S. Bonhoeffer. 2011. Population biological principles of drug-resistance evolution in infectious diseases. Lancet Infectious Diseases 11:236–247.

PART VI

CONSERVATION BIOLOGY
AND NATURAL RESOURCE MANAGEMENT

Collective Cooperation

From Ecological Communities to Global Governance and Back

Simon A. Levin

Collective cooperation within large groups has posed a fundamental puzzle in ecology and evolution at least since Darwin, especially to the degree that unequally distributed costs and benefits seem to be a part of the story. In the latter case, some individuals may appear to be altruistic; this created for Darwin a challenge to his theory of evolution through natural selection, a challenge he never fully resolved.

For the evolution of eusociality, Hamilton (1964) seemed to have shed useful light on the problem by working out the effects of relatedness and kin selection on promoting altruism; yet even that advance has been called into question (Nowak et al. 2010), in part because eusociality is not restricted to the haplodiploid insects. In any case, cooperation comes in many forms in nature, and clearly often arises among unrelated individuals. Such cooperation may aid individuals in the acquisition of resources or facilitate mating; often, however, cooperation is a mechanism that arises to bind individuals together into groups to provide advantages in conflict with other groups. Nowhere is this more evident than in human societies, making urgent an understanding of how to extend cooperation to all of humanity to deal with common threats, such as the deterioration of the biosphere.

The issue of individual incentives versus the collective good represents a dominant theme not only in evolutionary biology, but also in economic theory. Adam Smith (1776) wrote famously of "the invisible hand," which argued that, sometimes, individuals who pursued their own selfish agendas would produce outcomes that benefitted society. This argument is often used implicitly in arguing for financial markets relatively free of government control, yet Smith himself explicitly recognized that individual greed could undercut the public good. Similar holistic arguments persist, however, in fanciful dreams of Gaia (Kirchner 1991, Schneider et al. 1991, Kirchner 2002). We have learned from bitter experience that the invisible hand does not protect our economies; nor will the goddess Gaia protect our biosphere.

We do know, however, that individuals cooperate in collective activities that benefit the group as a whole. Group-selection arguments have been invoked to explain such behaviors (Wynne-Edwards 1962), but these are justified only under special conditions; more generally, one should seek explanations in actions that enhance individual fitness. Addressing the issue of how and when cooperation works presents deep theoretical challenges, with great relevance to how and whether we can achieve a sustainable future for humanity. In this article, I will highlight a few of the research areas that seem likely to bear fruit, especially through cross-fertilization across disciplines. Cooperation in dealing with global problems can benefit from an understanding of how evolution has solved such problems or failed to. In the latter cases, if we are to achieve a sustainable future, we must understand how to create the norms and institutions that bottom-up processes will not find.

As an example, Avinash Dixit, Daniel Rubenstein, and I examined the maintenance of insurance arrangements among east African herdsmen (Dixit et al. 2013). This is a problem of considerable interest in and of itself, but even more so as a model for cooperation in other spheres. In these situations, herdsmen engage in cooperative arrangements that allow those experiencing bad conditions to graze their cattle on others' ranges; it is a form of insurance, in that it relies on the unpredictability of when one or the other will experience good conditions. A first calculation computes the social optimum, that which maximizes the sum of the utilities of all participating herdsmen. Were there top-down controls, maintained by some higher authority, this would be the expected solution. However, if individual herdsmen are acting independently, such a social optimum may not represent a Nash equilibrium; that is, cooperation will collapse if individuals experiencing good years discount the future too heavily. Thus, a second calculation establishes what the critical discount rate is beyond which cooperation collapses. A final calculation then looks for *second-best solutions*, namely those that involve some sharing, and maximize total social welfare subject to the constraint that the arrangement can be sustained (is Nash) without the intervention of a higher authority. In unpublished work, we investigated the influence of *prosociality*, meaning that individuals incorporate others' welfare to some degree into their calculations (for example because others are kin); not surprisingly, this elevates the total social welfare that can be sustained. I will return to this theme in the next paragraph. The framework described above has obvious extensions to other problems of the commons, from fisheries to the global environment. But it also has the potential to help explain the evolution of cooperative arrangements in other species, or even among species, as well as the failure of evolution to achieve the collective optimum in many circumstances.

As mentioned above, *prosociality* obviously facilitates cooperation, and the phenomenon raises a number of deep questions: When will prosociality lead to the maintenance of a public good that would not have existed without it? Can social evolution lead to the maintenance of such prosociality? How has prosociality, especially among unrelated individuals, arisen through evolution? These questions have been addressed from multiple perspectives in diverse literature (Gintis 2003, Axelrod 2006, Akçay et al. 2009, Dixit 2009), and Dixit and I have been extending his basic results to more complex societal topologies, many involving environmental issues as with our work with Rubenstein (Dixit et al. 2013). The emergence of prosociality of course has long been a staple of evolutionary theory and remains among its most controversial. Marrying economic and evolutionary theory holds great potential to lead to enlightening synergisms for all these issues.

Even in the absence of prosociality, cooperation can arise through *reciprocal altruism;* in other words, through an effective contract involving two or more players. The example of the herdsmen is a case in point; but, more generally of course, such contracts are a staple of political theory (Axelrod 1997) and are crucial in the development of international agreements for such challenges as biodiversity loss or climate change (Barrett 2003). In any negotiations, implicit or explicit, leading to possible cooperation, each player must weigh the costs and benefits of particular actions; but the situation is more complicated, because the costs and benefits of other parties must be estimated to determine just how much one should yield in a negotiation. A framework for thinking about such issues involves assessing one's own likely benefits from cooperation and from outside options and estimating the same for potential partners (Akçay et al. 2012). Of course, information in such circumstances is asymmetric in the sense that another's costs and benefits (particularly involved so-called "outside actions") are not known with the same certainty as one's own, and this makes the decision as to whether to cooperate more complicated. Furthermore, the decision as to whether to cooperate is only part of the negotiation, which also must involve deciding how to divide the excess goods that have been achieved through cooperation. The theory of mechanism design, borrowed from economics (Maskin 2008, Myerson 2008), can be applied in an effort to determine pathways to cooperation; however, application of that theory demonstrates that potential collaborators who would benefit from cooperation may fail to realize this. Though the reasons for failure are not the same as for the prisoner's dilemma (Nash 1951), the bottom line is the same: Agents that should cooperate do not.

Cooperation can fall short of ideal in other ways, as in the second-best solution given earlier. The prisoner's dilemma only admits one Nash equilibrium (defect–defect) and is a noncooperative game. More generally, in international negotiations as well as in nature, we may be dealing with *cooperative games,* in which there may be multiple Nash equilibria (essentially multiple stable solutions), some better than others; there, individuals may be cooperating, but not as effectively as they might. Evolution may well lock in on inferior equilibria; but in international agreements, we should strive to find pathways to better solutions.

Cooperation among small numbers of individuals raises issues that also arise when many individuals are involved, though the analytical techniques relevant to large groups may be quite different. Still, just as in celestial mechanics, where much attention is given to two-body problems because they are tractable, the overwhelming number of game-theoretic examples that appear in the literature or in textbooks involve only two players. The challenge of scaling up to large ensembles hence is one of the most crucial areas for such research. In the game-theoretic literature, that has sparked interest in what are called *mean-field games,* namely games involving large numbers of interacting individuals (Huang et al. 2007, Lasry and Lions 2007, Nourian et al. 2012, Nourian et al. 2014); such advances have not yet had an impact on ecological investigations, although they should. Again, interdisciplinary transfers of insights and perspectives hold the potential for synergistic advances.

Of course, mean-field games in ecology have been explored to some extent under other names. For example, collective motion of bird flocks and fish schools and insect swarms has attracted much attention from ecologists and modelers alike (Hamilton 1971, Couzin et al. 2005, Cucker and Smale 2007, Bialek et al. 2012); indeed, few ecological topics have attracted attention across such a wide spectrum of researchers. Considerable work has focused on the dynamics of large aggregations, and on scaling from individuals to collectives; much less, however, has dealt with the public-goods dimensions of these ensembles, and the conflict between the decisions that individuals make and the good of the group. Information is perhaps the ultimate public good, and its production raises issues captured in the producer/scrounger dichotomy (Barnard and Sibly 1981, Vickery et al. 1991), a staple of behavioral ecology. In collective movement, some individuals pay an extra price for producing information, and others basically steal their information from the producers (Couzin et al. 2005, Guttal and Couzin 2011). Knowing the rules individuals use in determining their movement patterns, and what the consequences of such rules are for collective dynamics, leaves unanswered the question of whether such rules lead to behaviors that are best for all. Solutions that emerge from individual selection in such complex adaptive systems should not be expected to maximize total social welfare.

Public goods (and common pool–resource) problems are everywhere in the ecological world (Levin 2014), from bacteria and slime molds to nutrient cycling in ecosystems. John Tyler Bonner (1959) highlighted the social conflicts in the cellular slime mold long ago, and more recent work highlights the public goods dimensions of bacterial biofilms (Drescher et al. 2014). Nitrogen fixation, chelation, and a number of other fundamental ecosystem processes present similar challenges.

Such problems are also at the core in dealing with global environmental problems and are my candidates for the major unsolved problems in ecology. In the natural world, such studies will require scaling from individual actions to system consequences, within an evolutionary context that asks whether the social optimum in achieved. In achieving sustainability in our societies, however, we must go further, as in the case of the east African herdsmen, and even beyond. Garrett Hardin (1968), building on ideas of William Forster Lloyd (1833), famously brought to the attention of the ecological community the concept of the tragedy of the commons, in which individuals, by following their own selfish agendas, fail to sustain a resource essential to all. He concluded that the solution to such problems was in "mutual coercion, agreed upon," which for him meant societal "arrangements that create coercion of some sort." Elinor Ostrom, however, demonstrated that in small societies, like local fisheries, such mutual coercion could arise from bottom-up cooperative arrangements among individuals, essentially through the establishment of social norms (Ostrom 1990). She then built on such ideas to argue that the pathway to global cooperation, where the common heritages and objectives that exist in small societies are absent, is through *polycentric* approaches that built on local agreements (Ostrom 2009). Indeed, such ideas hold great promise for the most refractory problems of international cooperation (Hannam 2017). Some great unsolved problems facing us, in my view, are in understanding when and how cooperation in commons situations emerges in natural systems, and how we can overcome limitations to cooperation in dealing with the global commons in order to build a sustainable future.

References

Akçay, E., A. Meirowitz, K. Ramsay, and S. A. Levin. 2012. Evolution of cooperation and skew under imperfect information. Proceedings of the National Academy of Sciences of the United States of America 109:14936–14941.

Akçay, E., J. Van Cleve, M. W. Feldman, and J. Roughgarden. 2009. A theory for the evolution of other-regard integrating proximate and ultimate perspectives. Proceedings of the National Academy of Sciences of the United States of America 106:19061–19066.

Axelrod, R. 1997. The complexity of cooperation: Agent-based models of competition and collaboration. Princeton, N.J.: Princeton University Press.

Axelrod, R. 2006. The evolution of cooperation. New York: Basic Books.

Barnard, C. J., and R. M. Sibly. 1981. Producers and scroungers: a general model and its application to captive flocks of house sparrows. Animal Behaviour 29:543–550.

Barrett, S. 2003. Environment and statecraft: The strategy of environmental treaty-making. Oxford: Oxford University Press.

Bialek, W., A. Cavagna, I. Giardina, T. Mora, E. Silvestri, M. Viale, and A. M. Walczak. 2012. Statistical mechanics for natural flocks of birds. Proceedings of the National Academy of Sciences of the United States of America 109:4786–4791.

Bonner, J. T. 1959. Differentiation in social amoebae. Scientific American 201:152–162.

Couzin, I. D., J. Krause, N. R. Franks, and S. A. Levin. 2005. Effective leadership and decision making in animal groups on the move. Nature 433:513–516.

Cucker, F., and S. Smale. 2007. Emergent behavior in flocks. IEEE Transactions on Automatic Control 52:852–862.

Dixit, A. 2009. Governance institutions and economic activity. American Economic Review 99:5–24.

Dixit, A., S. A. Levin, and D. I. Rubenstein. 2013. Reciprocal insurance among Kenyan pastoralists. Theoretical Ecology 6:173–187.

Drescher, K., C. D. Nadell, H. A. Stone, N. S. Wingreen, and B. L. Bassler. 2014. Solutions to the public goods dilemma in bacterial biofilms. Current Biology 24:50–55.

Gintis, H. 2003. Solving the puzzle of prosociality. Rationality and Society 15:155–187.

Guttal, V., I. D. Couzin. 2011. Leadership, collective motion and the evolution of migratory strategies. Communicative & Integrative Biology 4:294–298.

Hamilton, W. D. 1964. The genetical evolution of social behavior: I and II. Journal of Theoretical Biology 7:1–52.

Hamilton, W. D. 1971. Geometry for the selfish herd. Journal of Theoretical Biology 31:295–311.

Hannam, P. M., V. V. Vasconcelos, S. A. Levin, and J. M. Pacheco. 2017. Incompletecooperation and co-benefits: deepening climate cooperation with aproliferation of small agreements. Climatic Change 144:65–69.

Hardin, G. 1968. The tragedy of the commons. Science 162:1243–1248.

Huang, M., P. E. Caines, and R. P. Malhame. 2007. An invariance principle in large population stochastic dynamic games. Journal of Systems Science & Complexity 20:162–172.

Kirchner, J. W. 1991. The Gaia hypotheses: are they testable? Are they useful? In: S. H. Schneider and P. J. Boston, eds. Scientists on Gaia. Cambridge, Mass.: MIT Press, 38–46.

Kirchner, J. W. 2002. The Gaia hypothesis: Fact, theory, and wishful thinking. Climatic Change 52:391–408.

Lasry, J. M., and P. L. Lions. 2007. Mean field games. Japanese Journal of Mathematics 2:229–260.

Levin, S. A. 2014. Public goods in relation to competition, cooperation, and spite. Proceedings of the National Academy of Sciences of the United States of America 111(Suppl. 3):10838–10845.

Lloyd, W. F. 1833. Two lectures on the checks to population. Oxford: J. H. Parker.

Maskin, E. S. 2008. Mechanism design: How to implement social goals. American Economic Review 98:567–576.

Myerson, R. B. 2008. Perspectives on mechanism design in economic theory. American Economic Review 98:586–603.

Nash, J. F. 1951. Non-cooperative games. Annals of Mathematics 54:286–295.

Nourian, M., Caines, P. E., and R. P. Malhame. 2014. A mean field game synthesis of initial mean consensus problems: A continuum approach for non-Gaussian behavior. IEEE Transactions on Automatic Control 59:449–455.

Nourian, M., Caines, P. E., Malhame, R. P., and M. Huang. 2012. Mean field LQG control in leader-follower stochastic multi-agent systems: likelihood ratio based adaptation. IEEE Transactions on Automatic Control 57:2801–2816.

Nowak, M., C. Tarnita, and E. Wilson 2010. The evolution of eusociality. Nature 466:1057–1062, doi:10.1038/nature09205.

Ostrom, E. 1990. Governing the commons: The evolution of institutions for collective action. Cambridge: Cambridge University Press.

Ostrom, E. 2009. A polycentric approach for coping with climate change. Policy Research Working Paper 5095. Washington, D.C.: World Bank.

Schneider, S. H., and P. J. Boston, eds. 1991. Scientists on Gaia. Cambridge, Mass.: MIT Press.

Smith, A. 1776. An inquiry into the nature and causes of the wealth of nations. Reprint, 1976. Chicago: University of Chicago Press.

Vickery, W. L., L.-A. Giraldeau, J. J. Templeton, D. L. Kramer, and C. A. Chapman. 1991. Producers, scroungers, and group foraging. American Naturalist 137:847–863.

Wynne-Edwards, V. C. 1962. Animal dispersion in relation to social behaviour. Edinburgh: Oliver and Boyd.

Keeping the Faith

The Case for Very-Large Terrestrial and Marine Protected Areas

Tim Caro

Very large strictly protected marine and terrestrial areas are the best long-term method of conserving intact natural ecological processes that are central to answering questions in ecology. Over the last half-century, anthropogenic pressures on protected areas have changed and multiplied yet very large areas are still better able to counter these challenges than smaller reserves, and full protection can assuage some of them entirely. The next decade may be the last opportunity to expand the extent of very-large terrestrial protected areas through enlarging existing reserves, creating new ones, and establishing buffer zones and corridors; in most cases international funding will be critical to success. Currently, we are making good progress in increasing the number of very-large marine protected areas. Ultimately the conservation of ecosystem processes will depend on these huge areas with minimal human influence which act as important baselines for answering unsolved questions in ecology.

Introduction

Functioning natural ecosystems enable us to investigate ecological processes and establish baseline data against which ecological perturbations can be compared (Sinclair et al. 2002). Future ecological discoveries will therefore depend enormously on the preservation of fully functioning ecosystems. In the 1960s when many national parks were first established it was simply accepted that large unexploited areas under strict protection were the best conservation strategy for maintaining intact ecological communities for perpetuity. These early protected areas constitute some of the world's most venerable and famous national parks such as Serengeti National Park, Chitwan National Park, and Yellowstone National Park. Similar sentiments have been reached using meta-analyses of marine systems which show that marine national parks (MPAs) under strict protection are the most effective way to conserve fish stocks (Edgar et al. 2014).

In this chapter, I arbitrarily define functioning terrestrial ecosystems as areas inhabited by large mammals (those with weight greater than 50 kg) including their predators, and functioning marine ecosystems as areas used by apex predatory fish or large marine mammals. I emphasize predators because their loss reduces cross-system connectivity, ecosystem stability, alters biogeochemical cycling, results in mesopredator release, and reshapes plant recruitment (Estes et al. 2011), although other ecological players such as large migratory herds certainly contribute to ecosystem integrity (Soule and Terborgh 1999). I define very-large terrestrial protected areas (VLTPAs) as those greater than 25,000 km^2 in area (Cantu-Salazar and Gaston 2010), and very-large marine protected areas (VLMPAs) as those greater 10^5 km^2 (Toonen et al. 2013). In line with others, I consider strict protection as occurring in International Union for Conservation of Nature (IUCN) protected area categories Ia (strict nature reserves managed mainly for science), Ib (wilderness areas managed mainly for wilderness protection), II (national parks managed mainly for ecosystem protection and recreation), III (natural monuments managed mainly for conservation of specific natural features), and IV (habitat/species management areas managed mainly for conservation through management intervention).

Despite its pedigree, the classic large protected area paradigm has been questioned repeatedly over the last 50 years. Antithetical views have included asking whether small reserves are equally good at conserving biodiversity (the "single large or several small" (SLoSS) debate), whether extractive reserves are better at conserving wildlife than no-take areas, whether top-down government control is really effective in protecting habitat, whether missing ecosystem components can be replaced by proxies, and even whether conserving natural habitats should be our primary conservation objective at all. Since 1975, these and other issues have diverted academics' attention, managers' money, journalists' priorities and governments' decisions away from what should be the crown jewel of conservation: very-large protected areas (VLPAs). Here, I try to reestablish why large unexploited areas coupled with strict top-down protection are the best strategy for conserving natural ecosystems, dismiss alternative propositions, discuss how large well-protected unexploited areas can weather many anthropogenic pressures, and summarize measures that can be used to broaden the very-large marine and terrestrial protected area project.

Extolling the importance of VLPAs is not new (Cantu-Salazar and Gaston 2010) and several massive MPAs have been gazetted recently (Jones and De Santo 2016) but here I stress their critical importance for *maintaining ecological processes* rather than conserving biodiversity. There are still opportunities to set aside VLTPAs and VLMPAs uninfluenced by human activities although these are diminishing rapidly, hence the urgency for reiterating the case for VLPAs.

Why Very-Large Protected Areas Work

Justification

VLTPAs are the most effective form of conservation for several reasons (Table 1). First, their size means that they often include intact species assemblages, including populations of relatively rare apex predators that can structure lower trophic levels (Terborgh and Estes 2010). They also contain robust populations of sufficient numbers of species so that ecological processes of competition, facilitation, succession, and scavenging can all play out. VLTPAs can also provide space for seasonal movements between spatially distinct food sources, and in some cases incorporate migratory routes. They can protect entire watersheds and aquatic ecosystem processes, and are likely to fare better in the face of climate change because of their greater range of elevation, climates, habitats, and latitude. Their low periphery-to-core ratio means that they are better buffered from anthropogenic influences around their edges (e.g., fire); proportionately, land-use conflict along terrestrial reserve edges is minimized; and in terrestrial reserves there will be proportionately less human–wildlife conflict because of fewer opportunities for wild animals to wander outside borders. Indeed, empirical data support the negative relationship between protected area size and intensity of human pressure (Jones et al. 2018). From a managerial perspective, on a per area basis, larger reserves are easier and less expensive to protect and maintain than smaller reserves (Peres 2005, Laurance 2005). Nonetheless, their size means that they are costly to purchase, patrol, and to monitor in absolute terms.

VLMPAs also protect intact species assemblages and whole ecosystems (Table 1) and provide baselines from which to compare many other marine areas that are exploited (Licuanan et al. 2017). They too can encompass animal seasonal movements (Block et al. 2011) and, potentially, all of the life history stages of fishes and invertebrates including larval dispersal. Whether fully protected or not, large MPAs have larger fish species, and higher fish biomass, particularly of sharks, than fished areas (Edgar et al. 2014). Their huge size means that they necessarily protect a variety of habitats. Nevertheless, their very-large size means that they are extremely costly to patrol, usually requiring aircraft or satellites.

Table 1. Some Advantages and Disadvantages of VLPAs

ADVANTAGES	DISADVANTAGES
Terrestrial	
Intact species assemblages	Costly to purchase land
Encompass large home ranges	Costly to survey and monitor
Encompass seasonal movements, sometimes migration	Costly to patrol
Incorporate several land cover types	
Habitat diversity may mitigate climate change	
Low periphery-to-core ratio proves insulation from external insult	
Proportionately easier to manage	
Marine	
Intact species assemblages	Costly to survey and monitor
Protect whole ecosystems.	Costly to patrol
Encompass seasonal movements, sometimes migration	
Protect dispersing larvae	
High fish biomass	
Protect variety of habitats	

Questioning the Paradigm

Although some would argue that it is axiomatic that large protected areas are our best conservation tool, ecologists, sociologists, and political scientists have questioned this paradigm in several orthogonal ways over the last 50 years. Early academic arguments about both the species richness and habitat heterogeneity benefits of small reserves (the SLoSS debate) no longer have practical relevance in a world where decisions about siting reserves are normally made on the basis of remoteness, land use, and political expediency. Nonetheless, small reserves are essential for conserving endemic species trapped in an area of human land use (e.g., giant pandas (*Ailuropoda melanoleuca*) in the Wolong National Nature Reserve, China), for capturing representative samples of otherwise degraded or destroyed biomes (e.g., tall grass prairie reserves in the United States), as stepping stones between larger protected areas (e.g., small forest reserves in Tanzania), as mountaintop reserves with unique flora and fauna (e.g., in Nicaragua), or where no other alternatives for conserving wildlife exist (e.g., many European protected areas), yet they are usually poor substitutes for conserving *ecosystem processes*.

Starting in the 1970s a movement to utilize wildlife—to make natural resources pay their way—was embraced by the IUCN, leading to a split in the conservation movement between protectionism and utilization. In Africa, for example, this is manifest as photographic versus hunting tourism or, at a broader scale, hunters against animal rights proponents. Limited data suggest that extractive conservation areas have different species' complements than strictly protected areas (Gardner et al. 2007) and lack species of conservation significance (Shahabuddin and Rao 2010). For those interested in maintaining functioning ecological communities, selected offtake of some species alters strengths of predator–prey interactions that can result in mesopredator release or prey eruptions (Myers et al. 2007), or changes in plant–herbivore dynamics. Nevertheless, some argue that well-managed exploitation carried out at low levels has a politically expedient role of making VLPAs more palatable, as seen in the Great Barrier Reef in Australia, or formerly, in the Selous Game Reserve in Tanzania, although overexploitation following reserve establishment must be guarded against.

It is easy to make prognoses about conserving large wilderness areas with a full component of biodiversity and functioning food webs but in reality there are increasingly few opportunities due to land conversion, population extirpations, and species extinctions. Consequently, species have been returned to areas where they were once present historically. These can be highly successful in restoring ecosystem processes as in wolf (*Canis lupus*) reintroduction into the Greater Yellowstone Ecosystem; other reintroductions of apex predators are planned (Wolf and Ripple 2018). Related subspecies are also introduced as ecosystem engineers. Such restoration projects are becoming more common and are hailed as successes, too; for example, heck cattle and Konik ponies as grazers in the Oostvaardersplassen, Netherlands (Stokstad 2015) although ecological reference points against which we can measure "success" are challenging. More controversially, plans to rewild areas using extant species to stand in for extinct species have been discussed although not yet implemented. These ideas involve using cheetahs (*Acinonyx jubatus*) to rewild areas of North America where extinct cheetah species once ran, or Asian elephants (*Elephas maximus*) to double as mammoths. Some academic and journalistic effort has been expended on justifying rewilding but logistical barriers including introduction costs, fences, disease prevention, and attitude change are all formidable, and these discussions carry the danger of diverting conservation effort away from protecting intact ecosystems (Caro 2007).

In the 1980s a debate arose in conservation science about the players involved in wildlife protection, centered on who should be responsible for protection—state or local communities—and who should benefit from protection practices. In the context of VLPAs being the most important conservation tool for conserving ecosystem processes, they can only be realistically policed by state-run personnel, whereas small no-take reserves can be protected by communities provided that resource spillover outside protected area boundaries or in buffer zones can be harvested. Admittedly, there are exceptions, such as large co-management areas in the Canadian Arctic, where strong conservation education and outreach programs are coupled with tightly limited resource extraction within reserves, but these examples are usually restricted to rich developed nations. As a rule of thumb, it is only state institutions that have the financial capability of managing areas greater than 10,000 km^2 in area. Management aside, profits from large reserves need to be apportioned appropriately to satisfy all stakeholders and win local support and this poses great challenges for nations with poor governance.

Recently, the very idea of conserving wilderness has been questioned. "Gardeners of nature" suggest that most, or all of nature is now subject to anthropogenic influence (Otto 2018) and our principal effort should be in conserving flora and fauna in agricultural landscapes (Kareiva et al. 2007). The origins of this approach are in Central America where human land use has left corridors of original vegetation or relatively benign crops connecting or being close to remnant forest patches. Yet many agricultural landscapes are far less benign; for example, monocultures of soybean and palm oil plantations and maize and wheat fields, areas where original vegetation is now absent. Such working landscapes are conceptually far from intact ecosystems so central to ecological research (Wuerthner et al. 2015).

Having suggested that very-large unexploited strictly protected areas are the principal strategy for maintaining functioning ecosystems with multiple trophic levels, I next enumerate a dozen important challenges that these areas face in 2020 (see also Geldmann et al. 2014). Many of these are associated with rapid world population growth over the last half-century (from 3.45 billion in 1967 to 7.68 billion in 2019) that has left an uneven human footprint across the terrestrial landscape (Venter et al. 2016).

Contemporary Challenges to Large Protected Areas

1. Human population pressure. In many parts of the world, human populations are growing quickly near reserve borders (Wittemyer et al. 2008). Sometimes this stems from migrants settling near reserves due to lack of resources elsewhere, especially arable land (Salerno et al. 2014), sometimes because people are attracted to features of the protected area, such as job opportunities in ecotourism, clean water, or the very resources that are being protected. Whatever the cause, increasing population pressure at terrestrial reserve borders isolates protected areas within the landscape and from each other. Larger reserves are *de facto* less impacted by neighboring land settlement.

2. Large projects. Worldwide, many large development projects are being implemented or planned close to or even inside protected areas. These comprise large-scale development corridors (Laurance et al. 2015), mining projects, and drilling operations within protected areas, together with associated access roads as seen in Murchison Falls National Park, Uganda. In the USA the possibility of oil exploitation in the federally owned section of the Arctic National Wildlife Refuge in Alaska continues to be debated while fossil fuel extraction in the lower forty-eight states has disrupted ungulate migration patterns and driven declines of the greater sage grouse (*Centrocercus urophasianus*). In the western Amazon rainforest, there are nearly 180 oil and gas blocks covering approximately 688,000 km²; many of these blocks overlap some of the most speciose areas on earth and are located within protected area boundaries (Finer et al. 2013). Yasuni National Park in Ecuador houses one of the most-biodiverse and least-degraded sectors of the Amazon but contains Ecuador's second-largest untapped oilfield.

Road proliferation drives deforestation and encourages development and exploitation at considerable distances from trunk roads (Ibisch et al. 2016). Animal mortalities, wildlife and timber exploitation, and species introductions all increase with these transport developments (Kleinschroth and Healey 2017). In general, larger reserves may be better able to tolerate these insults because they affect a relatively smaller proportion of the protected area.

3. Fragmentation and isolation. Habitat fragmentation between reserves reduces effective reserve size (Laurance et al. 2012) potentially lowering the genetic diversity of populations within reserves, slowing population growth rates, reducing trophic chain length of communities, altering species interactions, facilitating extinctions, and so ultimately lowering biodiversity (Rudnick et al. 2012, Crooks et al. 2017). A classic example comes from large carnivores with large home ranges that suffer human-induced mortality when wandering outside reserve borders (Woodroffe and Ginsberg 1998). Ecosystem effects such as changes in plant and consumer biomass and nutrient cycling and are often delayed following isolation but nonetheless appear later (Haddad et al. 2015). Obviously, very-large areas will be less influenced by reductions in effective reserve size than smaller areas.

4. Changing hydrology. Some protected areas serve as dry-season water sources for wildlife. New, excessive water demands outside reserves mean that major rivers may run dry during all or parts of the year. For example, overdrawing river water for irrigation and cattle watering outside Ruaha National Park, Tanzania, has caused the Great Ruaha River to become seasonally dry (Mtahiko et al. 2006). In Serengeti National Park, flow of the Mara River has decreased by 68% in the last 40 years due to increased water extraction upstream and deforestation of the Mau Forest in Kenya (Gereta et al. 2009). Agricultural demand on the rivers upstream of Kruger National Park, South Africa, has led to severe water shortages within the park (Du Toit et al. 2003). Even large reserves are not immune from hydrological pressures.

Runoff from both agriculture and mining activities pollutes watersheds and threatens survival of wildlife within protected areas. Widespread use of agricultural fertilizers, herbicides, and pesticides has negatively affected protected areas such as the Everglades National Park (Izuno et al. 1991), for instance. Similarly, pollution from agricultural systems can accelerate eutrophication in freshwater systems leading to freshwater species' declines; coastal protected areas can also be affected. VLTPAs are not immune from these affronts.

5. *Deforestation.* Selective logging results in forest degradation around and within protected areas. In protected areas already affected by logging, fuelwood extraction and charcoal production leads to further degradation. Many proximate factors (e.g., agriculture, wood extraction) and ultimate factors (e.g., economics and national policies) drive deforestation in protected areas; they are complex and often site- or region-specific so effects of deforestation are variable (Geist and Lambin 2002). Remoteness and difficulty of extracting resources from some protected areas may be responsible for lower rates of deforestation and degradation within reserves far from human settlement (*de facto* protection; Joppa et al. 2008) rather than reserve size *per se*.

6. *Wildlife exploitation.* Oceans were once heavily exploited for marine mammals but now are exploited for fish, with apex predators suffering badly. The principal function of MPAs is to halt or limit intense fishing pressure but to make MPAs palatable to stakeholders limited fishing is often allowed and very few MPAs are strictly protected (Costello and Ballantine 2015). Exceptions include new and very-large and remote Pacific and Indian Ocean MPAs (Devillers et al. 2015; Jones and De Santo 2016). In those that are legally protected, patrolling is a challenge.

In the terrestrial realm, there is a vast luxury trade in wildlife and wildlife products (Gross 2019) some of which comes illegally from protected areas. For example, African elephant (*Loxodonta africana*) populations in Tanzanian protected areas have recently declined by 60% due to ivory poaching (Tanzania Wildlife Research Institute 2015). Bushmeat hunting, often in protected areas where densities of animals are high, occurs on a massive scale: 78,000–110,000 wildebeest (*Connochaetes taurinus*) are poached in Serengeti National Park annually (Lindsey et al. 2013). Bushmeat can be a last-resort food source for poor communities but is often simply a source of additional income. In urban areas, bushmeat consumption is increasingly associated with greater wealth (Wilkie et al. 2016). There is also a huge illegal trade in threatened plants.

All types of extraction can lead to negative consequences for plant and animal communities and ecosystem processes within protected areas. For example, high levels of hunting in Amazonian forest sites have resulted in reduced densities of large-bodied game species (Peres 2000) several of which are important seed dispersers or predators. Respectively, exploitation can affect plant regeneration by decreasing seed dispersal, germination, and seed size, whereas removal of top predators negatively affects ecosystems by altering trophic cascades and reducing length of the food chain. Removal of ecosystem engineers, such as elephants, alters vegetation structure. Unfortunately, VLPAs are not immune from specific forms of wildlife exploitation (e.g., rhinoceros poaching) or from general political instability (Daskin and Pringle 2018).

7. Invasive Species. Exotic plant species are now present in many protected areas (Foxcroft et al. 2017) and exotic animals are increasing, too. The introduced Burmese python (*Python bivittatus)* has caused a dramatic decline in raccoons (*Procyon lotor),* and opossums (*Didelphis virginiana),* and a complete disappearance of once common rabbits (*Sylvilagus* spp). in the Florida Everglades National Park for example (Dorcas et al. 2012).

Introduced parasites and disease also have detrimental effects on native populations inside and outside protected areas. Avian malaria and avian poxvirus introduced to Hawaii in 1826 led to the probable extinction of at least 13 bird species. Domesticated animals can be vectors for the spread of invasive diseases into protected areas as in the case of domestic dogs carrying canine distemper virus to Serengeti lions (*Panthera leo*). This may be less of a problem in larger protected areas with lower periphery-to-core ratios. Invasive species can also outcompete native species in protected areas: in Yellowstone National Park and surrounding areas, the invasive plant *Linaria vulgaris* has dramatically reduced the cover of native plants (Pauchard et al. 2003).

Nonnative species, especially plants, are often difficult and costly to eradicate. Larger protected areas can limit the spread of nonnative plants, seen in Kruger National Park, where the number of nonnative plants inside the park drops off dramatically 1500 m from the border due to reduced disturbance (Foxcroft et al. 2011). In general, human density around protected areas is a major driver of invasions of both plants and animals and VLPAs are still affected by exotics.

8. Livestock–wildlife conflict. Incursions of livestock into reserves are common and livestock grazing has been implicated in environmental degradation, water shortages, and forage scarcity in and around protected areas; greater than 40% of 93 protected areas were ineffective at mitigating the impacts of grazing (Bruner et al. 2001). Although livestock may provide unexpected benefits to protected areas by promoting seed dispersal and increasing plant diversity, generally there is a negative competitive relationship between livestock density and wildlife density despite competition sometimes being seasonally dependent. In VLPAs, spatial partitioning of wildlife and livestock may lessen competition.

9. Fire. Fire shapes ecosystem structure and can maintain biodiversity, it can alter habitat structure, and can affect nutrient and particle content of soil, water, and air. In tropical forests, the natural fire return interval is long but human influence has shortened this through clearing land for agriculture. Fires set on nearby land can spread into protected areas as seen in the 2019 Brazilian amazon fires. In contrast, in subtropical and temperate forests, grasslands, and shrublands that are fire-adapted, fires are common but there human activities have suppressed fires leading to fuel buildup and woody species recruitment. When they do occur, effects can be dramatic as seen in recent severe fires in National Forests and other protected areas in the Northwest of the United States. Fires at all latitudes change ecosystem processes in protected areas of any size.

10. Recreation. Recreational activities in protected areas are accompanied by creation of roads and trails leading to animal mortalities and alterations to animal behavior. Litter and discarded food around recreational facilities and trails can result in locally increased animal populations and invasive species and diseases being introduced. Larger reserves will have a greater capacity to absorb such visitor pressure. In the vast majority of cases, tourist revenue fosters protected area establishment and management, and globally was worth $8.8 trillion in 2018 so we must learn to manage its adverse side effects.

11. PADDD. Protected area downgrading, downsizing, and degazettement (PADDD) is a growing threat to protected areas. Downgrading refers to a reduction in legal restrictions on human activities in protected areas, downsizing to a reduction in reserve area, and degazettement to a loss of legal protection for an entire protected area. PADDD usually occurs for extraction of natural resources or due to land pressure and claims. Since one-fifth of PADDD events occur more than once in the same protected area, large reserves may be better able to cope with these pressures than smaller reserves (Mascia et al. 2014).

12. Climate Change. Average global temperatures are predicted to increase by 1.1–6.4 °C this century leading to rising sea levels, increased ocean acidification, and greater frequency of extreme weather events. Climate change is a particular concern for coral reefs that have suffered severe bleaching events. Of course, these occur whether corals are under protection or not and forecasts suggest a 16%–46% reduction in coral reefs by 2100 (Descombes et al. 2015). Sea-water acidification is likely to compromise gastropods' abilities to form shells and to change fish behavior.

Climate change may alter water availability in terrestrial protected areas by increasing evaporation, altering snowpack melt, changing the quantity or form of precipitation, or modifying the seasonality of precipitation. Already, reductions in snowpack have led to decreases in seasonal water availability for protected areas throughout the western United States.

Meta-analyses have uncovered latitudinal shifts in terrestrial species' distributions of between 6.1 and 16.9 km per decade and altitudinal shifts of 6.1 to 11 m upwards due to climate change. While empirical data show that protected areas buffer detrimental effects of geographic range shifts in birds (Lehikoinen et al. 2019), there are concerns that species' geographic ranges will move out of protected areas, decreasing biodiversity inside and leaving species with inadequate protection. Clearly the probability of moving out of reserves is reduced when protected areas are very large, and altitudinal range shifts may be more possible in larger reserves with greater changes in elevation.

In both terrestrial and marine situations, differences in response rates to climate change may lead to genetic, phenological, population, and community changes with loss or addition of predators, prey, pollinators, or competitors (Dawson et al. 2011, Scheffers 2016). Alterations in interspecific interactions and formation of novel (nonanalog) communities will be important subjects of ecological study and pose challenges to managers, but VLPAs may buffer these effects by providing a greater number and variety of ecological refuges..

Policing

Given this alarming roster of contemporary threats, many of which involve people living around park borders, there is a need to maintain the sanctity of protected areas to stop encroachment and prevent exploitation of animal and plant resources (Zupan et al. 2018). In theory, economic benefits accruing to people living around reserve borders could be sufficiently large that people will avoid diminishing protected area resources. This can work well in marine protected areas where large fish stray outside into fishing nets although preliminary evidence is mixed in the case of VLMPAs (Ban et al. 2017). In the terrestrial realm this argument suffers from four major problems (Adams and Hutton 2007). First, it is very rare that there are sufficient economic returns emanating from a single protected area to satisfy economic demands of all neighboring households. Second, even where neighbors reap rewards such as construction of clinics or schools around park borders, it is still worthwhile for people to remove plants or animals from a protected area—a tragedy of the commons.

Third, changing market forces, often beyond national borders, can suddenly make certain natural resources extremely valuable (e.g., hardwood timber). Thus, compliance with reserve regulations at one point in time may flip to disregard. Fourth, international gangs or nationals living far away may want to exploit the protected area for items such as ivory whether or not local neighbors garner health care or economic benefits. Locals are ill-suited to police reserves unless paid and equipped properly. In essence, strictly community policing is only likely to work under a rather restricted set of circumstances (Agrawal and Gibson 1999): when the protected area is small and most areas can be easily reached on foot or by boat, when resources in the reserve are of little economic value on the open market, and when outsiders can be dissuaded from incursion. Empirical work indicates that strictly protected areas in IUCN categories I to IV are associated with higher levels of biodiversity than less-well-protected categories (Bradshaw et al. 2015) in both terrestrial and marine realms (Sciberras et al. 2013) lending some support to this idea.

Effective protection of VLPAs necessitates sufficient government funding, competent staff, large salaries, good equipment, vehicles, fuel, and maintenance, as well as wildlife monitoring (Edgar et al. 2014). For example, stopping illegal extraction requires law enforcement and negotiating with local communities; tackling invasive species requires prevention and removal techniques; and managing fire may require suppression or prescribed burning. Currently, perhaps the only effective framework for protected area conservation is adaptive management, where policies are adjusted based on frequent monitoring of biodiversity. Some people worry that this flexible strategy may lead to a lack of accountability and shifting benchmarks but adaptive management is well-suited to facing unpredictable threats like climate change.

Very-Large Protected Areas Are Needed Now

Natural functioning ecosystems are critical to the discipline of ecology because they enable us to explore unsolved ecological problems. I have made a case for very-large unexploited protected areas under strict protection being *the best approach for conserving functioning ecosystems* because they can often buffer the 12 challenges listed above although empirical verification for some of these assertions are still required. Assuming for the moment that I am correct, I now suggest four ways to enhance the VLPA agenda. Currently very few MPAs are fully protected (6%) (Costello and Ballantine 2015) and remarkably little terrestrial area is under strict protection (15.8%) (Jenkins and Joppa 2009): As a result, rather a low percentage of species enjoy this sort of conservation. Some species and regions are not protected at all; as of 2004, more than 1,400 terrestrial vertebrate species did not have ranges that overlapped protected areas (Rodrigues et al. 2004), and 50% of important sites for biodiversity conservation remain unprotected (Butchart et al. 2012).

Some of the problems facing protected areas are exacerbated by insufficient space to protect populations properly. Climate change may shift species' ranges outside of designated regions; fragmentation may divide habitat, making parts of the original habitat inaccessible; and resource use outside protected areas affects habitat within protected areas (Laurance et al. 2012). Addressing these threats will require increasing the areal extent under protection by making existing protected areas bigger, creating new protected areas, and making protected areas better insulated and more connected (Watson et al. 2018).

Enlarge Protected Areas

Possibly the easiest way to increase the number of VLPAs is to enlarge current large protected areas. When protected areas are established along international boundaries as transboundary conservation areas neighboring countries can create VLPAs without incurring all the costs themselves. As illustrations, the Kavango–Zambezi Transfrontier Conservation Area, spanning the borders of Angola, Botswana, Namibia, Zambia, and Zimbabwe, is set to cover an extraordinary 520,000 km^2, including 36 national parks, game reserves, community conservancies, and game management areas. Other terrestrial transboundary protected areas include the 1,008,470-km^2 Ellesmere–Greenland area in Canada and Greenland (Denmark), and the 67,855-km^2 Glaciares–Torres del Paine–O'Higgins Complex in Argentina and Chile.

Another method of increasing effective protected area size is to set aside adjacent protected areas for exploitation (IUCN categories V and VI) next to a strictly protected area. For example, Katavi National Park (4471 km^2) in western Tanzania is surrounded by Rukwa and Lukwati Game Reserves (4194 and 3566 km^2, respectively), Mlele Game Controlled Area (3544 km^2), and Lwafi Game Reserve–Nkamba Forest Reserve (3369 km^2) where tourist hunting and timber extraction occurs, a total of 19,144 km^2.

New Protected Areas

We also need to create VLPAs *de novo*. Big nongovernmental organizations have created algorithms for siting new protected areas by delineating wilderness areas (Mittermeier et al. 2003), through the Last of the Wild initiative (Sanderson et al. 2002b, Watson et al. 2016), by uncovering biodiversity hotspots (Myers et al. 2000), by identifying Global 200 ecoregions (Olson and Dinerstein 2002), and by protecting endemic vertebrates (Pimm et al. 2018). Although these protected areas are set to target specific areas of the globe, in reality protected areas are usually established on the basis of political expediency. Of 63 VLPAs only a very-low percentage corresponded to these global prioritization schemes showing that VLPAs have not actually been established with regard to global conservation concerns (Cantu-Salazar and Gaston 2010). The problem with algorithms produced by nongovernmental organization teams is that they are neither read much by governmental officials making decisions about protected areas, nor by local stakeholders. There is an extraordinary disconnect between academic large-scale analyses and decision-making at a national level.

Several huge new MPAs have been established in the last 15 years (see Toonen et al. 2013, Jones and De Santo 2016) including the Papahanaumokuakea Marine National Monument (362,074 km²), Phoenix Islands Protected Area (408,250 km²), and British Indian Ocean Territory Marine Reserve (640,000 km²). These are very optimistic developments. Other options include networks of interconnected marine reserves (Althaus et al. 2017).

Buffer Zones

Protected area effectiveness in maintaining biodiversity is directly related to the larger landscape context (Laurance et al. 2012). Buffer zones around protected areas are designed to insulate them from the negative impacts of anthropogenic activities. For example, intermediate-to-large-sized buffers are thought to reduce illegal extraction within a protected area core and can help prevent destruction of forest immediately bordering protected areas. Furthermore, buffer zones can help protect wide-ranging carnivores that wander outside reserves. They also support low-impact land-use wherein people can sustainably extract resources or even practice agriculture and in so doing foster political will. They are particularly useful in the context of MPAs.

Despite the recognized importance of buffer zones, only general guidelines exist for their development and management, and current understanding of the dynamics of anthropogenic pressures at park boundaries is shaky. Nevertheless, buffer zones remain an important protection strategy for helping to protect ecosystem processes and providing access for local people to utilize resources outside protected area boundaries.

Corridors

Wildlife corridors are a related concept aimed at enhancing the effective size of protected areas. They offer more area for species with large home ranges, provide connectivity between protected areas enabling species to disperse between them, maintain genetic variability within populations, and can rescue populations from local extinction (Belote et al. 2017; Tucker et al. 2018; Smith et al. 2019). Elephants, for instance, routinely move between protected areas using venerated corridors that cross unprotected land. Linking existing protected areas may be an important tool for mitigating the threat of climate change by enabling species to shift their distributions to climatically favorable areas. In a promising development, some countries are now formalizing national corridor plans (Riggio and Caro 2017).

Effective Management

Unless large protected areas are protected by governments, they will simply be "paper parks," there only in name. Thus, any protected area requires careful establishment, including boundary marking, adequate legislation, especially tenure resolution, community involvement, and law enforcement and funding (Leverington et al. 2010). In most tropical countries, where biodiversity is high, governments have many other pressing needs concerning education, infrastructure, health, energy, and the military. Funds for conservation are low on their priorities list, often very low, but there are situations where substantial improvements to park infrastructure and restoration have been successfully accomplished (Pringle 2017). Therefore, it is important that foreign nongovernmental organizations and foreign governments offer more aid in protecting these areas. This will be especially important in developing VLPAs, which are costly to establish (Table 1). If conservation of functioning ecosystems is important to western culture, we must be prepared to pay for it.

Big Thinking

There is abundant evidence that VLPAs are the most effective way of conserving fully functioning ecosystems, as well as biodiversity, although the minimum area required to achieve these goals is an empirical issue likely to vary on a case-by-case basis. However, reserves are still vulnerable to many human activities that they aimed to ameliorate at their time of establishment, and they are now facing a new generation of threats that were formerly unanticipated. Very-large reserves are in a commanding position to absorb some of these pressures. Fortunately, some conservation biologists continue to "think big" putting forward ambitious plans that link Yellowstone to the Yukon (1,200,000 km²) across the North American Rockies, and setting aside a jaguar corridor (2,562,378 km²) linking

182 areas between Mexico and northern Argentina. These and other bold initiatives are currently held in limbo at their planning stages, which is unfortunate since time is short: Land conversion is now at an advanced stage in many biodiverse countries and there are few remaining ecosystems that are uninfluenced by humans. For MPAs the prognosis is better: There are still openings to protect large areas of ocean to bolster fish stocks and to save multitrophic-level ecosystems. We still have opportunities to protect large scale ecological processes in both marine and terrestrial habitats and we must grasp them.

Acknowledgments

I thank Grace Charles, Dena Clink, Jason Riggio, Alexandra Weill, and Carolyn Whitesell for early discussions and help with the references and Bob Holt for comments.

References

Adams, W. M., and J. Hutton. 2007. People, parks, and poverty: Political ecology and biodiversity conservation. Conservation and Society 5:147–183.

Agrawal, A., and C. C. Gibson. 1999. Enchantment and disenchantment: The role of community in natural resource conservation. World Development 27:629–649.

Althaus, F., A. Williams, P. Alderslade, and T. A. Schlacher. 2017. Conservation of marine biodiversity on a very large deep continental margin: How representative is a very large offshore reserve network for deep-water octocorals? Diversity and Distributions 23:90–103.

Ban, N. C., T. E. Davies, S. E. Aguilera, C. Brooks, M. Cox, G. Epstein, et al. 2017. Social and ecological effectiveness of large marine protected areas. Global Environmental Change 43:82–91.

Belote, R. T., M. S. Dietz, C. N. Jenkins, P. S. McKinley, G. H. Irwin, T. J. Fullman, et al. 2017. Wild, connected, and diverse: building a more resilient system of protected areas. Ecological Applications 27:1050–1056.

Block, B. A., I. D. Jonsen, S. J. Jorgensen, A. J. Winship, S. A. Shaffer, S. J. Bograd, et al. 2011. Tracking apex marine predator movements in a dynamic ocean. Nature 475:86–90.

Bradshaw, C.J.A., I. Craigie, and W. F. Laurance. 2015. National emphasis on high-level protection reduces risk of biodiversity decline in tropical forest reserves. Biological Conservation 190:115–122.

Bruner, A. G., R. E. Gullison, R. E. Rice, and G. A. da Fonseca. 2001. Effectiveness of parks in protecting tropical biodiversity. Science 291:125–128.

Butchart, S.H.M., J.P.W. Scharlemann, M. I. Evans, S. Quader, S. Arico, J. Arinaitwe, et al. 2012. Protecting important sites for biodiversity contributes to meeting global conservation targets. PLOS One 7:e32529.

Cantu-Salazar, L., and K. J. Gaston. 2010. Very large protected areas and their contribution to terrestrial biological conservation. Bioscience 60:808–818.

Caro, T. 2007. The Pleistocene rewilding gambit. Trends in Ecology and Evolution 22: 281–283.

Costello, M. J., and B. Ballantine. 2015. Biodiversity conservation should focus on no-take Marine Reserves, 94% of Marine Protected Areas allow fishing. Trends in Ecology and Evolution 30:507–509.

Crooks, K. R., C. L. Burdett, D. M. Theobald, S.R.B. King, M. Di Marco, C. Rondinini, et al. 2017. Quantification of habitat fragmentation reveals extinction risk in terrestrial mammals. Proceedings of the National Academy of Sciences of the United States of America 114:7635–7460.

Daskin, J. H., and R. M. Pringle. 2018. Warfare and wildlife declines in Africa's protected areas. Nature 553:328.

Dawson, T. P., S. T. Jackson, J. I. House, I. C. Prentice, and G. M. Mace. 2011. Beyond predictions: Biodiversity conservation in a changing climate. Science 332:53–58.

Descombes, P., M. S. Wisz, F. Leprieur, V. Parravicini, C. Heine, S. M. Olsen, et al. 2015. Forecasting coral reef decline in marine biodiversity hotspots under climate change. Global Change Biology 21:2479–2487.

Devillers, R., R. L. Pressey, A. Grech, J. N. Kittinger, G. J. Edgar, T. Ward, and R. Watson. 2015. Reinventing residual reserves in the sea: are we favouring ease of establishment over need for protection? Aquatic Conservation: Marine and Freshwater Ecosystems 25:480–504.

Dorcas, M. E., J. D. Willson, R. N. Reed, R. W. Snow, M. R. Rochford, M. A. Miller, et al. 2012. Severe mammal declines coincide with proliferation of invasive Burmese pythons in Everglades National Park. Proceedings of the National Academy of Sciences of the United States of America 109: 2418–2422.

Du Toit, J., K. Rogers, and H. Biggs. 2003. The Kruger experience: Ecology and management of savanna heterogeneity. Washington, D.C.: Island Press.

Edgar, G. J., R. D., Stuart-Smith, T. J. Willis, S. Kininmonth, S. C. Baker, S. Banks, et al. 2014. Global conservation outcomes depend on marine protected areas with five key features. Nature 506:216–220.

Estes, J. A.., J. Terborgh, J. S. Brashares, M. E. Power. J. Bergerr, W. J. Bond, S. R. Carpenter, T. E. Essington, R. D. Holt, and J. B. Jackson. 2011. Trophic downgrading of planet Earth. Science 333:301–306.

Finer, M., C. N. Jenkins, and B. Powers. 2013. Potential of best practice to reduce impacts from oil and gas projects in the Amazon. PLOS One 8:e63022.

Foxcroft, L. C., V. Jarosik, P. Pysek, D. M. Richardson, and M. Rouget. 2011. Protected-area boundaries as filters of plant invasions. Conservation Biology 25:400–405.

Foxcroft, L. C., P. Pysek, D. M. Richardson, P. Genovesi, and S. MacFadyen. 2017. Plant invasion science in protected areas: Progress and priorities. Biological Invasions 19:1353–1378.

Gardner, T., T. Caro, E. Fitzherbert, T. Banda, and P. Lalbhai. 2007. Conservation value of multiple use areas in East Africa. Conservation Biology 21:1516–1525.

Geist, H. J., and E. F. Lambin. 2002. Proximate causes and underlying driving forces of tropical deforestation. BioScience 52:143–150.

Geldmann, J., L. N. Joppa, and N. D. Burgess. 2014. Mapping change in human pressure globally on land and within protected areas. Conservation Biology 28:1604–1616.

Gereta, E., E. Mwangomo, and E. Wolanski. 2009. Ecohydrology as a tool for the survival of the threatened Serengeti ecosystem. Ecohydrology & Hydrobiology 9:115–124.

Gross, M. 2019. Hunting wildlife to extinction. Current Biology 29:R551–R554.

Haddad, N. M., L. A. Brudvig, J. Clobert, K. F. Davies, A. Gonzalez, et al. 2015. Habitat fragmentation and its lasting impact on Earth's ecosystems. Science Advances 1:p. e1500052.

Ibisch, P. L., M. T. Hoffmann, S. Kreft, G. Pe'er, V. Kati, L. B. Freudenberger, et al. 2016. A global map of roadless areas and their conservation status. Science 354:1423–1427.

Izuno, F., C. Sanchez, F. Coale, A. Bottcher, and D. Jones. 1991. Phosphorus concentrations in drainage water in the Everglades agricultural area. Journal of Environmental Quality 20:608–619.

Jenkins, C. N., and L. Joppa. 2009. Expansion of the global terrestrial protected area system. Biological Conservation 142:2166–2174.

Jones, K. R., O. Venter, R. A. Fuller, J. R. Allan, S. L. Maxwell, P. J. Negret, et al. 2018. One-third of global protected land is under intense human pressure. Science 360:788–791.

Jones, P. J. S., and E. M. De Santo. 2016. Is the race for remote, very large marine protected areas (VLMPAs) taking us down the wrong track? Marine Policy 73:231–234.

Joppa, L. N., S. R. Loarie, and S. L. Pimm. 2008. On the protection of "protected areas". Proceedings of the National Academy of Sciences of the United States of America 105:6673–6678.

Kareiva, P., S. Watts, R. McDonald, and T. Boucher. 2007. Domesticated nature: Shaping landscapes and ecosystems for human welfare. Science 316:1866–1869.

Kleinschroth, F., and J. R. Healey. 2017. Impacts of logging roads on tropical forests. Biotropica 49:620–635.

Laurance, W. F. 2005. When bigger is better: The need for Amazonian mega-reserves. Trends in Ecology & Evolution 20:645–648.

Laurance, W. F., S. Sloan, L. Weng, and J. A. Sayer. 2015. Estimating the environmental costs of Africa's massive "development corridors". Current Biology 25:1–7.

Laurance, W. F., D. C. Useche, J. Rendeiro, M. Kalka, C. J. Bradshaw, S. P. Sloan, et al. 2012. Averting biodiversity collapse in tropical forest protected areas. Nature 489:290.

Lehikoinen, P., A. Santangeli, K. Haatinen, and A. Lehikoinen. 2019. Protected areas act as buffer against detrimental effects of climate change—evidence from large-scale, long-term abundance data. Global Change Biology 25:304–315.

Leverington, F., K. L. Costa, H. Pavese, A. Lisle, and M. Hockings. 2010. A global analysis of protected area management effectiveness. Environmental Management 46:685–698.

Licuanan, W. Y., R. Robles, M. Dygico, A. Songco, and R. van Woesik. 2017. Coral benchmarks in the center of biodiversity. Marine Pollution Bulletin 114:1135–1140.

Lindsey, P. A., G. Balme, M. Becker, C. Begg, C. Bento, C. Bocchino, et al. 2013. The bushmeat trade in African savannas: Impacts, drivers, and possible solutions. Biological Conservation 160:80–96.

Mascia, M. B., S. Pailler, R. Krithivasan, V. Roschanka, D. Burns, M. C. Mlotha, et al. 2014. Protected area downgrading, downsizing, and degazettment (PADDD) in Africa, Asia, and Latin America and the Caribbean, 1900–2010. Biological Conservation 169:355–361.

Mittermeier, R., C. G. Mittermeier, T. M. Brooks, J. D. Pilgrim, W. R. Konstant, G. Da Fonseca, and C. Kormos. 2003. Wilderness and biodiversity conservation. Proceedings of the National Academy of Sciences of the United States of America 100:10309–10313.

Mtahiko, M., E. Gereta, A. Kajuni, E. Chiombola, G. Ng'umbi, P. Coppolillo, et al. 2006. Towards an ecohydrology-based restoration of the Usangu wetlands and the Great Ruaha River, Tanzania. Wetlands Ecology and Management 14:489–503.

Myers, N., R. A. Mittermeier, C. G. Mittermeier, G. A. Da Fonseca, and J. Kent. 2000. Biodiversity hotspots for conservation priorities. Nature 403:853–858.

Myers, R. A., J. K. Baum, T. D. Shepherd, S. P. Powers, C. H. Peterson. 2007. Cascading effects of the loss of apex predatory sharks from a coastal ocean. Science 315:1846–1850.

Olson, D. M., and E. Dinerstein. 2002. The Global 200: priority ecoregions for global conservation. Annals of the Missouri Botanical Garden 89:199–224.

Otto, S. P. 2018. Adaptation, speciation and extinction in the Anthropocene. Proceedings of the Royal Society of London B 285:20182047.

Pauchard, A., P. B. Alaback, and E. G. Edlund. 2003. Plant invasions in protected areas at multiple scales: *Linaria vulgaris* (Scrophulariaceae) in the West Yellowstone area. Western North American Naturalist 63:416–428.

Peres, C. A. 2000. Effects of subsistence hunting on vertebrate community structure in Amazonian forests. Conservation Biology 14:240–253.

Peres, C. A. 2005. Why we need megareserves in Amazonia. Conservation Biology 19:728–733.

Pimm, S. L., C. N. Jenkins, and B. V. Li. 2018. How to protect half of Earth to ensure it protects sufficient biodiversity. Science Advances 4:p.eaat2616.

Pringle, R. M. 2017. Upgrading protected areas to conserve biodiversity. Nature 546:91–99.

Riggio, J., and T. Caro. 2017. Structural connectivity at a national scale: wildlife corridors in Tanzania. PLOS One 12:p.e0187407.

Rodrigues, A. S., S. J. Andelman, M. I. Bakarr, L. Boitani, T. M. Brooks, R. M. Cowling, et al. 2004. Effectiveness of the global protected area network in representing species diversity. Nature 428:640–643.

Rudnick, D. A., P. Beier, S. Cushman, F. Dieffenback, and C. Epps. 2012. The role of landscape connectivity in planning and implementing conservation and restoration priorities. Issues in Ecology 16:1–20.

Salerno, J. D., M. Borgerhoff Mulder, and S. C. Kefauver. 2014. Human migration, protected areas, and conservation outreach in Tanzania. Conservation Biology 28:841–850.

Sanderson, E. W., M. Jaiteh, M. A. Levy, K. H. Redford, A. V. Wannebo, and G. Woolmer. 2002. The human footprint and the last of the wild: the human footprint is a global map of human influence on the land surface, which suggests that human beings are stewards of nature, whether we like it or not. BioScience 52:891–904.

Scheffers, B. R., L. De Meester, T. C. Bridge, A. A. Hoffmann, J. M. Pandolfi, R. T. Corlett, et al. 2016. The broad footprint of climate change from genes to biomes to people. Science 354:p.aaf7671.

Sciberras, M., S. R. Jenkins, M. J. Kaiser, S. J. Hawkins, and A. S. Pullin. 2013. Evaluating the biological effectiveness of fully and partially protected marine areas. Environmental Evidence 2:4.

Shahabuddin, G., and M. Rao. 2010. Do community-conserved areas effectively conserve biological diversity? Global insights and the Indian context. Biological Conservation 143:2926–2936.

Sinclair, A. R., S. A. Mduma, and P. Arcese. 2002. Protected areas as biodiversity benchmarks for human impact: agriculture and the Serengeti avifauna. Proceedings of the Royal Society of London B: Biological Sciences 269:2401–2405.

Soule, M. E., and J. Terborgh. 1999. Conserving nature at regional and continental scales: a scientific program for North America. BioScience 49:809–817.

Smith, J. A., T. P. Duane, and C. C. Wilmers. 2019. Moving through the matrix: promoting permeability for large carnivores in a human-dominated landscape. Landscape and Urban Planning 183:50-58.

Stokstad, E. 2015. Bringing back the aurochs. Science 350:1144–1147.

Tanzania Wildlife Research Institute. 2015. Wildlife Status Report. Population status of the African Elephant in Tanzania. Dry and wet season 2014. Unpublished report. Arusha, Tanzania: Tanzania Wildlife Research Institute (TAWIRI).

Terborgh, J., and J. A. Estes, eds. 2010. Trophic cascades: Predators, prey, and the changing dynamics of nature. Washington D.C.: Island Press.

Toonen, R. J., T. A. Wilhelm, S. A. Maxwell, D. Wagner, B. W. Bowen, C. R. C. Sheppard, et al. 2013. One size does not fit all: the emerging frontier in large-scale marine conservation. Marine Pollution Bulletin 77:7–10.

Tucker, M. A., K. Böhning-Gaese, W. F. Fagan, J. M. Fryxell, B. Van Moorter, S. C. Alberts et al. 2018. Moving in the Anthropocene: global reductions in terrestrial mammalian movements. Science 359:466-469.,7

Venter, O., E. W. Sanderson, A. Magrach, J. R. Allan, J. Beher, K. R. Jones, et al. 2016. Sixteen years of change in the global terrestrial human footprint and implications for biodiversity conservation. Nature Communications 7:12558.

Watson, J. E., D. F. Shanahan, M. Di Marco, J. Allan, W. F. Laurance, E. W. Sanderson, et al. 2016. Catastrophic declines in wilderness areas undermine global environment targets. Current Biology 26:2929–2934.

Watson, J.E.M., O. Venter, J. Lee, K. R. Jones, J. G. Robinson, H. P. Possingham, and J. R. Allan. 2018. Protect the last of the wild. Nature 563:27–30.

Wilkie, D. S., M. Wieland, H. Boulet, S. Le Bel, N. van Vliet, D. Cornelis, et al. 2016. Eating and conserving bushmeat in Africa. African Journal of Ecology 52:402–414.

Wittemyer, G., P. Elsen, W. T. Bean, A. C. O. Burton, and J. S. Brashares. 2008. Accelerated human population growth at protected area edges. Science 321:123–126.

Wolf, C., and W. J. Ripple. 2018. Rewilding the world's large carnivores. Royal Society Open Science 5:1722.

Woodroffe, R., and J. R. Ginsberg. 1998. Edge effects and the extinction of populations inside protected areas. Science 280:2126–2128.

Wuerthner, G., E. Crist, and T. Butler, eds. 2015. Protecting the wild: Parks and wilderness, the foundation for conservation. Washington D.C.: Island Press.

Zupan, M., E. Fragkopoulou, J. Claudet, K. Erzini, B. Horta e Costa, and E. J. Gonçalves. 2018. Marine partially protected areas: Drivers of ecological effectiveness. Frontiers in Ecology and the Environment 16:381–387.

How Does Biodiversity Relate to Ecosystem Functioning in Natural Ecosystems?

Rachael Winfree

The accelerating loss of biodiversity worldwide generates many unsolved problems for the twenty-first century. Primary among these, from a human perspective, is how biodiversity loss affects ecosystem functions and services,[1] many of which are essential to human life. For example, as we lose plant species, what will happen to carbon storage; or as we lose bee species, what will happen to pollination? Based on the results of biodiversity-function experiments, ecologists expect ecosystem function to be strongly dependent on the number of species present (Cardinale et al. 2012, Tilman et al. 2014). On this basis, maintaining ecosystem services has become a leading argument for biodiversity preservation. For example, The Nature Conservancy, which spends more on conservation efforts than the next eight international conservation organizations combined, now focuses its biodiversity conservation strategies on maintaining ecosystem services.

However, ecologists actually know rather little about how the biodiversity–function relationship works in real-world ecosystems. In contrast to the consensus of more than 600 studies finding that species richness is important to function within an experimental context, ecologists don't know to what extent real-world ecosystem functions are dependent on the number of species (species richness[2]), on particular species, or merely on the abundance of individuals, regardless of species (Cardinale et al. 2012; Winfree et al. 2018). The answer determines whether arguments based on ecosystem services can motivate the conservation of all species, or only the subset, perhaps a small one, of functionally important species.

In this essay I propose five unsolved research problems, the answers to which would help expand our understanding of the biodiversity–ecosystem function relationship from laboratory and field experiments (generally done at scales of a few liters or square meters) to real-world ecosystems (generally measured in hundreds of square kilometers). Part of my aim is to point out systematic differences between experiments and real-world ecosystems and suggest ways in which our thinking might need to differ systematically between scales as well. First, though, I briefly summarize what is already known at each scale.

What We Know from Biodiversity–Ecosystem Function Experiments

There is clear consensus on several key ways that biodiversity affects ecosystem function at small scales (Cardinale et al. 2012, Tilman et al. 2014). First, across many taxa and types of ecosystem functions, greater species richness leads to greater function. Second, the increase in function with increasing richness often starts out roughly linear and then approaches an asymptote, often in the range of five to ten species (Hooper et al. 2005). Third, greater species richness increases the stability of function over time. Essentially this happens through compensatory fluctuations among species. For example, in multispecies communities, species can have differential responses to environmental variation (response diversity), or species that compete for the same resources can covary negatively as a result of competition (density compensation).

What We Know from Real-World Biodiversity–Ecosystem Function Studies

A major challenge in real-world research is separating effects of species richness from other aspects of communities, such as species composition or the total abundance of individuals. These factors are separated by design in experiments but tend to be intercorrelated in nature. The few studies that have experimentally separated the roles of richness, composition, and abundance in natural systems found richness to be less important to function than the identity or abundance of particular high-functioning species (Dangles and Malmqvist 2004, Smith and Knapp 2003). At very large scales, manipulative experiments on natural communities are not feasible, but some researchers have tried to isolate the effect of richness statistically or analytically. The results range from positive, hump-shaped relationships to no relationship between richness and function (Hoehn et al. 2008, Paquette and Messier 2011, Gamfeldt et al. 2012, Garibaldi et al. 2013, Winfree et al. 2015).

A final challenge in real-world systems is ruling out the possibility that causality operates in the opposite direction, such that high levels of ecosystem function lead to higher species richness. This reverse causality is predicted by coexistence theory, and found for some ecosystem functions, such as plant productivity (Chesson et al. 2000, Fraser et al. 2015), which unfortunately has been widely used as the outcome variable in the biodiversity–ecosystem function literature. This reverse causality is logically impossible for some other ecosystem services, however, such as crop pollination (because increased pollination does not lead to increased floral resource availability for pollinators, and thus potentially more pollinator species, in human-managed plant populations).

Biological mechanisms through which biodiversity *could* increase function have been found in real-world systems, however. These include functional complementarity among species (Hoehn et al. 2008, Chagnon et al. 1993, Tylianakis et al. 2015) and facilitation between species (Brittain et al. 2013). Likewise, real-world studies have documented mechanisms through which biodiversity could stabilize ecosystem function. For example, species that provide the same function respond differently to environmental variation (Rader et al. 2013, Bartomeus et al. 2013, Brittain et al. 2012, Cariveau et al. 2013), which suggests (although it does not show) that communities with more species should provide more stable function as environmental conditions vary.

Five Unsolved Problems

The typical biodiversity–function experiment is 3 m^2 in size and lasts less than one organismal generation (Cardinale et al. 2009, 2011), whereas real-world ecosystem functions are delivered over hundreds of kilometers and many years. This increase in scale has been predicted to make biodiversity both more (Cardinale et al. 2012, Tilman et al. 2014, Duffy 2009) and less (Jiang et al. 2009) important to ecosystem function, but we lack systematic guidelines for determining how and why the two scales differ. In the remainder of this essay, I propose four research questions as a place to start. A fifth question asks how the results of this research are relevant to biodiversity conservation.

1. Which mechanisms found in experiments might increase the mean, and decrease the variability, of ecosystem function in the real world? Species richness increases *mean* function in experiments in two ways. The first mechanism is complementarity among species. This can occur through niche-related complementarity in resource use, or in how the different species provide the function. With complementarity, a diverse community can exceed the maximum function achievable by a monoculture of any one species. Complementarity is the mechanism that accounts for most of the positive richness-function relationships found in aquatic experiments, and about half of those found in terrestrial experiments (Cardinale et al. 2011).

The second mechanism is known as the selection (or sampling) effect and is more statistical than biological. It arises when the species that provides the most function per individual also comes to dominate the experimental unit, whether through competition or for other reasons. Thus, in the context of a random community assembly experiment, function can increase with richness simply because the probability of selecting the highest-functioning species for inclusion in the experimental community increases with richness. With selection effects, the function provided by the community will never exceed the maximum function provided by any one species, but it will attain that maximum level more frequently under higher richness treatments. Selection effects account for approximately half of the positive biodiversity-function relationships found in terrestrial experiments, although few of those found in aquatic experiments (Cardinale et al. 2011); the reason for this difference is not known (Cardinale, personal communication).

Selection effects might occur less often in real-world ecosystems than in experiments, and therefore, we might find fewer positive biodiversity–function relationships there. First, for most real-world functions, there is no reason to expect that the species providing the most function per individual will also be dominate the system. Experiments may have found selection effects unusually often because they have focused on the functions of biomass production and nutrient acquisition (Jiang et al. 2008). These two functions have strong pre-existing links between functional and competitive dominance. For example, large plant species compete well for space and light, and also produce more biomass per capita. Likewise, R* theory predicts that the species that reduces nutrients to the lowest level outcompetes other species. But for many important ecosystem functions that have rarely been studied in experiments, there is no such expectation. For example, there is no reason why the bee species that outcompetes other bee species should also be the best pollinator, especially when one considers that leaving pollen on female flowers is largely a mistake from a bees' point of view.

Selection effects are even less likely to make biodiversity important to function when multiple functions are considered simultaneously. This is because a single competitively dominant species is unlikely to make the greatest per individual contribution to each of many different functions. But many functions need to be considered in a real-world setting. Overall, then, it seems doubtful that selection effects will link species richness to function in real-world ecosystems. Because selection effects account for about half of the contribution richness makes to function in terrestrial experiments, we might expect the importance of richness to decline roughly by half when studies are done in the real world (Cardinale et al. 2011).

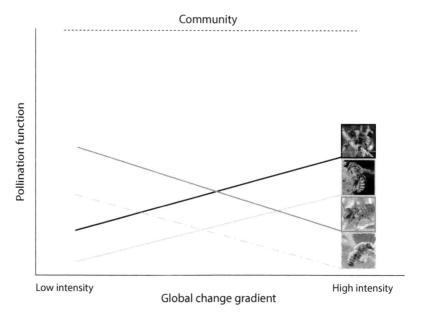

FIGURE 1. Schematic representing how response diversity among the species in a community (diagonal lines) can result in stable function at the community level across a gradient. After Cariveau et al. 2013.

The second mechanism through which species richness increases function, complementarity, should show a trend opposite to that of selection effects: More species should become important as the number of functions increases. This is because different species are needed for different functions (Isbell et al. 2011). What will the net change be in the importance of species to mean function as scale expands—smaller due to missing selection effects, or greater due to increased complementarity? At present, we don't know.

Greater species richness decreases the *variability* of function when compensatory dynamics among species cancel each other to make community-level aggregate function more stable. One way this can happen is through differences in how different species respond to environmental conditions, or *response diversity* (Fig. 1). Response diversity is probably underestimated by experiments, which tend to control environmental conditions (but see Leary and Petchey 2009, Tilman and Downing 1994). For the same reason, response diversity might be far more important in real-world settings than is predicted by experiments. In fact, insofar as species have environmental niches, we should expect different species to be functionally important as environmental conditions fluctuate (Amarasekare 2003).

In principle, response diversity could even provide a way for a kind of selection effect to work in real-world, non–randomly assembled communities. Specifically, communities with more species will be more likely to contain a species that does well under a particular environmental condition, such as drought (Tilman and Downing 1994). Thus far, this parallels how the selection effect operates to increase the mean function under controlled environmental conditions. However, the second requirement of the selection effect, that the species providing the most function per individual also becomes the competitive dominant (or becomes more abundant for some other reason), is more likely when there is response diversity. Response diversity provides a lurking variable—namely, the new environmental condition—that increases both per individual function and competitive ability for the species that is adapted to that condition. For example, a drought-tolerant plant can both produce more biomass than, and also outcompete, other species that do not survive drought well. A classic example of this phenomenon, although it was not referred to as response diversity at the time, is provided by the work of Tilman and Downing (1994).

Researchers have only recently begun to look for, and find, evidence of response diversity in real-world ecosystems. For example, the pollination provided by different pollinator species shows diverse responses to changes in temperature (Rader et al. 2013), land use (Cariveau et al. 2013), and wind speed (Brittain et al. 2012). Simulation models suggest that the compensatory effect of this variation should increase with the number of species present (Bartomeus et al. 2013). However, the two key components of response diversity are yet to be shown in any real-world system. These components are, first, that increased biodiversity leads to increased response diversity, and second, that increased response diversity stabilizes ecosystem services.

Competition is a second mechanism that can stabilize community-level function over time. This works when one species' decline allows other species, with which it competes for resources, to increase (Gonzalez and Loreau 2009). Such compensatory effects have often been found in experiments. However, I expect they will turn out to be less important in real-world systems. Experimental designs generally create a zero-sum game among species by standardizing resource availability within and across replicate communities. All else being equal, such a resource cap should increase the probability of compensatory dynamics occurring due to competitive release (Winfree 2013). In contrast, real-world resource availability varies over both space and time, and in a global-change context, across disturbance gradients. I predict that species' abundances might be negatively correlated over time *within* relatively undisturbed communities (either experimental or natural), but positively correlated *across*

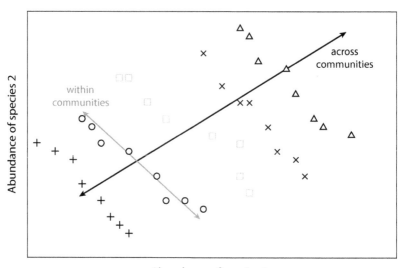

FIGURE 2. A schematic for the proposed hypothesis about density compensation, showing negative correlations between species within communities (same symbol), but positive correlations across communities (different symbols). Within communities, resources vary less, leading to competitive interactions and negatively correlated abundances. Across communities, resources vary more, with some sites being better for all species. Note that although competition could still generate compensatory interactions across communities in this situation, it would be a weaker effect that is less likely to stabilize ecosystem function against powerful drivers of species loss such as land use change. Figure modified from Winfree (2013); conceptually, Shea and Chesson (2002) made an analogous argument for scale-dependence in species invasions.

more disturbed real-world communities, because some locations or time periods simply have more resources for all species (Fig. 2). Note that this doesn't preclude compensatory effects from happening; it just means they will be effectively drowned out by noncompensatory effects (Gonzalez and Loreau 2009). In this case, competition might not be an important force through which biodiversity stabilizes real-world communities against global change.

2. Is biodiversity less important to ecosystem function in the real world because of the species–abundance distribution? Biodiversity–function experiments generally use even species–abundance distributions, such that each species is represented by a similar number of individuals (Dangles and Malmqvist 2004, Kirwan et al. 2007, Wilsey and Polley 2004). However, ecological communities universally have a few common species and

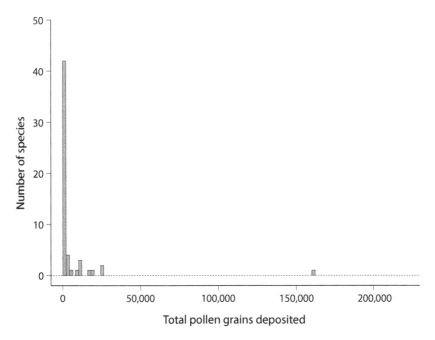

FIGURE 3. The species-function distribution for the wild-bee community pollinating watermelon flowers over a 5400-km^2 region and 3 years. This species-function distribution is driven by the underlying species-abundance distribution: the correlation between the total number of individuals collected for a bee species, and the total number of pollen grains deposited by that species, was $r = 0.95$. (After Winfree et al. 2015.)

many rare ones (McGill et al. 2007). In cases where the function a species provides is proportional to its abundance, most of the function will be contributed by a few species, with the large number of rare species contributing little (Fig. 3). In this situation we expect species richness to matter less to function than the abundance of a few common species, as found by several studies that have measured the distribution of function provided per species in natural communities (Winfree et al. 2015, Balvanera et al. 2006, Kleijn et al. 2015) (Fig. 3). How general might this pattern be? We don't know, because very few functions have been analyzed in this way. However, Vázquez et al. (2005) proposed the following unifying argument. Function per species is the product of the number of individuals and function per individual. The variance of the former (i.e., of the species–abundance distribution) greatly exceeds the variance of the latter (i.e., the distribution of per-individual function across species), at least for the functions that have been analyzed. Thus, of the two factors, abundance is the more predictive of the product. Ecologists need to test this idea by measuring both distributions for a variety of ecosystem functions, including those functions for which the variance in per-individual function might be high. For example, carbon storage by tropical forests relies on tree species that vary over orders of magnitude in per-individual biomass. Thus, carbon storage per tree, not the abundance of trees, might be the most important factor in carbon storage.

The real-world species–abundance distribution could change the importance of species richness to function in a second way: by changing the order in which species are lost as human disturbance increases. In general, if the functionally important species are lost first, function will decline rapidly as disturbance increases and richness declines (and conversely if unimportant species are lost first). Several studies have found that the functionally important species are most sensitive to disturbance, resulting in a rapid loss of function with increasing human disturbance (Larsen et al. 2005, Wolf and Zavaleta 2015). However, there is no reason to expect a general relationship between disturbance sensitivity and functional importance, across taxa, disturbances, or functions. In contrast, there may be a general relationship between the probability of a species' loss from the community and its functional importance if both are driven by rarity. Specifically, if rare species are both less important to function (because there aren't enough individuals to make a significant contribution), and more likely to be lost from communities (because the loss of few individuals would mean the loss of the species), then function might be relatively robust to species loss. We found this was the case for one ecosystem function, pollination (Winfree et al. 2015), but the idea needs to be tested with other functions. This idea could have wide generality given the ubiquity of rare species in real-world communities.

3. Do more species become important to function as spatial, temporal, and environmental scale expands? The fact that the biodiversity–function relationship often asymptotes at a small number of species in experiments could be an artifact of experiments being conducted at small spatial and temporal scales (Cardinale et al. 2012, Isbell et al. 2011). In the real world, species composition changes across space, and this species turnover should make more species important to a single function when it is required over a larger area (Fig. 4). This idea, although perhaps a bit obvious, has only recently been tested (Winfree et al. 2018).

Interestingly, even very conservative tests that use identical experimental communities in multiple locations and thus control, rather than measure, species turnover, find that many more species become important to function as spatial scale increases (Isbell et al. 2011). Similarly, within a given experimental community, the cumulative number of species that are important to function in at least one year increases with the number of

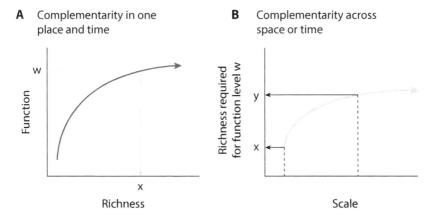

FIGURE 4. (A) A typical plot from biodiversity–function experiments, showing that the richness-function relationship is increasing and asymptotic (curved line), and that in order to achieve function at level *w*, *x* species are needed. (B) A plot of how the number of species required for a particular level of function changes with spatial scale. The curve represents the number of species required to achieve function at level *w*; *x* species are required at a smaller scale (as in (A)), but *y* species are required at a larger scale. This curve would, however, likely lie far below the complete species–area curve that would include all species present, not only those required for function. After Winfree et al. 2018.

years examined (Isbell et al. 2011). Thus, in both cases the biodiversity–function relationship becomes less saturating when it is evaluated at larger scales (Isbell et al. 2011). Presumably this is due to different species being important to function in different environmental contexts, i.e., to response diversity across space and time. A related study with a very different interpretation measured the form of the biodiversity–function relationship *within* each year of a long-term experiment and found that the function became less asymptotic as the experimental communities matured (Reich et al. 2012). Although the mechanism here is not known, it is unlikely to be response diversity.

Thus far we have strong predictions that more species will become important to function at larger spatial and temporal scales. However, the species–abundance distribution could counter this effect. Species that are abundant are likely to be widespread (Gaston et al. 2000). If most of the function is provided by abundant species, as the discussion above suggests, then a small set of common species might suffice to provide ecosystem function even across large geographical scales. Of course, species could still turn over dramatically as spatial scale expands, but the functional consequences of species turnover would be small if the turnover were driven by rare and functionally less-important species (Winfree et al. 2015). Graphically, this means that the widespread occurrence of the dominant species would flatten the slope of the curve shown in Figure 4B (Winfree et al. 2013). In sum, species turnover and dominance will likely have opposing effects on the role of biodiversity in ecosystem function as spatial scale increases.

4. How can we measure biodiversity meaningfully in the real world? All of the biodiversity–function analyses discussed thus far require an empirical measurement of biodiversity. Experiments have predominantly used species richness as their biodiversity metric, and this makes sense insofar as the species in an experimental unit can be completely counted. However, data collected from real-world systems are almost always samples, and species richness is highly sensitive to the extent of sampling. To compare richness among real-world communities, sampling needs to be standardized such that equivalent information is obtained from each community. The problem is how to do this standardization. If sampling effort is equalized, the more diverse communities will be under-sampled relative to others. If the amount of data collected is standardized (e.g., each sample is rarefied to the same number of individuals), the higher-abundance sites will be relatively under-sampled. The best solution is to equalize by coverage, or the proportion of the not-yet-sampled individuals in the community that belong to species not yet detected; this equalizes the information obtained across communities (Chao and Jost 2012). However, there are no clear guidelines about the level of coverage, and thus the sampling effort, that is necessary in order for the richness calculated from the sample to accurately represent the true community richness (Roswell et al., in press). Thus, a key unsolved problem for biodiversity research is how to equalize samples prior to measuring biodiversity.

One solution to this problem would be to use a metric that is more robust to the extent of sampling than is species richness. For example, the Shannon and Simpson diversities both stabilize more quickly than richness does with increased sampling. This is because they are less sensitive to rare species, which are hard to measure, and because they include information on relative abundance, which is easier to measure accurately at small sample sizes (Roswell et al., forthcoming). For the same reasons, though, compared with species richness, they upweight the common species and downweight the rarest species. Thus, their use seems counterintuitive to ecologists accustomed to thinking about richness, which is often driven by rare species, and to conservationists, who often seek to protect rare species.

Lastly, real-world sampling problems are exacerbated when one needs to measure mobile organisms. The vast majority of biodiversity–function experiments have used immobile organisms, such as plants, or immobilized ones, such as containerized microbes. Yet many important real-world ecosystem services, such as pollination, pest control, and seed dispersal, are provided by highly mobile organisms (Kremen et al. 2007). Biodiversity measurement is qualitatively different in this situation. For example, all of the plant species in a plot of a few square meters can be counted in a few hours, and the addition of new species would occur on a time scale of months or years, not hours. In contrast, the number of bee species pollinating those same plants will continue increasing with the number of hours that the plot is observed. Generally speaking, for mobile organisms, the longer one samples, the more species are detected, and this will change the biodiversity–function relationship. This general issue is explored by work on the species–time relationship (Adler et al. 2005), but we are far from having guidelines about how to measure richness of mobile organisms in a meaningful way for biodiversity–function studies.

5. *Will preserving real-world ecosystem services preserve real-world biodiversity, and vice versa?* This is an important question because leading conservation organizations are now using ecosystem services as a primary motivation for preserving biodiversity. The answer to this question is not known but could be approached in several ways. First, ecologists could determine to what extent biodiversity is needed for the provision of the ecosystem service itself at the present time (or over an empirically investigable time period). This approach is the focus of this article.

Second, it could be assumed that over longer time periods than are generally studied (e.g., decades), the identity of the key ecosystem service providing species will change, as formerly functionally dominant species are replaced by new dominants (the insurance hypothesis; Yachi and Loreau 1999). Thus, because we cannot predict the identity of the future dominants, it is pragmatic to conserve all species. Turnover in the identity of the key species providing a service is very likely to occur at small scales even on short time frames. For example, different bee species pollinated the same crop fields across two years (Kremen et al. 2002).

Third, even if many rare species in need of protection are not important contributors to ecosystem services, the same local-scale habitat conservation or restoration methods might benefit both rare species and ecosystem service–providing species. For example, there are plant species that are preferred by both crop-pollinating and regionally rare bee species, and these plants could be used in pollinator restorations (MacLeod et al., in press). For this approach to work, the rare species would need to exist in the same general place where ecosystem services are required. This might not be true for some services; for example, rare bee species are largely absent from the intensively agricultural landscapes where most crop pollination takes place (Kleijn et al. 2015). However, it could readily work for ecosystem services that operate at a global scale, such as carbon storage.

Fourth, even if ecosystem service providers and rare species do not in general live in the same place, particular areas that are hotspots for both ecosystem services and biodiversity could be identified a priori, and reserves situated there (Chan et al. 2006).

Fifth, rare species might benefit from our current inability to target the conservation of particular service-providing species. The convention in biodiversity–function research is to answer the question of "how many species are required for function?" by listing species in descending order of importance and choosing the minimum set of species that provides some threshold level of function. However, this convention does not really apply in the real world. For most kinds of ecosystem services, ecologists and land managers have no robust methods for conserving populations of some species in an ecosystem but not others. Therefore, if protected areas are the only practical way to conserve ecosystem service providers, then these species might serve as umbrellas for the protection of the rest of their ecosystem.

Acknowledgments

James R. Reilly, Mark A. Genung, Tina Harrison, and Michael E. Roswell contributed significantly to the development of the ideas in this article. Helene Muller-Landau's and Bob Holt's insightful comments greatly improved the final paper.

Notes

1. Ecosystem functions are simply the ecological processes that contribute to the functioning of ecosystems (for example, nutrient cycling or decomposition). Ecosystem services are ecosystem functions that are useful to people (Cardinale et al. 2012).
2. For simplicity, I focus on species as the units of biodiversity, but many of the ideas discussed for species could be applied to other forms of biodiversity such as genetic, functional, or trait diversity. Likewise, although there are many metrics of species-level biodiversity, I focus on species richness. I do this not because it is a good metric (it isn't! (Hooper et al. 2005) but simply because it is the main metric used in the biodiversity–ecosystem function literature thus far (Tilman et al. 2004). Alternative biodiversity metrics, as well as the limitations of richness, are discussed under the section *"How can we measure biodiversity meaningfully in the real world?"*

References

Adler, P. B., et al. 2005. Evidence for a general species-time-area relationship. Ecology 86:2032–2039.
Amarasekare, P. 2003. Competitive coexistence in spatially structured environments: A synthesis. Ecology Letters 6:1109–1122.

Balvanera, P., C. Kremen, and M. Martinez-Ramos. 2005. Applying community structure analysis to ecosystem function: examples from pollination and carbon storage. Ecological Applications 15:360–375.

Bartomeus, I., M. Park, J. Gibbs, B. N. Danforth, and R. Winfree. 2013. Biodiversity ensures plant-pollinator phenological synchrony against climate change. Ecology Letters 16:1331–1338.

Brittain, C., C. Kremen, and A. M. Klein. 2012. Biodiversity buffers pollination from changes in environmental conditions. Global Change Biology 19:540–547.

Brittain, C., N. M. Williams, C. Kremen, and A. M. Klein. 2013. Synergistic effects of non-Apis bees and honey bees for pollination services. Proceedings of the Royal Society B: Biological Sciences 280:20122767.

Cardinale, B. J., E. Duffy, D. Srivastava, M. Loreau, M. Thomas, and M. Emmerson. 2009. Towards a food web perspective on biodiversity and ecosystem functioning. In: S. Naeem, D. E. Bunker, A. Hector, M. Loreau, and C. Perrings, eds. Biodiversity, Ecosystem Functioning, and Human Wellbeing. Oxford: Oxford University Press, 105–120.

Cardinale, B. J., J. E. Duffy, A. Gonzalez, D. U. Hooper, C. Perrings, P. Venail, et al. 2012. Biodiversity loss and its impact on humanity. Nature 486:59–67.

Cardinale, B. J., K. L. Matulich, D. U. Hooper, J. E. Byrnes, E. Duffy, L. Gamfeldt, P. Balvanera, M. I. O'Connor, and A. Gonzalez. 2011. The functional role of producer diversity in ecosystems. American Journal of Botany 98:572–592.

Cariveau, D. P., N. M. Williams, F. Benjamin, and R. Winfree. 2013. Response diversity to land use occurs but does not necessarily stabilize ecosystem services provided by native pollinators. Ecology Letters 16:903–911.

Chagnon, M., J. Gingras, and D. de Oliveira. 1993. Complementary aspects of strawberry pollination by honey and indigenous bees (Hymenoptera). Ecology and Behavior 86:416–420.

Chan, K.M.A., R. Shaw, D. R. Cameron, E. C. Underwood, and G. C. Daily. 2006. Conservation planning for ecosystem services. PLOS Biology 4:2138–2152.

Chao, and A., L. Jost. 2012. Coverage-based rarefaction and extrapolation: standardizing samples by completeness rather than size. Ecology 93:2533–2547.

Chesson, P. 2000. Mechanisms of maintenance of species diversity. Annual Review of Ecology, Evolution, and Systematics 31:343–366.

Dangles, O., and B. Malmqvist. 2004. Species richness–decomposition relationships depend on species dominance. Ecology Letters 7:395–402.

Duffy, J. E. 2009. Why biodiversity is important to the functioning of real-world ecosystems. Frontiers in Ecology and the Environment 7:437–444.

Fraser, L. H., J. Pither, A. Jentsch, M. Sternberg, M. Zobel, D. Askarizadeh, et al. 2015. Worldwide evidence of a unimodal relationship between productivity and plant species richness. Science 349:302–305.

Gamfeldt, L., T. Snall, R. Bagchi, M. Jonsson, L. Gustafsson, P. Kjellander, et al. 2012. Higher levels of multiple ecosystem services are found in forests with more tree species. Nature Communications 4(1340):1–8.

Garibaldi, L. A., I. Steffan-Dewenter, R. Winfree, M. A. Aizen, R. Bommarco, S. A. Cunningham, et al. 2013. Wild pollinators enhance fruit set of crops regardless of honey bee abundance. Science 339:1608–1611.

Gaston, K. J., T. M. Blackburn, J. J. D. Greenwood, R. D. Gregory, R. M. Quinn, and J. H. Lawton. 2000. Abundance-occupancy relationships. Journal of Applied Ecology 37(s1):39–59.

Gonzalez, A., and M. Loreau. 2009. The causes and consequences of compensatory dynamics in ecological communities. Annual Review of Ecology, Evolution and Systematics 40:393–414.

Hoehn, P., T. Tscharntke, J. M. Tylianakis, and I. Steffan-Dewenter. 2008. Functional group diversity of bee pollinators increases crop yield. Proceedings of the Royal Society B: Biological Sciences 275:2283–2291.

Hooper, D. U., F. S. Chapin III, J. J. Ewel, A. Hector, P. Inchausti, S. Lavorel, et al. 2005. Effects of biodiversity on ecosystem functioning: a consensus of current knowledge. Ecological Monographs 75:3–35.

Isbell, F., V. Calcagno, A. Hector, J. Connolly, W. S. Harpole, P. B. Reich, et al. 2011. High plant diversity is needed to maintain ecosystem services. Nature 477:199–202.

Jiang, L., Z. Pu, and D. R. Nemergut. 2008. On the importance of the negative selection effect for the relationship between biodiversity and ecosystem functioning. Oikos 117:488–493.

Jiang, L., S. Q. Wan, and L. H. Li. 2009. Species diversity and productivity: why do results of diversity-manipulation experiments differ from natural patterns? Journal of Ecology 97:603–608.

Kirwan, L., A. Lüscher, M. T. Sebastià, J. A. Finn, R. P. Collins, C. Porqueddu, et al. 2007. Evenness drives consistent diversity effects in intensive grassland systems across 28 European sites. Journal of Ecology 95:530–539.

Kleijn, D., R. Winfree, I. Bartomeus, L. G. Carvalheiro, M. Henry, R. Isaacs, et al. 2015. Delivery of crop pollination services is an insufficient argument for wild pollinator conservation. Nature Communications 6:7414.

Kremen, C., N. M. Williams, M. A. Aizen, B. Gemmill-Herren, G. LeBuhn, R. Minckley, et al. 2007. Pollination and other ecosystem services produced by mobile organisms: a conceptual framework for the effects of land use change. Ecology Letters 10:299–314.

Kremen, C., N. M. Williams, and R. W. Thorp. 2002. Crop pollination from native bees at risk from agricultural intensification. Proceedings of the National Academy of Sciences of the United States of America 99:16812–16816.

Larsen, T. H., N. W. Williams, and C. Kremen. 2005. Extinction order and altered community structure rapidly disrupt ecosystem functioning. Ecology Letters 8:538–547.

Leary, D. J., and O. L. Petchey. 2009. Testing a biological mechanism of the insurance hypothesis in experimental aquatic communities. Journal of Animal Ecology 78:1143–1151.

McGill, B. J., R. S. Etienne, J. S. Gray, D. Alonso, M. J. Anderson, H. K. Benecha, et al. 2007. Species abundance distributions: moving beyond single prediction theories to integration within an ecological framework. Ecology Letters 10:995–1015.

MacLeod, M., J. Reilly, D. Cariveau, M. Genung, M. Roswell, J. Gibbs, and R. Winfree. In press. How much do rare and crop-pollinating bees overlap in identity and flower preferences? Journal of Applied Ecology.

Paquette, A., and C. Messier. 2011. The effect of biodiversity on tree productivity: from temperate to boreal forests. Global Ecology and Biogeography 20:170–180.

Rader, R., J. Reilly, I. Bartomeus, and R. Winfree. 2013. Native bees buffer the negative impact of climate warming on honey bee pollination of watermelon crops. Global Change Biology 19:3103–3110.

Reich, P. B., D. Tilman, F. Isbell, K. Mueller, S. E. Hobbie, D.F.B. Flynn, et al. 2012. Impacts of biodiversity loss escalate through time as redundancy fades. Science 336:589–592.

Roswell, M. E., J. Dushoff, and R. Winfree. In press. A conceptual guide to measuring species diversity. Oikos.

Shea, K., and P. Chesson. 2002. Community ecology theory as a framework for biological invasions. Trends in Ecology and Evolution 17:170–176.

Smith, M. D., and A. K. Knapp. 2003. Dominant species maintain ecosystem function with non-random species loss. Ecology Letters 6:509–517.

Tilman, D., and J. Downing. 1994. Biodiversity and stability in grasslands. Nature 367:363–365.

Tilman, D., F. Isbell, and J. M. Cowles. 2014. Biodiversity and ecosystem functioning. Annual Review of Ecology and Systematics 45:471–493.

Tuomisto, H. 2010. A consistent terminology for quantifying species diversity? Yes, it does exist. Oecologia 164:853–860.

Tylianakis, J. M., T. A. Rand, A. Kahmen, A.-M. Klein, N. Buchmann, J. Perner, and T. Tscharntke. 2008. Resource heterogeneity moderates the biodiversity-function relationship in real world ecosystems. PLOS Biology 6:947–956.

Wilsey, B. J., and H. W. Polley. 2004. Realistically low species evenness does not alter grassland species-richness-productivity relationships. Ecology 85:2693–2700.

Winfree, R. 2013. Invited view: Global environmental change, biodiversity, and ecosystem services: what can we learn from studies of pollination. Basic and Applied Ecology 14:453–460.

Winfree, R., J. W. Fox, N. M. Williams, J. Reilly, and D. Cariveau. 2015. Abundance of common species, not species richness, drives delivery of a real-world ecosystem service. Ecology Letters 18:626–635.

Winfree, R., J. R. Reilly, I. Bartomeus, N. M. Williams, D. Cariveau, and J. Gibbs. 2018. Species turnover promotes the importance of bee diversity for crop pollination at regional scales. Science 359: 791-793.

Wolf, A. A., and E. S. Zavaleta. 2015. Species traits outweigh nested structure in driving the effects of realistic biodiversity loss on productivity. Ecology 96:90–98.

Yachi, S., and M. Loreau. 1999. Biodiversity and ecosystem productivity in a fluctuating environment: the insurance hypothesis. Proceedings of the National Academy of Sciences of the United States of America 96:1463–1468.

PART VII

FINAL THOUGHTS

A Science Business Model for Answering Important Questions

Kevin D. Lafferty

Perhaps the biggest question in science is how to do better science. Many ecologists, including this book's editors and authors, have succeeded under the current science "business model" and, from our perspective, the status quo works well enough. But science business models are under increased scrutiny. For instance, since 2012, at least nine papers have critiqued government-sponsored biomedical research, with the most-suggested (self-serving) solution being to spend more government funds on science (Pickett et al. 2015). To get more funding, scientists might consider first improving their return on investment. To increase return on investment, ecologists (and scientists in general) could rethink training programs, reproducibility, funding distribution, synthesis, publication models, and evaluation metrics.

Training

Unlike a professional school, like law, veterinary, or medicine, where students pay tuition in exchange for a degree by exam, science students are usually apprentices who learn while training under a mentor. Although most PhD students train for a career in research, just one in five ecology PhDs will stay as practicing ecologists, presumably because there are fewer opportunities to do ecological research than there are ecologists wishing to do research (Kennedy et al. 2004). Such culling might maintain high quality in research professions, but when supply exceeds demand, wages stagnate and career-focused students follow other paths (Alberts et al. 2014), leading to a poor return on the considerable investment made in ecology graduate training programs. In addition, with more students comes less support and attention per student, which makes it hard to determine the extent to which competition increases overall professional quality. In comparison, medicine's pipeline is sound (~1% leakage) because once a person gets a medical degree, there are medical jobs (Ryten et al.

1998). This is not to say that medicine is less competitive (medical school acceptance rates are far lower, at ~20%), only that competition filters prospective doctors out at the beginning, whereas science filters many out at the end (Ryten et al. 1994). Depending on the details, PhD programs might benefit from putting the competition on the front end (such as by requiring a master's degree), with fewer students, with better prospects, and more individual training (Kennedy et al. 2004). Cost savings could help support more postdocs and staff scientists, helping to fix the opportunity deficit for new PhDs. Although, that would mean fewer yoga instructors would have a PhD.

To balance graduate student supply and demand, some have suggested funders eliminate research assistants (RAs) on grants and instead fund students with fellowships (Alberts et al. 2014). Competitive fellowships weed out graduate students at the application stage, rather than at graduation. Furthermore, fellowship winners become free agents for which faculty must compete by offering better mentoring, facilities, and exciting projects. From the taxpayer's perspective, fellowships create opportunities for a nation's residents, whether to be trained at home or abroad. Fellowships also have the potential to add diversity to the student pool through affirmative action. Losing in this proposed model are faculty at undesirable institutions with labor-intensive projects, and foreign students seeking to work through graduate school. More fellowships could also create barriers to students who don't have the test scores, experience, and letters used to evaluate fellowship applicants, reduce the vibrancy of large graduate programs, and, by creating a teaching assistant shortage, put more pressure on faculty to (gasp) teach.

The science career pipeline disproportionately leaks women. Although women increasingly dominate among graduate students, gender career disparity remains. For example, 55% of Ecological Society of America (ESA) members are women, but only 28% of 2011 ecology authors (and only 21% of the authors of the most-cited papers) were female, compared with 44% of persons acknowledged for assistance (Beck et al. 2014). Similarly, only 30% of National Science Foundation (NSF) ecology funding went to female investigators (Beck et al. 2014). Although in 2019 53% of 2018 ecology hires are women, 40% of associate professors are women, and 28% of full professors are women (Fox, J. October 22, 2019 Dynamic Ecology Blog). These shifts likely represent both recent increases in female job applicants and higher attrition rates for female faculty. Female scientists face more barriers, including increased teaching loads, reduced resource allocation, and few role models (Beck et al. 2014). Unconscious bias (by men and women) is pervasive in science, such as the higher valuations given to identical work products, depending on whether the first name is masculine or feminine (Moss-Racusin et al. 2012). Furthermore, mixed-gender fieldwork exposes women to sexual harassment outside the lab. Although such obstacles make research careers challenging for women, they do not explain why so many women with PhDs do not start a faculty position. Ecology is impoverished when women are lost from the profession, and when so few under-represented groups are attracted to the profession in the first place. Ecologists should embrace workforce diversity as much as species diversity.

Recent studies find the main challenge for new female faculty is the disproportionate societal, cultural, and biological demands that parenting puts on women in their late 20s and early 30s when career competition is most intense (O'Brien and Hapgood 2012). This is not just because parenting is hard. Medicine does not see a drop in women doctors after medical school (Adamo 2013). In medicine, doctors slide into a career track earlier, and this makes it easier to establish a career before establishing a family. Young physicians can also earn more, which makes it easier to trade income for inconvenience. Most importantly, most female physicians do not need to compete with men for limited positions during motherhood (Adamo 2013). This is not to say that women are less competitive than men; it relates to how women view competition. From a hundred female post docs at the University of California, 46 start out thinking they want a career in science; by the end, only 11 want one, mostly due to how a career would affect their families (Goulden et al. 2011). Losing these female postdocs means that fewer women apply for faculty positions. For married US faculty with children, mothers were almost one-third less likely to get tenure than fathers or single women (Goulden et al. 2011). But in Canada, where competitive funding rates do not differ by gender, faculty retention does not differ between women and men (Adamo 2013). One way by which women could be retained in ecology (and other disciplines) would be to reduce professional competition when women are starting families and to decrease demands associated with parenting (Young *2015*).

Hillary Young, a recent faculty member at UC Santa Barbara, explains why (Young 2015):

> It is an unfortunate biological reality that child-bearing age for women coincides nearly perfectly with the age when scientists' research productivity needs to be the highest. Yet most institutions have woeful basic childcare available, no support for urgent or extended childcare during work related obligations, inadequate support for maternity needs, financial disincentives to PIs for hiring postdoctoral fellows with children needing insurance coverage or those who are likely to need maternity leave, no insurance coverage for fertility treatment for women who choose to delay reproduction

for career reasons, and little way to account for delays in productivity due to time spent in childcare in the faculty hiring process.

One would think that with most ecology students being women, departments would adopt female-friendly policies to attract the best students, postdocs, and faculty. Existing policies do include reducing institutionalized discrimination, training that defines harassment, and encouraging more diverse role models (Beck et al. 2014). There is also an effort to shift the ecology culture from success through competition to fostering community building and empowering all our early-career women scientists (Horner-Devine et al. 2016). Simple steps, like double-blind proposal and manuscript review would make it more difficult for subconscious biases to favor some groups over others. Less clear is whether there is a willingness for institutions to address Young's suggestion that more–parent friendly policies, like free onsite child care and after-school programs, could reduce the parent–career tradeoff for women. Furthermore, adding competition to graduate school admission and reducing it at the faculty stage, would minimize intense competition during the period when women are at the greatest disadvantage (Adamo 2013). No person in this modern age should have to choose between a career and a family; right now, only women do (Mason and Goulden 2004).

Reproducibility

We don't know the extent to which ecological results are reproducible but concerns about reproducibility from other disciplines suggest this is a topic that ecologists should think about (Alberts et al. 2014). Whereas economics, psychology, and biomedical research study humans and a few model organisms, ecologists study biodiversity in its entirety. For this reason, ecologists expect that a single study might not be general, and it is only after amassing many studies from many researchers on many systems do ecologists consider whether support for a hypothesis is general. Ecology is, by its nature, often not reproducible, and there is a tradeoff between ecologists replicating specific studies versus gaining insight from doing similar studies in different contexts (Nakagawa and Parker 2015). And that might be why progress in ecology sometimes seems like a random walk more than a stable attractor.

Although the goal for ecology might not be reproducibility, ecologists should at least strive for transparent and unbiased data interpretation (Oberbillig et al. 2014). Unfortunately, complex modern statistical analyses allow multiple interpretations, leaving it up to ecologists which results to report and emphasize. Increasing ambiguity is revealed by lower R^2

values and higher *p*-values per paper (Oberbillig et al. 2014). The desire to report something significant can lead authors to subconsciously report significant outcomes from multiple tests without controlling for multiple comparisons (*p*-hacking) (Head et al. 2015) and ecologists report more significant findings when they gather data with a preconceived hypothesis (Parker et al. 2016). On the other hand, the joy in reporting an unexpected finding leads to HARKing (hypothesizing after the results are known) (Kerr 1998), which is encouraged by high-impact journals that require authors to emphasize novelty and importance (Alberts et al. 2014). Furthermore, under the current biodiversity crisis, it is harder to remain neutral and dispassionate about the systems ecologists study. Ecologists' personal concerns for the environment can emphasize catastrophes, collapses, and crises that attract readership, provoking calls for more careful analyses and sober interpretations (Connell 2012). The best way for ecologists to help solve environmental problems is to generate unbiased information that policy makers can trust.

Ecologists increasingly acknowledge that reproducibility is important (Parker 2016), and there is already a move among journals for transparency and openness guidelines that could help foster reproducibility by having authors adhere to citation standards, data transparency, code archiving, materials archiving, design transparency, preregistering hypotheses and analytical methods, and replicating past studies (Nosek et al. 2015). Some have argued that research institutions should implement good institutional practices (i.e., rules, standards, documentation, transparency, blind assessment) (Begley et al. 2015), but ecologists do not often follow such practices even when they would be easy to implement (Parker et al. 2016). For instance, blind assessment helps researchers avoid bias, and is standard in clinical trials, but is not common in ecology (Kardish et al. 2015). Although no journal has adopted all transparency and openness guidelines, several have their own lists. For example, *Nature* has an 18-point checklist for good institutional practices in its instructions to authors. Independent assessment could be extended into several other publication steps with the aim to reduce bias, increase specialization, and foster critical thinking. For instance, basing publication acceptance on sound hypotheses and methods rather than the significance or findings (as per the Public Library of Science (PLOS) journals scope) makes it possible to publish negative results, which helps reduce the file-drawer problem. In addition, a few journals embrace reproducibility by inviting repeat studies (e.g., F1000Research). To push the reproducibility envelope, in Box 1, I propose a publishing model called Collaborative Independent Review, which increases reproducibility, but at the cost of time, expense, creativity and investigator control.

Box 1: Collaborative independent review

If taxpayers were to realize scientists spend the public's money on irreproducible results, their logical response should be to either withhold funds or demand a new process that emphasizes reproducibility. Here is how funders, journals, and scientists could implement Collaborative Independent Review whereby four independent teams and an editor author a paper together.

The first step is for a Principal Investigator (PI) to propose the questions, hypotheses, predictions, and methods (including proposed analyses). The preregistered proposal includes an Introduction and Methods, and suggests a target budget for the methods and analyses (Parker et al. 2016). Proposals receive double-blind panel review based on expected return on investment. Competitive proposals are revised according to panel review and then put out for bids for a lead technician (who can be the PI) and a lead analyst (not associated with the PI). The technician receives half the funds upfront to implement the methods and report the data. The analyst, who maintains independence by remaining anonymous until publication, blindly tests the a priori predictions and writes and illustrates the results, including appendices describing the analyses in detail. The analyst sends draft results to the PI and technicians for review. Once the three parties agree on the results, the PI submits the Introduction, Methods, and Results to an editor who sends the sections to outside referees. In response to the referee reports, the PI, technician, and analyst revise the Introduction, Methods, and Results. The referees then write a collaborative Discussion about the Results, at which point the funder pays the award balance to all authors (PI, technician, analyst, referees, and editor). All data produced in the project become available to the public at the publication date so others can repeat the analyses.

A drawback is that Collaborative Independent Review could discourage the scientific creativity that generates new ideas and hypotheses when unexpected results occur. It is probably best suited for controversial or politicized topics that could benefit from impartiality or when attempting to repeat important or unusual findings.

Funding

Funders assume that scientists convert funding into knowledge. Yet funding rates predict only 30% of an ecologist's science impact, even in the low-variance Canadian system (Aarssen and Lortie 2010). Regardless, financial investment in science is limited (Staff 2017). The default funding mechanism in the United States is the big investigator-driven original research proposal. Unlimited submissions coupled with low success create a positive feedback that increases submissions to the point that some NSF funding rates are less than 5% (according to the NSF Division of Environmental Biology, 2013–2015 funding rates were 7.6%) (Staff 2017). This is near the 6% "not worth it" breakpoint where the effort spent writing proposals has a value equal to the expected funds gained (von Hippel and von Hippel 2015). Low funding rates mean risky, creative, or ambitious proposals cannot make the cut, demoralizing reviewers and applicants alike. And the need to get unanimous positive reviews stifles researchers from taking the risks needed to make major breakthroughs. When twenty proposals are written and reviewed for each one that is funded, the US model seems like an inefficient way to fund science (as says anyone except the day they land a big grant).

Funders could increase funding rates for quality proposals by reducing award size. In stark contrast to the NSF sweepstakes model, Canada's Natural Sciences and Engineering Research Council program (NSERC) reduces the need to write proposals by funding people more than projects, leading to a high return on investment (Aarssen and Lortie 2010). A similar approach is Brazil's Coordination for the Improvement of Higher Level Personnel (CAPES), which gives research productivity grants to ecologists who generate high-impact science (Loyola et al. 2012). Such funding models provide positive feedbacks whereby productive researchers gain more funding, which they are encouraged to convert into more productive research. Although this rewards productivity, positive feedbacks might also funnel funding allocations into bigger and bigger labs, which by some measure, might become inefficient at doing science as more time is spent on managing money and addressing derivative questions. This might explain why productivity per dollar declines with award size (Fortin and Currie 2013). Evidently, when you have to spend time counting money, there is less time to count species.

Funders could increase proposal success by reducing submission rates. To reduce full proposals, the NSF Division of Environmental Biology has new submission limits and uses preproposals to filter out uncompetitive applicants. And eliminating submission deadlines reduces submissions by half (Hand 2016). Another way to reduce submission rates and reduce demands on outside reviews is to couple submission with reviewing. NSF experimented with this by asking some submitting investigators to help with the review burden (Mervis 2014). Time-strapped investigators would be less likely to submit 20 proposals per year if it meant reviewing 140 proposals per year.

If universities stopped rewarding investigators for the overhead they bring in, faculty might be less likely to propose expensive projects. A different incentive, one based on research efficiency, could flow from funders to universities directly. Direct flow from funders to universities already happens in some countries, like the United Kingdom, where funders reward institutions based on cumulative research output (Atkinson 2014). But my favorite example is Mexico, which rewards mentors when their students author a paper.

Data Streams and Synthesis

Although ecologists traditionally collected their own data, new data streams ranging from satellite imagery to climate information to various sensor arrays have increasingly high value for ecological synthesis. Furthermore, real-time global species distributions are emerging from crowdsourced data. Worldwide programs such as eBird, iNaturalist, and Reef Environmental Education Foundation (REEF) compile species accounts daily across the globe, and have a culture of conservative identification. Such programs (which are often initiated by museums) have incomparable scope and effort, but their design could benefit from more input from ecologists skilled in sampling design and data interpretation. Furthermore, field ecologists could consider contributing their data to these platforms. The public already collects far more species distribution data than ecologists do, suggesting citizens are now our main collaborators.

More data sounds great, but by the 1990s, NSF realized that there were more ecological data than ecological analyses. NSF imagined that finding general answers to big questions in ecology required evaluating many studies and massive data sets, something that was increasingly in reach due to the internet. NSF's radical idea was to fund the National Center for Ecological Analysis and Synthesis (NCEAS) in 1994. Ecologists were more than willing to spend a week with their colleagues in sunny Santa Barbara. Rather than give PowerPoint presentations to each other, 5,000 individuals from 70 countries discussed and synthesized. This approach was so productive that NSF renewed NCEAS funding twice (spending roughly $30 million over 15 years), during which center working groups and postdocs published more than 2,000 papers, many being among the most highly cited in ecology (and a bargain at $15,000 per paper), with an institutional h-index of 242. This vast and influential intellectual output changed ecology from a lonely endeavor to a team sport. Although NSF discontinued funding for NCEAS (despite its incredible success), its spirit limps along with nearly 20 synthesis centers modeled after NCEAS. In simplified terms, it is surprisingly simple to break down the barriers to collaboration: Cover travel to an enjoyable destination, add some big ideas, and the best minds will give up their days off to synthesize together.

Publication

If you cannot make a good living from doing science, you can from publishing it. Elsevier (Amsterdam, the Netherlands), which publishes several ecological journals, has an enviable 40% profit margin (DeutscheBank 2005). The ESA took some by surprise when, instead of switching ecology to open access as some members suggested, it contracted with the publishing company Wiley (Hoboken, N.J.) as a way to underwrite the society's bills without having to run a journal (Inouye and McCarter 2015). Wiley and the ESA profit because ecologists donate their content for free in exchange for professional validation and packaging. You might not see many Gucci tote bags at ESA, but ecologists, like most scientists, give credence to the branding that respected journals offer.

Journals count on scientists to donate their labor. Even though authors gripe about how long it takes reviewers to comment on their papers, they gripe even more about being asked to review papers, refusing 70% of requests (Hochberg et al. 2009). This is the tragedy of the reviewer commons, whereby benefits accrue most to researchers who have their papers reviewed, but don't accept requests to review themselves. Some have suggested that authors solicit their own non-anonymous peer reviews (Aarssen and Lortie 2010). Others, worried about cronyism, have suggested a cooperative points system for reviewers, whereby authors need to show that they have reviewed papers before they can submit papers (see Box 2).

Many have questioned the ethics and legality of publishers profiting from taxpayer-funded research that remains behind a paywall. This is the motivation behind several institutions canceling subscriptions to some for-profit journals, and the controversial Plan S (funders forcing authors to make their papers open access by 2020) endorsed by the Bill and Melinda Gates Foundation, several European research funding agencies, and gaining momentum in China and India. The gold open access model is where the author pays publication costs upfront (Hochschild 2016), freeing content for readers. Although gold open access still privatizes a public good, it should drive publishing fees closer to publishing costs because it allows authors to shop around for where to submit in contrast to libraries that cannot shop around for which journals to subscribe. Unfortunately, gold open access has created a market for predatory stand-alone "open access" journals, which, by skimping on peer review and editorial oversight, provide a pay-to-publish service akin to a vanity press for amateur poets. In the popular green open access model, authors do not pay publication costs, but post their articles' PDFs on their websites or repositories like Research Gate. But the fastest growing model is what I call black open access, where scientists download 150,000 articles a day, at the Robin Hood–like server

---

Box 2: Brand name cooperative diamond open access

Hypothetically, submitting a cooperative diamond open access (CDOA) article costs four tradeable credits per submission. Researchers that need credits log on to the server and, if the handling editor approves their experience and expertise, they can choose from paper topics they want to evaluate. In exchange for a credit, they are assigned the paper in their topic(s) that has been in the queue the longest. Authors who have written several quality peer reviews can be assigned as handling editor. Once a reviewer completes a paper, review credit is banked after being weighted (0–1) by the editor's assessment of that review (incentivizing quality reviews). The time to review is also noted. After sufficient back and forth between reviewers and authors, editors (who also receive a credit per paper) determine when the paper is suitable for publication. With four banked credits, authors can submit a paper, which is reviewed and edited as above. Once accepted, the publication date is embargoed by the average time to review associated with the author's cashed credits (incentivizing speedy reviews). CDOA is therefore rigorous, fast, and equitable, and the product is easy to make open, free to readers and authors, and affordable to subsidizing institutions. To create branding, societies like ESA could establish their own portals to set standards, vet reviewers, and customize citations. For example, accepted articles could carry a citation linking them to the society (or its paper journals) that handled the paper. Long-term success for CDOA would require dedicated subsidies for servers, such as would become available if libraries invest less in subscriptions to expensive commercial journals and institutions invest less in author charges.

Sci-Hub (*34*). Black and green publishing could disrupt the publishing industry the same way free downloads disrupted the music industry.

Then there is diamond open access, in which neither authors nor readers pay a fee (Fuchs and Sandoval 2013). A familiar example is how Cornell University and member institutions subsidize the arXiv preprint server. A longstanding publishing model in physics, arXiv has become popular enough across other science fields that it posts 10,000 preprints per month. These preprints are filtered by moderators and an endorsement system; however, because most scientists distinguish this process from traditional peer review, authors usually also submit their preprints to conventional journals. To build on this model, a group of researchers from public research institutions added a peer-review layer in 2017 called "Peer Community in . . ." (with parallel sections for ecology, evolution, and paleontology). This project has been supported by some scientific councils of research institutions (Institut National de la Recherche Agronomique, Centre National de la Recherche Scientifique in particular), scientific societies, journals, and laboratories. Authors deposit preprints in a preprint server like arXiv and then submit that preprint to PCI Ecology (at no cost). Named volunteers play the role of editors and invite other volunteers (named or anonymous) to review submitted preprints. Reviews and revisions are uploaded to the preprint server where they remain public. If accepted, the final version is labeled by PCI as a "peer-reviewed" article, and the editor adds a short recommendation. However, because a peer-reviewed preprint stored on a file server lacks the brand value that motivates authors and readers to affiliate with particular journals, it remains to be seen whether authors will send PCI their best work. For this reason, accepted versions and associated reviews can also be submitted to traditional journals (which can then charge various fees). An ideal solution to the current publishing model would make publicly financed content available for free without profiting from author labor, have a rigorous and equitable peer-reviewing system, and provide the branding that authors want to associate with their work. It would not take much to make a platform like PCI Ecology meet all three needs (Box 2). I call this brand-name cooperative diamond open access.

For-profit journals could still play an important role in the future science-publishing model. The top science magazines, funded by advertising revenue, could publish news, invited reviews, and commentaries on notable papers. Someday having an open-access paper highlighted in *Science* might have the same caché that publishing a *Science* paper has today.

Science Metrics

Discovery motivates scientists, but so does recognition for discoveries. The science enterprise now ranks and compares science success for journals, scientists, universities, and even countries. Ideally, evaluating individual success would involve inspecting a scientist's record, including reading their publications, as often happens during tenure reviews and job searches. But shortcuts include many standardized metrics, grouped as inputs (grants received, operating costs), outputs (papers published, students graduated) and outcomes (student job success, policy influence). Many metrics are easy to calculate with online citation information. For instance, the website Publish or Perish computes a dozen citation metrics (Harzing 2017). When evaluating research institutions, the United Kingdom's Natural Environment Research Council impact report considers 28 inputs, 12 outputs, and 8 outcomes (Goff 2015). When ranking university departments (for instance, to help graduate students decide where to apply), the US National Research Council uses the publications per faculty member, citations per publication, grant support, faculty and student diversity, student exam scores, graduate student funding, number of graduate students, PhD completion percentage, time to degree, academic plans of graduating students, student work space, student health insurance, and student activities (Ostriker et al. 2001). The Leiden Rankings look at total publications (with optional partial credit for multi-institutional papers), high-impact publications, and fraction of high-impact publications per university (emphasizing either total output or output quality rather than per-capita output) (Centre for Science and Technology Studies 2017). Like most short cuts, simple metrics can have shortfalls, but it seems inevitable that scientists and institutions will change their behavior in response to how the system grades them.

To the extent that institutions reward scientists based on metrics and scientists respond to those awards, metrics could alter, for better or worse, how scientists prioritize their time. Science moves fastest when researchers take creative approaches to difficult questions. Not long ago, committees simply counted a scientist's published papers. Rewarding the number of papers published presumably steered some researchers to write many uninteresting or trivial papers. Eventually, citation indexing made it easy to separate well-cited from forgotten papers. Accumulating citation counts, on the other hand, favors one-hit wonders over productivity and conflates popularity with quality. These loopholes inspired the now ubiquitous h-index (count of papers (N) by an author cited N times or more), which combines information on cumulative popularity and quantity into a single number. However, the h-index can reward self-citation (though often this has a trivial effect), values review papers over primary literature, and gives overlapping credit to group authors. Easy solutions are to share a paper's citations among its authors (i.e., an individual h-index) (Box 3), discount self-citations, and calculate separate metrics for primary and review literature (Box 4). Regardless, praise or disdain for a metric often lies with the extent that it flatters you.

A citation metric assigned to a paper seems logical (e.g., a well-cited versus a poorly cited paper), but the same metrics assigned to individual scientists are often illogical because they imply a per-person impact but are not calculated as such. Now that most papers are group efforts, the ninth author on a 40-author paper (often contributing the equivalent of a single paragraph) gets the same credit in their h-index as if they had written the paper by themselves. To make individual metrics satisfy basic logic, some have advocated adjusting individual metrics for papers written by groups (Schreiber 2008). Such individual metrics are easy to calculate and help share credit done by groups. This is particularly important when comparing the impacts of ecologists (who often work in small

---

Box 3: Individual metrics

Individual metrics should share credit among authors. Ideally, authors would indicate proportional credit in their published paper. In lieu of this, others have suggested only counting a scientist's first-authored publications, or dividing citations per paper, C, by the number of authors on a paper, as in Schreiber's multiauthored h-index (Schreiber 2008). As a slight improvement to either approach, I propose a simple default algorithm that weights credit to first authors more than secondary authors. Here, a solo author ($N = 1$) obviously merits all C citations to that paper. The lead author of a multiauthor paper ($N > 1$) claims half ($C/2$), and nonlead authors divide the remaining half equally ($C/2)/(N − 1$). Because this simple algorithm is only a best guess, journals might start allowing authors to specify proportional credit at publication.

---

groups) with others such as physicists (who often work in huge groups). And although the average scientist might be wise enough not to compare metrics across fields, it might not be beyond the average dean.

Most science impact metrics fail to meet the key societal expectation that scientists should wisely use the public's money to make important discoveries. In part, this is because the public does not reward scientists. Instead reward comes from research institutions that depend on overhead from research grants to subsidize their operations. Given that institutions reward investigators for overhead received, scientists can succeed by emphasizing grant writing over paper writing. In 2016, US science and engineering researchers spent $72 billion to publish 409,000 papers, or about $176,000 per paper (National Science Board 2018). Yet many of the top-funded researchers have moderate scientific impact (Fortin and Currie 2013). Because institutions will continue to reward overhead generation, it is up to funders to reward return on investment. This could include free-market ideas like funding completed rather than proposed science.

One way to improve the science business model in the public's interest could be to recognize return on investment. A current metric for return on investment is the publication impact efficiency (PIE) index, which divides citations by research dollars (e.g., accumulated grant funding or grant funding plus salary) (Aarssen and Lortie 2010) (Box 4). A more abstract metric for return on investment would be the individual h-index–to–dollar ratio. As the ratio of two integrals over time, return on investment metrics are relatively independent of seniority, although they can be extreme (e.g., near zero or infinity) for young scientists. Such ratios can be easier to imagine as cost per impact (e.g., dollars per citation or

---

Box 4: Individual return on investment

What is the return on investment for an anonymous author chosen from this volume? Web of Science (WoS) tabulates their publication count (~190), citation count (~17,000) and *h*-index (59). Over their career, this author and their colleagues received about $14 million in extramural funding (not counting salaries). Their PIE, therefore, is 0.001 (or $823 per WoS citation). However, the author's adjusted individual citation count (6,000), individual *h*-index (37), and grant share ($3 million) are all lower, so that the individual PIE changes to 0.002 (or $500 per WoS citation). Filtering out review papers in WoS further reduces the individual citation count (4,500), and individual *h*-index (28). Overall their individual empirical PIE would be 0.0015 (or $667 per empirical WoS citation), or $107,000 per individual *h*-index. Metrics that account for research costs and individual rather than group impact might make it easier to determine if this researcher has been a good investment, or if they just spend time at the beach.

---

per *h*-index). To better reflect the empirical discoveries that funding agencies expect, return on investment metrics can be estimated after excluding review papers. If used to assess individual performance, return on investment metrics could motivate researchers to spend more time conducting and writing up science and less time writing long-shot proposals to land huge grants, resulting in more research productivity and lower variance in funding among researchers.

Conclusions

The science business model has evolved from patron-sponsored intellectuals, to university-funded faculty, to taxpayer-funded scientists. Taxpayers might expect scientists to spend funds wisely on valid, public knowledge. To give the public what they expect might require different models for funding science, publishing science, and rewarding scientists. Although I have focused on ecologists in North America, most of these issues apply to other fields and other countries, especially in Europe. To efficiently train the brightest and most creative scientists might require changing the training model from a rear-filtered, family-unfriendly, apprentice model, to a front-filtered, family-friendly, free-agent model. Funding graduate students with competitive fellowships rather than research assistantships would reduce student entry rates and help funders improve diversity. Fewer students would increase student quality, improve mentoring, and increase return on investment in training, particularly if funding fewer students frees funds for permanent science staff or postdoctoral opportunities. Furthermore, shifting the filter earlier would level the playing field for women and men. And having childcare as a standard benefit for scientists would make a science career more appealing to women. Funding could generate a greater return on investment if scientists write fewer proposals for less money. One model is NSERC, which encourages scientists to produce rather than propose. Furthermore, given the spectacular success of NCEAS, funders could encourage creative incentives for new data streams and synthesis. Scientists might eventually be able to choose a publishing model like collaborative independent review, that emphasizes reproducibility over novelty. One solution for making science products more available to the public is the cooperative diamond open access model, which shares the editorial and reviewing tasks needed for peer review, frees publishing and reading, and gives libraries a way to keep subscription costs down. Finally, a business model for ecology must differentiate success from failure. Metrics for success should share credit among collaborators and emphasize return on investment, where "return" refers to empirical discoveries rather than well-cited review papers or opinionated book chapters like this.

College deans, funding agency program officers, and journal editors belong to the current science business model. Ecologists, however, can help create the new science business model imagined here. Certainly the skills needed to study behavior, demographics, and consumer-resource interactions are not all that different from the skills needed to create business models. If successful, a new science business model will improve training, increase reproducibility, distribute funding efficiently, synthesize new data streams, liberate journal publications, and evaluate success with appropriate metrics. With such changes, ecologists will be ready to solve the challenging problems posed by this book.

Any use of trade, product, or firm names in this publication is for descriptive purposes only and does not imply endorsement by the US government. Thanks to Hillary Young, Andy Dobson, Dave Tilman, Thomas Guillemaud, Denis Bourguet, Ben Halpern, and my lab group for advice and comments.

References

Aarssen, L., and C. J. Lortie. 2010. Ideas for judging merit in manuscripts and authors. Ideas in Ecology and Evolution 3:28–34.

Adamo, S. A. 2013. Attrition of women in the biological sciences: Workload, motherhood, and other explanations revisited. Bioscience 63:43–48.

Alberts, B., M. W. Kirschner, S. Tilghman, and H. Varmus. 2014. Rescuing US biomedical research from its systemic flaws. Proceedings of the National Academy of Sciences of the United States of America 111:5773–5777.

Atkinson., P. M. 2014. Assess the real cost of research assessment. Nature 516:145–145.

Beck, C., K. Boersma, C. S. Tysor, and G. Middendorf. 2014. Diversity at 100: Women and underrepresented minorities in the ESA. Frontiers in Ecology and the Environment 12:434–436.

Begley, C. G., A. M. Buchan, and U. Dirnagl. 2015. Robust research: institutions must do their part for reproducibility. Nature 525:25–27.

Bohannon, J. 2016. Who's downloading pirated papers? Everyone. Science 352:508–512.

Centre for Science and Technology Studies at Leiden University. 2017. Leiden, The Netherlands.

Connell, S. 2012. F1000Prime Recommendation of [Dayton PK, Ecol Process Coast Mar Syst 1979, Plenum Press, New York:3–18]. F1000Prime.

Deutsche Bank. 2005. Reed Elsevier: Moving the supertanker. Company focus: Global equity research report. Berlin.

Fortin, J.-M., and D. J. Currie. 2013. Big science vs. little science: How scientific impact scales with funding. PLOS One 8:e65263.

Fuchs, C., and M. Sandoval. The diamond model of open access publishing: Why policy makers, scholars, universities, libraries, labour unions and the publishing world need to take non-commercial, non-profit open access serious. TripleC 13:428–443.

Goff, F. 2015. NERC Impact Report. Natural Environment Research Council, Swindon, United Kingdom.

Goulden, M., M. A. Mason, and K. Frasch. 2011. Keeping women in the science pipeline. Annals of the American Academy of Political and Social Science 638:141–162.

Hand, E. 2016. Science. April 15, 2016.

Harzing, A. W. 2017. Harzing.com, vol. 2017.

Head, M. L., L. Holman, R. Lanfear, A. T. Kahn, and M. D. Jennions. 2015. The extent and consequences of p-hacking in science. PLOS Biology 13:e1002106.

Hochberg, M. E., J. M. Chase, N. J. Gotelli, A. Hastings, and S. Naeem. 2009. The tragedy of the reviewer commons. Ecology Letters 12:2–4.

Hochschild, J. 2016. Redistributive implications of open access. European Political Science 15:168–176.

Horner-Devine, M. C., J. W. Yen, P. N. Mody-Pan, C. Margherio, and S. Forde. 2016. Beyond traditional scientific training: the importance of community and empowerment for women in ecology and evolutionary biology. Frontiers in Ecology and Evolution 4:119.

Inouye, D. W., and K. S. McCarter. 2015. The next century of ESA publications. Frontiers in Ecology and the Environment 13:67–67.

Kardish, M. R., U. G. Mueller, S. Amador-Vargas, E. I. Dietrich, R. Ma, B. Barrett, and C.-C. Fang. 2015. Blind trust in unblinded observation in ecology, evolution and behavior. Frontiers in Ecology and Evolution 3.

Kennedy, D., J. Austin, K. Urquhart, and C. Taylor. 2004. Supply without demand. Science 303:1105–1105.

Kerr, N. L. 1998. HARKing: Hypothesizing after the results are known. Personality and social psychology review: an official journal of the Society for Personality and Social Psychology, Inc 2:196–217.

Loyola, R. D., J. A. F. Diniz-Filho, and L. M. Bini. 2012. Obsession with quantity: a view from the south. Trends in Ecology & Evolution 27:585–585.

Mason, M. A., and M. Goulden. 2004. Marriage and baby blues: redefining gender equity in the academy. Annals of the American Academy of Political and Social Science 596:86–103.

Mervis, J. 2014. Research grants: A radical change in peer review. Science 345:248–249.

Moss-Racusin, C. A., J. F. Dovidio, V. L. Brescoll, M. J. Graham, and J. Handelsman. 2012. Science faculty's subtle gender biases favor male students. Proceedings of the National Academy of Sciences of the United States of America 109:16474–16479.

Nakagawa, S., and T. H. Parker. 2015. Replicating research in ecology and evolution: feasibility, incentives, and the cost–benefit conundrum. BMC Biology 13:88.

National Science Board. 2018. Science and engineering indicators 2018. NSB-2018-1. Alexandria, Va.: National Science Foundation.

Nosek, B. A., G. Alter, G. C. Banks, D. Borsboom, S. D. Bowman, S. J. Breckler, et al. 2015. Promoting an open research culture: Author guidelines for journals could help to promote transparency, openness, and reproducibility. Science 348:422.

Oberbillig, D., D. C. Randle, G. Middendorf, and C. L. Cardelus. 2014. Outdoor learning in formal ecological education: Looking to the future. Frontiers in Ecology and the Environment 12:419–420.

O'Brien, K. R., and K. P. Hapgood. 2012. The academic jungle: ecosystem modelling reveals why women are driven out of research. Oikos 121:999–1004.

Ostriker, J. P., C. V. Kuh, and J. A. Voytuk, eds. 2011. A data-based assessment of research-doctorate programs in the United States. Washington, D.C.: National Academies Press.

Parker, T. H., W. Forstmeier, J. Koricheva, F. Fidler, J. D. Hadfield, J. E. Chee, et al. 2016. Transparency in ecology and evolution: real problems, real solutions. Trends in Ecology and Evolution 31.

Pickett, C. L., B. W. Corb, C. R. Matthews, W. I. Sundquist, and J. M. Berg. 2015. Toward a sustainable biomedical research enterprise: Finding consensus and implementing recommendations. Proceedings of the National Academy of Sciences of the United States of America 112:10832–10836.

Ryten, E., A. D. Thurber, and L. Buske. 1998. The class of 1989 and physician supply in Canada. Canadian Medical Association Journal 158:723–728.

Schreiber, M. 2008. To share the fame in a fair way, h_m modifies h for multi-authored manuscripts. New Journal of Physics 10.

Staff, D. S. 2015. Division of Environmental Biology, NSF, vol. 2017.

von Hippel, T., and C. von Hippel. 2015. To apply or not to apply: A survey analysis of grant writing costs and benefits. PloS One 10:e0118494.

Young, H. S. 2015. Sexism discussion misses the point. Science 349:390–391.

Going Big

Stefano Allesina

Many biological systems are large: the gene-regulatory network of *Drosophila melanogaster* is composed of more than 15,000 genes interacting through activation and inhibition; over a thirty-year period, Charles Robertson documented more than 15,000 distinct interactions between 456 plants and their 1420 pollinators in a single location in western Illinois (Robertson 1929; Memmott and Waser 2002); the spread of influenza in Chicago and surrounding areas is mediated by a large social network comprising about ten million nodes, connected by billions of interactions.

These large biological systems can be represented as networks of interactions between agents (genes, individuals, populations): empirical biological networks invariably contain many agents with few interactions, and few with very many. This type of network structure makes it difficult to describe the systems using mean-field approximations, in which every agent is approximated by a typical, average agent. Empirical network structures are also very different from what we would expect under simple models (such as the Erdős–Rényi random graph, in which every node has the same probability of connecting to every other)—nodes in biological networks have been coevolving for long periods of time, and thus network structure is believed to reflect the underlying biological processes of the system.

An underappreciated aspect of these large networks is that they are quite variable: When measuring an arctic pollination network in Greenland, Olesen et al. (2008) found that the number of interactions between plants and pollinators varied dramatically from day to day. Similarly, were we to build the food webs of two lakes a few kilometers apart, we would find substantial differences. However, the dynamical processes occurring on these networks seem to be fairly robust to their ever-changing nature: Every day in Chicago somebody is born, somebody dies, somebody breaks up with their boyfriend—yet we do not expect these events to dramatically alter the number of cases of influenza in a given year. This suggests that, although we can only measure *a* network, rather than *the* network of interactions, all sampled network structures for a given system belong to a well-defined statistical ensemble, in which some unknown but important quantities are kept constant.

Historically, theoretical ecologists have devoted much effort to the study of small dynamical systems. This year marks the 94th anniversary of the work of Volterra (1926) on predator–prey interactions (the same set of equations had been published by Lotka (1925) the year before). These seminal contributions led to the birth of the field of theoretical ecology and spurred the development of ever more refined sets of equations describing the interaction between two populations. Interestingly, after almost a century, there is still debate over which variants of these equations can best model how species interact (Arditi and Ginzburg 2012). Besides the great influence of the pioneering work by Lotka and Volterra, the field concentrated on small systems also out of mathematical and empirical convenience. From a mathematical point of view, studying even the simplest system of nonlinear, autonomous differential equations describing the growth and decrease of interacting populations becomes prohibitively difficult when we include more than three populations. For larger systems, we can rely on numerical integration on a computer, but again it is difficult to understand these systems well enough to know what to look for in the rich output of such simulations. In the context of empirical studies, measuring interaction strengths is very complex and quite debated (Berlow et al. 2004); accurately measuring many interactions among many species seems to be outside the realm of possibility at the moment.

The study of small ecological systems has produced many beautiful and obviously useful results. I am thinking especially of the successes in modeling tightly coupled systems, such as pathogen–host dynamics and laboratory experiments (e.g., Costantino et al. 1997, Yoshida et al. 2003, Dai et al. 2012, among many). Small, treatable systems such as these are also great for developing intuition about the behavior of complex systems in general and can be seen as building blocks of larger systems. However, I sometimes feel that the focus on small systems could have serious side effects for the discipline as a whole. For example, much of our understanding of competitive interactions descends from the careful study of two interacting populations, and yet the beautifully simple result that intraspecific competition must exceed interspecific competition for populations to coexist does not extend to more than two competitors (Barabás et al. 2016). In addition, some documented dynamics leading to coexistence (such as intransitive competition) cannot even be studied for less than three competitors (Sinervo and Lively 1996, Allesina and Levine 2011), and focusing on two species at a time does not allow for the study of the community structure that emerges when pooling all interactions together (Barabás et al. 2016).

Summarizing, much progress in ecology has been made by taking the limit of $n \to 2$ for communities of interacting populations. Being somewhat of a contrarian, I cannot help but wonder what ecology would look like today, had we instead taken the limit $n \to \infty$, starting from infinitely many species and working our way down.

This might seem a strange idea, but it has a strong parallel in the development of physics. The celebrated equations by Newton can accurately describe the motion of the Earth circling the Sun. Take three bodies along with their mass, position, and velocity, however, and you will end up with a system of equations that cannot be solved analytically. This three-body problem has important consequences. For example, although many "proofs" of the stability of our solar system have appeared in the literature, we currently believe the motion of the planets to be chaotic, with large-scale computations being the only means to produce accurate results (Laskar 2013). However, when physicists are confronted with very many particles, for example gas molecules in a room, they do not write equations of motions for each and every one of them and integrate the equations in a computer. Rather, they rely on a completely different theory that, by describing the system statistically, provides important insights on its dynamics. In this way, one can show that fundamentally statistical quantities, such as temperature and pressure, play a key role in determining the system's behavior.

Continuing the analogy with physics, one might ask what the equivalent of temperature and pressure for ecological systems would be—and especially what kind of a toolbox one would need to be able to determine which quantities have the largest effect on the fate of large ecological dynamical systems.

In my explorations around these themes, I stumbled upon the theory of random matrices. This branch of mathematics is concerned with the characterization of the distribution of eigenvalues and eigenvectors of (typically infinitely) large matrices whose coefficients, rather than being fixed numbers, are random variables (Bai and Silverstein 2010). Several characteristics make results in random matrices directly applicable to the study of biological systems. First, this is a theory for large matrices, and networks and matrices are two ways of representing the same problem. Second, although it is difficult to precisely measure interactions in biological systems (and these numbers would be subject to change anyway), one might hope that describing them with distributions would be simpler and more natural. Third, and most importantly, many results in random matrix theory are *universal* (Tao and Vu 2010): the exact distributions for the coefficient do not matter, provided that some quantities (e.g., mean, variance, and correlation) are kept constant. Hence, one can take a very complex network problem and use the machinery of random matrix theory to determine which quantities have the largest influence on eigenvalue distribution. Furthermore, the universality property guarantees that these conclusions will be robust.

Many ecological problems (notably, local asymptotic stability (May 1972), metapopulation capacity (Hanski and Ovaskainen 2000), and spread of infectious diseases (Van Mieghem et al. 2009) can be turned into problems regarding the eigenvalues of certain matrices, so that one can directly translate mathematical results into ecological insights. Though the first application of random matrix theory in ecology dates back to the seminal work by May (1972) on complexity and stability, ecologists have rarely used these tools in the past forty years. Right now, the timing for an exploration of these topics and their potential for biology could not be more perfect: In mathematics, random matrix theory is experiencing a phase of exponential growth, with many active researchers producing new, fundamental results and methods at a very high rate (e.g., Rogers 2010, O'Rourke and Renfrew 2014, Aljadeff et al. 2015). Thanks to these advances, one can analyze ever more complex, structured random matrices, unveiling which broad-scale properties of ecological systems strongly influence dynamics (Allesina and Tang 2015, Allesina et al. 2015, Grilli et al. 2015). Ecologists should engage with the mathematical community, proposing new challenges and classes of random matrices that are of ecological interest. This dialogue between mathematicians and ecologists could both further the study of random matrices and produce beautiful results in the realm of ecology.

Besides the opportunity granted by the development of this theory in mathematics, the availability of new, high-throughput ecological data should spur the development of a toolbox capable of harnessing the information contained in these datasets of unprecedented quality and size. It is not surprising that a random matrix approach immediately appealed to the community of scientists working on microbial communities and metagenomics (Coyte et al. 2015), as our ability to deal with communities containing hundreds of "species" using more traditional approaches is quite limited.

In summary, ecological systems are large, network-structured, and variable, and we need to build a toolbox that can handle these complex systems in a simple and natural way. The theory of random matrices is a promising research avenue, with many active researchers, and fundamental, new results published every year. Using this and similar techniques, we can pinpoint important quantities that largely drive the dynamics of ecological systems.

The need for good ecological theories capable of dealing with the complexity of natural communities is growing every day, be it for managing ecosystems, or to understanding how to preserve or restore essential ecosystems services. With these growing needs in mind, I maintain that in the next century ecology should go big or go home.

References

Aljadeff, J., D. Renfrew, and M. Stern. 2015. Eigenvalues of block structured asymmetric random matrices. Journal of Mathematical Physics 56:103502.

Allesina, S., J. Grilli, G. Barabás, S. Tang, J. Aljadeff, A. Maritan, and A. 2015. Predicting the stability of large structured food webs. Nature Communications 6(7842).

Allesina, S., and J. M. Levine, 2011. A competitive network theory of species diversity. Proceedings of the National Academy of Sciences of the United States of America 108:5638–5642.

Allesina, S., S. Tang, 2015. The stability–complexity relationship at age 40: A random matrix perspective. Population Ecology 57:63–75.

Arditi, R., and L. R. Ginzburg. 2012. How species interact: Altering the standard view on trophic ecology. New York: Oxford University Press.

Bai, Z., and J. W. Silverstein. 2010. Spectral analysis of large dimensional random matrices. Berlin: Springer.

Barabás, G., M. S. Michalska-Smith, and S. Allesina. 2016. The effect of intra- and interspecific competition on coexistence in multispecies communities. American Naturalist, in press.

Berlow, E. L., A.-M. Neutel, J. E. Cohen, P. C. De Ruiter, B. Ebenman, M. Emmerson, et al. 2004. Interaction strengths in food webs: issues and opportunities. Journal of Animal Ecology 73:585–598.

Costantino, R. F., R. A. Desharnais, J. M. Cushing, and B. Dennis. 1997. Chaotic dynamics in an insect population. Science 275:389–391.

Coyte, K. Z., J. Schluter, and K. R. Foster. 2015. The ecology of the microbiome: Networks, competition, and stability. Science 350:663–666.

Dai, L., D. Vorselen, K. S. Korolev, and J. Gore. 2012. Generic indicators for loss of resilience before a tipping point leading to population collapse. Science 336:1175–1177.

Grilli, J., G. Barabás, and S. Allesina. 2015. Metapopulation persistence in random fragmented landscapes. PLOS Computational Biology 11:e1004251.

Hanski, I., and O. Ovaskainen. 2000. The metapopulation capacity of a fragmented landscape. Nature 404:755–758.

Laskar, J. 2013. Is the solar system stable? In: B. Duplantier, S. Nonnenmacher, and V. Rivasseau, eds. Progress in mathematical physics. Vol. 66. Chaos. Berlin: Springer, 239–270.

Lotka, A. J. 1925. Elements of physical biology. Baltimore: Williams and Wilkins.

May, R. M. 1972. Will a large complex system be stable? Nature 238:413–414.

Memmott, J., and N. M. Waser. 2002. Integration of alien plants into a native flower-pollinator visitation web. Proceedings of the Royal Society of London B: Biological Sciences 269:2395–2399.

Olesen, J.M., J. Bascompte, H. Elberling, and P. Jordano. 2008. Temporal dynamics in a pollinator network. Ecology 89:1573–1582.

O'Rourke, S., and D. Renfrew. 2014. Low rank perturbations of large elliptic random matrices. Electronic Journal of Probability 19:1–65.

Robertson, C. 1929. Flowers and insects: Lists of visitors to four hundred and fifty-three flowers. Carlinville, IL: C. Robertson.

Rogers, T. 2010. Universal sum and product rules for random matrices. Journal of Mathematical Physics 51(9):093304.

Sinervo, B., and C. M. Lively. 1996. The rock-paper-scissors game and the evolution of alternative male strategies. Nature 380:240–243.

Van Mieghem, P., J. Omic, and R. Kooij. 2009. Virus spread in networks. IEEE/ACM Transactions on Networking 7:1–14.

Volterra, V. 1926. Fluctuations in the abundance of a species considered mathematically. Nature 118:558–560.

Tao, T., and V. Vu. 2010. Random matrices: universality of ESDs and the circular law. Annals of Probability 38:2023–2065.

Yoshida, T., L. E. Jones, S. P. Ellner, G. F. Fussmann, and N. G. Hairston. 2003. Rapid evolution drives ecological dynamics in a predator–prey system. Nature 424:303–306.

Index

Milton Keynes UK
Ingram Content Group UK Ltd.
UKHW020815041123
431884UK00007B/240